# 2D MATERIALS
## Characterization, Production and Applications

T0133558

*Editors*

**Craig E. Banks**
Faculty of Science and Engineering
Manchester Metropolitan University
John Dalton Building
Chester St., Manchester
UK

**Dale A. C. Brownson**
Faculty of Science and Engineering
Manchester Metropolitan University
John Dalton Building
Chester St., Manchester
UK

**CRC Press**
Taylor & Francis Group
Boca Raton London New York

CRC Press is an imprint of the
Taylor & Francis Group, an **informa** business

A SCIENCE PUBLISHERS BOOK

*Cover credit:*

Front image inset adapted from the publication Materials Today, 20, P. Dadhich, S. Dhara, "Calcium phosphate flowers A bone filler substitute", 657–658, Copyright (2017), with permission from Elsevier. The second front image inset is adapted from the publication Science, 353, K. S. Novoselov, A. Mishchenko, A. Carvalho, A. H. Castro Neto, "2D materials and van der Waals heterostructures", aac9439, Copyright (2016), with permission from The American Association for the Advancement of Science.

CRC Press
Taylor & Francis Group
6000 Broken Sound Parkway NW, Suite 300
Boca Raton, FL 33487-2742

First issued in paperback 2021

© 2018 by Taylor & Francis Group, LLC
CRC Press is an imprint of Taylor & Francis Group, an Informa business

No claim to original U.S. Government works

Version Date: 20180322

ISBN-13: 978-0-367-78108-8 (pbk)
ISBN-13: 978-1-4987-4739-4 (hbk)

| **Library of Congress Cataloging-in-Publication Data** |
| --- |
| Names: Banks, Craig E., editor. \| Brownson, Dale A. C., editor. |
| Title: 2D materials : characterization, production, and applications / editors, Craig E. Banks (Faculty of Science and Engineering, Manchester Metropolitan University, Manchester, UK), Dale A.C. Brownson (Faculty of Science and Engineering, Manchester Metropolitan University, Manchester, UK). |
| Other titles: Two-dimensional materials |
| Description: Boca Raton, FL : CRC Press, Taylor & Francis Group, 2018. \| "A science publishers book." \| Includes bibliographical references and index. |
| Identifiers: LCCN 2018000483 \| ISBN 9781498747394 (hardback) |
| Subjects: LCSH: Graphene. \| Thin films. \| Chemical vapor deposition. \| Layer structure (Solids) \| Electrochemistry. |
| Classification: LCC TA455.G65 T86 2018 \| DDC 620.1/15--dc23 |
| LC record available at https://lccn.loc.gov/2018000483 |

Visit the Taylor & Francis Web site at
http://www.taylorandfrancis.com

and the CRC Press Web site at
http://www.crcpress.com

# Preface

Since the re-discovery of Graphene in 2004/2005 there has been a surge of interest into isolating and utilising other novel 2D materials. Such 2D materials (*Graphene, h-Boron Nitride and various Transition Metal Dichalcogenides such as Molybdenum Disulphide or Tungsten Diselenide to name a few*) have been reported to possess a range of unique and exclusive properties and have consequently been explored in a plethora of scientific disciplines.

There is a global pursuit to find new 'industrial scale' methodologies for the facile fabrication of advanced 2D materials and related structures are being extensively incorporated into an ever diversifying range of applications across numerous fields in the search for greatly improved device performance, ranging from sensing through to energy storage and generation and carbon-based molecular electronics. Chapters herein explore state-of-the-art synthesis techniques, biomedical applications and related biological interactions/responses to such nano-materials, associated antimicrobial and chemical/mechanical properties for surface coating and engineering, advances in biomolecule detection and sequencing, electrochemical sensing and catalysis including energy applications and Hydrogen generation from water splitting.

This book aims to provide readers with a fundamental introduction into 2D materials, bringing together research into the various 2D materials across the sciences such that, with time, a *real encyclopaedia* materialises, allowing one to be able to design and interpret experiments in their field/area of interest whilst utilising such novel materials given the valuable and transferrable insights gained across the disciplines. Due to the range of currently available 2D nano-materials, their reported properties, diversity in fabrication and wide range of potential applications, the study of 2D materials and their scope is truly fascinating.

**March 2018**                                                    **Dale A. C. Brownson**
**Craig E. Banks**

# Contents

# CHAPTER 1

# CVD Synthesis of Graphene and Advanced 2D Materials

*Thomas Hardisty Bointon*

## Introduction

In the last decade, the technological advances in materials and fabrication techniques have vastly changed the face of consumer electronics. The introduction of lithium ion batteries and indium tin oxide (ITO) have provided the increase in energy density and transparent conductors required to produce ultra-portable phones, tablets and laptops. However, there is ever more drive for smaller, faster and longer battery life devices which can only be satisfied by the next-generation of nanomaterials.

In this chapter the CVD synthesis of the following materials are covered:

### Graphene

An atomically thin allotrope of carbon, was first isolated in 2004 by means of mechanical exfoliation of natural graphite.[2] Monolayer and few layer graphene demonstrate a combination of unique properties. For monolayer graphene there is no band gap between the valence and conduction band, due to the hexagonal arrangement of the carbon atoms in the lattice, shown in Fig. 1. This allows for electrical conduction regardless of the fermi energy of the material, hence graphene can be utilized as a conduit between components for the next generation ultra-portable processors and batteries.[3]

The University of Manchester, National Graphene Institute, Booth Street East, Manchester, M13 9PL.
Email: Tom@Bointon.com Corresponding author.

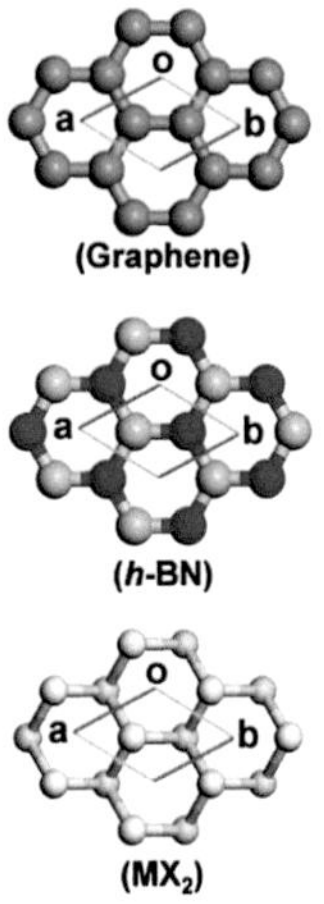

| 2D crystals | a=b (Å) | Lattice mismatch (%) |
|---|---|---|
| Graphene | 2.464 | 0 |
| *h*-BN | 2.504 | 1.6 |
| MoS$_2$ | 3.148 | 27.8 |
| WS$_2$ | 3.154 | 28.0 |
| VS$_2$ | 3.221 | 30.7 |
| MoSe$_2$ | 3.289 | 33.5 |
| WSe$_2$ | 3.286 | 33.4 |
| SnS$_2$ | 3.645 | 47.9 |

(Graphene)

(*h*-BN)

(MX$_2$)

**Fig. 1.** Demonstrating the structure of Graphene, H-BN and a selection of TMDCs with a table showing the lattice spacing of each of the materials. Reproduced from Ref. 1 with permission from The Royal Society of Chemistry.

The tightly bound covalent structure of graphene lends to a tensile strength greater than that of steel, a high resilience to chemical reactions and as a physical barrier to oxygen. The application of graphene to a substrate can protect from corrosion or oxidation.[4,5] The atomically thin nature of graphene allows for a the high transmission light for a large range of optical wavelengths. Monolayer graphene has an optical transparency of 97%, while in the few layer limit increasing the number of layers monotonically reduces the transparency to > 80% for 5 layers of graphene.[6] Although graphene is a good conductor, the sheet resistance is far higher than that of the current generation of materials.

Various functionalizations of graphene have been implemented to tailor the properties to exceed the properties of current industrial materials. A notable functionalization of graphene is achieved through FeCl$_3$ intercalation. When functionalized with FeCl$_3$,[6-10] few layer graphene exhibits a resistivity as low as 8Ω, while having an optical transparency similar to that of commercialised materials such as ITO and FTO.

## Hexagonal Boron Nitride (H-BN)

A 2D crystal formed from atoms of alternating boron and nitrogen atoms. The atoms are arranged in a hexagonal lattice and is closely matched to that of graphene, as shown in Fig. 1, with a lattice constant ~ 1.8% larger than that of graphene.[11,12] H-BN has similar properties to graphene in terms of tensile strength and integrity as a barrier layer, due to the covalently bonded hexagonal structure. The similar properties and structure have earned H-BN the moniker of 'White Graphene'.

In bulk form, H-BN is used an oxygen compatible, high temperature ceramic. This property is inherited in the 2D allotrope, where H-BN shows a remarkable thermal stability and chemical inertness.[13] The covalent bonds between B and N have a high bond energy ($\sim$ 4eV) even exceeding the bond energy of the CC bonds in graphene ($\sim$ 3.7eV).[14] This combined with the absence of dangling bonds, gives rise to the high thermal stability and oxidation resistance of H-BN at high temperatures. In high quality H-BN, there are no available energy states in the intrinsic fermi energy. This manifests in a the high chemical stability of H-BN.

In contrast to graphene with a continuous band structure, H-BN is an excellent insulator with a wide band gap of 5-6eV.[15] Thin layers of H-BN make an ideal insulating protective layer to prevent electrical leakage of sensitive electrical devices without impacting their inherent properties. When H-BN is reduced to ultra-thin crystals it can be used as a tunnel barrier for photodetectors and photovoltaics. Thicker layers of H-BN make an ideal atomically flat substrate. The intrinsic insulating properties, low levels of defects, and similar structure to that of graphene allow for H-BN to be used as a substrate for graphene to observe exotic physics such as the Hofstadter butterfly[16] and ballistic superconducting graphene hybrid devices.[17]

When used in combination with other 2D materials, the tensile strength and stability of H-BN has been utilised as a protective layer to prevent oxidation. Formed heterostructures using H-BN allows for the tailoring of the properties of unstable 2D materials.[18,19]

## Transition Metal Dichalcogenides (TMDC)

A family of 2D crystals typically composed of a transition metal paired with a chalcogen in a similar lamella structure to that of graphite. Figure 1 shows some examples of the typical transition metals are molybdenum and tungsten whereas the chalcogens are typically sulphur and selenium.[20]

Microprocessors and CPU consist of billions of transistors, a reduction in the size of transistors allow for a higher density and processing power. However, increasing the density of transistors also increases the power requirement and the heat generated by traditional transistors. There is a fundamental limitation to the processing power for high end CPU, as the significant increase in the power requirements leads to diminishing returns. On the other end of the scale ultra-portable devices such as phones and tablets have a limited power capacity. Therefore there is a requirement for high efficiency processors.

This is where there is a requirement for next generation materials such as semiconducting TMDCs. The structure and composition of some monolayer TMDC give a direct band gap. This is an ideal property for use as an ultra-thin transistor. The reduced dimensions of transistor devices made using monolayer and few layer $MoS_2$, $WS_2$, $MoSe_2$ and $WSe_2$ TMDCs allow for lower power dissipation and hence higher efficiency.

## Fabrication

Small samples for the next-generation 2D materials are typically produced using micro-mechanical cleavage of natural or synthetic bulk crystals.[21] The resulting materials are high quality 2D crystals and the simplicity of fabrication has been the primary drive for research into these next generation materials. However, micro-mechanical cleavage produces a random array of 2D crystals with a variety of thickness, which limits the reproducibility and scalability for industrial fabrication. To produce large area, high quality crystals in a reproducible manner requires an alternative production process.

Chemical vapour deposition (CVD) is a process used in the semiconductor industry to deposit thin high quality films onto a substrate. It is the primary method used for the fabrication of gallium arsenide, silicon nitride and silicon dioxide thin films. Each material produced via CVD requires a suitable recipe consisting of; a compatible substrate; the correct combination of precursor gases to facilitate a surface chemical reaction; a suitable temperature for the reaction to occur; and a pressure at which the chemical reaction occurs at.

In the following sections of this chapter the typical CVD procedures for the fabrication of 2D crystals are outlined.

# CVD Growth of Graphene

There are two dominant approaches to the growth of graphene through CVD processes, monolayer growth on a copper substrate[22–24] and multilayer growth using a nickel substrate.[25–27] Both processes require an experimental setup control the temperature, atmosphere and pressure, for this purpose a hot walled vacuum tube furnace or a cold wall CVD reactor are typically used. This primary focus of this section is for the growth of monolayer graphene on copper substrates.

## Monolayer Graphene Growth

The growth of monolayer graphene was first demonstrated in 2009 by reducing a carbon feedstock with a hydrogen atmosphere and a copper catalyst substrate.[22] Monolayer graphene is produced on the copper foil and can be transferred onto an arbitrary substrate.[28]

There are several steps required to produce large-area continuous graphene on a copper substrate. The copper substrate is annealed at high temperature under a hydrogen atmosphere. The purpose of which is to clean the substrate to facilitate graphene growth. The combination of high temperature and hydrogen atmosphere act to reduce the native oxide,[29] help remove any organic impurities and to enlarge the crystalline domains in the copper substrate.[23,30,31] Each of these factors increase the uniformity and quality of the grown graphene.

To grow graphene on the surface of the substrate, a carbon feedstock is required, the most popular of which is methane. The growth of graphene on the copper substrate can be separated into two steps, the nucleation stage and growth stage. As the methane enters the furnace and impinges on the copper substrate it is catalysed into carbon monomers.[32] There are three possible actions for the monomers on the surface of the copper; nucleate a graphene island, bind to the edge of a graphene island or to leave the surface of the copper substrate. Each action has an activation energy determined by the pressure, surface purity and roughness, and the temperature, all of which influence the density and size of the formed graphene islands.[33] At high enough feedstock partial pressure the graphene islands continue to grow domains and coalesce into a near continuous film.[34] Each graphene island/domain has a random crystalline orientation. Where there is a mismatch between two adjacent domains a disordered grain boundary forms with missing carbon atoms and dangling bonds.[35,36]

Once a continuous film of graphene is produced, the furnace is cooled and the copper substrate is removed to be further processed. The resulting copper substrates have a lighter appearance as the annealing stage of the growth reduces any native copper oxide and the graphene acts as an oxygen barrier.[22] The graphene can be separated using a wet etch approach, which then allows the film to be transferred to an arbitrary substrate.[28]

Since the original experiment was performed there has been driven research to optimise the graphene growth process which includes the effects of annealing conditions, the effect of the growth pressure and ratio of hydrogen to the carbon feedstock, and process temperatures.

## Effects of Annealing and Substrate Preparation

As received graphene is coated in a native oxide, predominantly $CuO$ and $CuO_2$ which has been shown to inhibit the catalytic activity required to grow graphene.[37,38] Before the annealing stage the oxide can be partially removed via wet etching in acetic acid, which predominantly removes $CuO_2$ from the copper substrate.[37,39]

The surface roughness of the copper substrate affects the rate of nucleation and resulting quality.[40] Structural defects resulting from the manufacture of thin copper foils leave surface grooves. These areas reduce the mobility of carbon monomers on the surface increasing the likelihood of nucleation. The benefit of this is the generation of a large amount of sites for the graphene to grow from. However, the high density of sites limits the maximum size the graphene islands/domains can grow to. When grown to a continuous film this also has the effect of increasing the number of boundaries between the graphene domains. The grain boundaries have an adverse affect on the quality of the graphene as they cause discontinuities in the band structure and have dangling bonds which introduce doping to the graphene.[41] The surface roughness can be reduced using electro-polishing prior to processing[42] or by taking the copper foil to a temperature close to the bulk melting point during the annealing process, where the surface partially melts to form a smooth surface.[43]

Copper substrates have a grain structure with domains, each with a different crystalline orientation. Annealing the substrate modifies the morphology, increasing the size of the crystalline domains and changing the orientation of the crystalline domains.[29] It has been demonstrated that different crystalline orientations of the domains on the copper exhibit different rates of graphene growth, nucleation and edge structure of the formed graphene islands.[44] Reducing the number of differently orientated domains helps produce a uniform growth rate upon the substrate, while increasing the size of the copper domains reduces the presence of edge defects that occur between the graphene that coalesce at the copper domain boundaries.

Each of these factors discussed have an effect on the resulting quality of the grown graphene films, however there is a trade off between rate of growth and the quality. For a rougher copper substrate there is a higher amount of graphene island nucleation. This provides a higher growth rate and faster coalescence into a continuous film, but, the resulting small graphene domains are connected by a high density of domain boundaries. The defects introduced to the graphene film directly affect the band-structure, intrinsic doping and act to dilute the some of the incredible properties of graphene. For a smooth defect free copper substrate the contrary is true, there is a lower nucleation of graphene islands. With a greater distance between the graphene islands, the time required to coalesce into a continuous films is increased. The resulting islands form larger crystalline domains with a reduced density of grain boundaries and produce a higher quality film.

## Effect of Growth Pressure

The pressure during the growth stage has a significant effect on the structure of the formed graphene islands.

At low pressure ($\sim 10^{-2}$ mBar), graphene islands grow in a dendritic manner, which resemble snowflakes. As carbon is added to the nucleation site it spurs off as spears. The graphene spears grow and spur off forming many more arms which eventually coalesce with each-other and adjacent graphene islands.[45,46]

At higher pressures ($\sim$ atmospheric pressure), graphene islands grow in a hexagonal manner. The edges of the island grow outward from the nucleation site and maintain this shape and structure over hundreds and microns.[46]

During the growth stage by combining a standard methane feedstock with a methane feedstock containing the $^{13}$C isotope, the direct observation of each growth type has been observed.[47,48] Alternating the two feed stocks during the growth process there are regions in the graphene islands populated either $^{12}$C and $^{13}$C atoms and creates a set of tree rings that show the chronological growth of graphene. The $^{13}$C atoms in the grown graphene are identified by a shift in the observed Raman spectra compared to that of $^{12}$C. By performing a Raman map of the graphene island the exact growth morphology can be observed. There is a difference in the quality of the graphene produced at low pressure and high pressure due to the intrinsic growth structure.

In the low pressure regime as the spears grow and spur they coalesce creating grain boundaries within the graphene island. The islands continue to grow and coalesce with adjacent graphene islands creating more grain boundaries with highly disordered interfaces. The increased density of grain boundaries and disordered edges has an adverse effect on the electronic properties of the graphene. Raman spectroscopy has been used to reveal the disorder of the grain boundaries within the graphene inland with the presence of the D-band.[48]

In the atmospheric regime, the hexagonal structure grows outward from the nucleation site. Growth within the island has the same 2D crystalline orientation, each island is considered a single crystal of graphene.[49] The lack of intersecting growth trajectories, due to the uniform island boundary growth, minimises the occurrence of grain boundaries within the islands. However, grain boundaries are formed as the islands grow and coalesce with other different orientation islands. The resulting graphene has large highly ordered domains which yield exceptionally high mobilities when compared to low pressure growth.[50]

## Effect of Ratio of Hydrogen to Carbon Feedstock

The rate of graphene nucleation and growth can be controlled through the ratio of hydrogen gas and the carbon feedstock during the graphene growth stage. A commonly used ratio for methane gas as the feedstock ranges between 2:7 and 2:35 (H2:CH4).[22,23,51] However, there is evidence that increasing the ratio of hydrogen to that of the feedstock or carrier gas has an effect on the structure of the graphene island growth.

With a low ratio between hydrogen and methane, there is a lower nucleation density and a lower growth rate.[34,51] The direct consequence is that the graphene islands have a larger separation. During the growth the islands increase in size, however, if the amount of carbon feedstock is too low, the island size saturates and no continuous film is formed. Increasing growth time has no effect on increasing graphene island size or reaching a point where the islands coalesce giving a poor coverage of graphene on the copper substrate.

In the case or a high ratio between hydrogen and methane a higher nucleation density, higher growth rate and film coalescence is observed.[52] As the nucleation density is higher there is less separation between the graphene islands. A high growth rate of the islands allow for the rapid expansion of the graphene. The film coalesces, resulting in many small graphene domains with a high density of grain boundaries and defects. Furthermore, high feedstock concentrations also cause the formation of multilayered graphene[52,53] islands within the monolayer film.

To improve the quality of the grown graphene, it has been shown that a combination of growth stages can optimise the growth of the graphene while maintaining a high growth rate and quality.[34] Starting with a low ratio between hydrogen and methane, the low nucleation density of the graphene islands is a achieved. Once saturation occurs, the ratio is increased. It has been shown that at there is only minimal increase in new island nucleation, while there is a significant increase in the growth rate of the islands. The result is the coalescence of the low density graphene islands leading to a higher quality film.[34,51]

The structure of the grown graphene have been shown to be influenced not only by the pressure at which the growth occurs, but the hydrogen partial pressure. In the high pressure regime, varying the quantity of hydrogen has been shown to promote dendritic growth at low concentrations and hexagonal growth at high concentrations[54] as shown in Fig. 2. For hydrogen partial pressures between these bounds there is a hybrid structure which inherits both the dendritic structure and the hexagonal structure. Understanding the role of hydrogen in the growth process gives another variable to control the resulting quality of the grown graphene films.

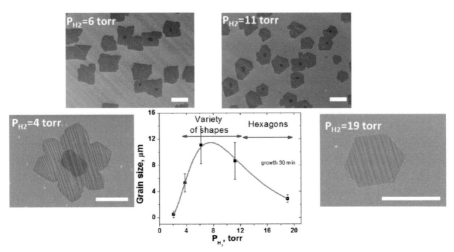

**Fig. 2.** A demonstration of the effect of different hydrogen concentrations on the growth shape of CVD graphene. For low hydrogen concentrations there is a combination of dendritic and rhomboid shaped graphene islands. At higher hydrogen concentrations hexagonal graphene islands are formed. Scale bars are 10 μm (top two images) and 3 μm (bottom two images). Reprinted with permission from Ref. 54. Copyright 2011 American Chemical Society.

## Effect of Growth Temperature

The temperature at which graphene is grown at is critical, as it affects the quality of the film, growth rate and suitability of catalytic substrate.[33] Low temperature growth allows for better integration of graphene with plastics and organic materials, but the minimum temperature required is governed by the catalyst. The limitation of low temperature growth is the activation of the copper catalyst substrate, whereas the limit for high temperature growth is the melting point of the substrate.

The growth temperature is correlated with the nucleation density of the graphene islands that form.[33,51] At low temperature < 950°C, there is a high nucleation density. Increasing the temperature reduces the nucleation density, as shown in Fig. 3. The temperatures effect on the nucleation density can be quantified by the activation energy for the three possible actions of the carbon monomers generated on the surface of the substrate; nucleation, addition to existing graphene islands, and leaving the substrate surface. For lower temperatures the process is dominated by nucleation. The probability that a carbon monomer seeds a nucleation site is greater than that of addition to an island and leaving the surface, due to a lower nucleation activation energy, as shown in the sketch in Fig. 4. As the growth temperature is increased the probability of seeding a nucleation site and the probability of addition to a graphene island are reduced, while the probability that carbon monomers leave the surface. This means for the same hydrogen/methane ratio and pressure the nucleation density reduces when the temperature is increased.[33]

One of the primary objectives for the CVD graphene community is to reduce the growth temperatures required to produce graphene. However, as discussed previously the catalytic activity of the substrate limits the generation of carbon monomers and hence graphene growth. In order to reduce the required growth temperatures,

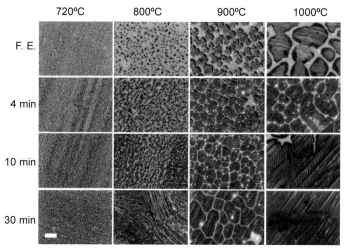

**Fig. 3.** A demonstration of the effect of growth temperature on the nucleation density of graphene on copper substrates. SEM micrographs have been used to capture the evolution of graphene growth for 720°C, 800°C, 900°C and 1000°C for a range of growth times. At low temperatures high nucleation density is observed. The effect of increasing growth temperatures shows a reduction in the nucleation density. Scale bars are 1 μm. Reprinted with permission from Ref. 33. Copyright 2012 American Chemical Society.

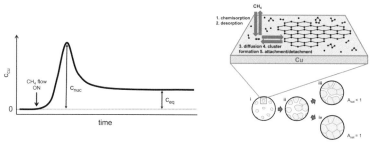

**Fig. 4.** A sketch demonstrating the role of the three processes influencing the growth of graphene on a copper substrate, graphene island nucleation, graphene island growth and desorption from the substrate surface. Adapted with permission from Ref. 33. Copyright 2012 American Chemical Society.

researchers have been combining traditional growth methods with RF plasma.[55] The high energies present in the plasma ionise the hydrogen and the carbon feedstock generate reactive carbon monomers at a far lower temperatures than on the catalyst surface. Graphene growth has been reported using plasma-enhanced techniques, but the method still needs refinement to compete in quality and repeatability of higher temperature growth methods.

## The Production of High Quality Single Crystal Graphene

Understanding of the factors that dominate the characteristics of grown graphene allow for the optimisation of growth parameters to grow large-area single crystals. Recent scientific developments have shown with the correct growth conditions, single crystals of graphene with areas in the order of $mm^2$ and even $cm^2$. The primary methodology used grow graphene in large single crystals is to minimise nucleation. In a typical graphene growth the nucleation density is of the order of 1 island for every 10 μm². [51,56] For large-area single crystal growth the density must be reduced to 1 island per mm² or even cm². To achieve this each stage of the growth has been optimised.

The anneal stage of the process is designed to reduce any factor that encourages nucleation. This is primarily achieved through the minimising the surface roughness, by melting the copper substrate under atmospheric pressures.[57] The liquid surface of the substrate is ultra smooth. By gently lowering the temperature prevents excessive evaporation of the copper substrate and allows for the recrystallisation of the solid ultra-smooth surface. There are also reports where hydrogen is gas is substituted for Argon. The effect leaves much of the native surface oxide intact which has been shown to suppress the nucleation of graphene.[58]

Low pressure growth of single crystal graphene was first achieved using the envelope technique. By folding the copper foil substrate into a sealed pocket carbon must diffuse through the copper.[48] This reduces the amount of carbon with respect to hydrogen inside the pocket allowing for a low nucleation density and large graphene crystals to form. The graphene morphology is affected by the low pressure leading to a dendritic formation. The grown crystals reach a maximum size of 0.5 mm in width. Electrical measurements of the grown graphene reveal a charge carrier mobility of 4000 cm²V⁻¹S⁻¹, limited by the multiple grain boundaries that exist between the coalesced spurs.

Growing the graphene at a higher pressure results in a hexagonal growth pattern with less intrinsic grain boundaries. In one case the growth has been performed at a near atmospheric pressure, where using a ratio of hydrogen to methane of 470:1 and a pressure of 144 mBar[50]. The authors report an 18 μm per minute growth rates using this technique yielding highly ordered single crystals of with widths of up to 2.3 mm were produced, as shown in Fig. 5. Measuring the electrical quality of the resulting graphene demonstrates high carrier mobility of 11,000 cm²V⁻¹S⁻¹.

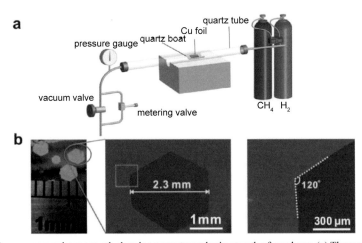

**Fig. 5.** Large area graphene growth showing near atmospheric growth of graphene. (a) The experimental setup, with copper foils loaded into a hot walled tube furnace and flow control of the growth gases. (b) Images showing large are single crystals of graphene with widths up to 2.3 mm. Reprinted with permission from Ref. 50. Copyright 2012 American Chemical Society.

The largest graphene crystals that have been grown by increasing the growth pressure to atmospheric.[59] By growing at atmospheric pressure and with a ratio of hydrogen to methane up to 8000:1, the density of nucleation has been reduced to as low as 4 sites per $cm^2$. The hexagonal grown single crystals have widths up to 5 mm and have a reported mobility of up to 16000 $cm^2V^{-1}S^{-1}$.

The primary draw back associated with large area single crystal graphene growth is the limited substrate coverage and the long time required to grow the graphene. A novel method has been developed to increase the rate of graphene growth.[60] By using a hybrid copper nickel alloy and directing the growth gases directly at the surface of the substrate the rate of growth has been reduced from in excess of 13 hours to produce 5 mm to 2.5 hours to produce crystals as large as 38 mm.

## Low Energy Graphene Production

The cost of producing graphene is an important factor the effects the viability when compared to traditional materials where the properties required are similar. Hot walled furnaces that are used for typical graphene growth have a high energy expenditure,[51] approximately 6 kW. Furthermore, growth in a hot walled furnace has an intrinsic time cost in addition to the time required for graphene growth. Heating a hot walled furnace to 1000°C requires between 20 and 40 minutes.[22] The combination of long growth times and a high energy cost leads to a high cost of graphene production.[51]

Reducing the energy and time required to produce graphene drastically increases the viability of graphene. A cold walled furnace uses significantly less energy to raise the substrate temperature to 1000°C.[51] Due to the small thermal mass that is heated, the growth temperature can be reached in as little as 90 seconds. This lowers the intrinsic energy cost of producing graphene.

Rapid cold walled growth of monolayer graphene has been reported by optimising the nucleation density and growth rate. With the selection of a suitable growth temperature, hydrogen to methane ratio and implementing the two stage growth technique the total time required for continuous graphene growth has been reduced to less than 15 minutes. Combining the rapid growth time and low intrinsic energy cost of operating a cold walled furnace the authors report a reduction in cost of graphene growth by 99%.[51]

# CVD Growth of Hexagonal Boron Nitride

Chemical vapour deposition is seen as a viable route for the production of large-area monolayer hexagonal boron nitride. The formation of graphene requires only the restructuring of carbon to form a film while Hexagonal boron nitride has a more complex structure as two different species (boron and nitrogen) are required to form in a specific order.

There are many suitable substrates that can be used for the growth of H-BN including Ir, Rb, Rh, Ag, Pd and Pt.[61–66] However, the primary focus of this section is the growth of H-BN on copper and copper alloy substrates, as these methods have been demonstrated to produce large-area H-BN and use relatively inexpensive catalysts.[67]

As with the growth of graphene, the growth process involves an annealing stage, where structural defects in the substrate are minimised to increase the quality of the resulting film.

## Effects of Annealing and Substrate Preparation

The first stage required to grow H-BN is to anneal the substrate. As with graphene growth substrate preparation has a significant effect on the quality of the produced H-BN film. The copper foils are annealed under a hydrogen atmosphere in order to clean the native oxides and reduce surface roughness. It has been shown that surface roughness has a significant effect on the growth, where a rough surface drastically increases the nucleation density of the H-BN islands.[68,69] The increased grain boundaries and dangling bond present after coalescence of H-BN islands gives rise to a high density of structural defects in the resulting film. For example, a high density of dangling bonds increases the charged impurities present in the film reducing the materials performance in electronic devices.[70]

To significantly reduce the surface roughness of the substrate, copper foils are melted on a tungsten substrate during the annealing process. This produces an ultra smooth substrate that is required to minimise the nucleation density of H-BN islands. However, this approach is limited by a maximum domain size of approximately 50 μm,[67] due in part to the high density of nucleation that occurs in the early stages of H-BN growth.

## Effects of the Growth Precursor and Concentrations

The precursor for H-BN growth is typically Borazine, which has a high concentrations of boron and nitrogen.[71] The precursor is carried by a mixture of Argon and Hydrogen, by bubbling the carrier gas through a Borazine solution. This carries the borazine feedstock to the reactor hot zone where it is decomposed into monomers through thermal decomposition and catalytic action of the copper foil. The partial pressure of borazine has been demonstrated to modify the thickness of the resulting H-BN film. For low partial pressures of borazine, the H-BN film grows in a thin layer, while increasing the partial pressure results in a thicker film. It was shown that for low pressure growth thickness was limited to 10 nm,[71] where the catalytic action of the substrate is suppressed.

H-BN growth using Borazine is associated with a high density of nucleation, due in part to the rapid decomposition that occurs on the catalyst surface during the early stages of H-BN growth.[72] In order to minimise the nucleation density and increase the quality of the produced film, H-BN growth with alternative precursors has been performed, Ammonia Borane is an ideal candidate. When compared to Borazine, Ammonia Borane has a lower quantity of Nitrogen and Boron. During the vital early growth stages, the lower amounts of Nitrogen and Boron present on the catalyst surface and different chemical composition reduces the rate of nucleation of H-BN islands. Substituting Borazine for Ammonia Borane, it has been shown that H-BN islands as large as 20 μm in width have been produced.[73] Further growths involving

Ammonia Borane are discussed later in this section which demonstrate exceptionally large domains.

Additional growth precursors can also be utilised to dope the grown H-BN. Modifying the doping of the produced H-BN films allows for band gap engineering of the work function and dielectric parameters required for LED, photo-detectors and tunnel current devices. Growing H-BN on InP substrates and adding Phosphine to the precursor mix has been demonstrated to hole dope the resulting film,[74] where it was then used as a high-K dielectric for an electrostatic gate.

## High Quality H-BN Synthesis

The production of continuous films H-BN with large crystalline domains can be achieved through careful selection of growth parameters such as the growth substrate, annealing treatments and precursors to control H-BN island nucleation. In 2015 it was demonstrated that large area H-BN islands can be produced with widths up to 72 μm,[75,76] which coalesce into a continuous film. Using nitric acid treated copper foils and annealing on a tungsten substrate above the melting point of copper, produced an ultra smooth substrate on which H-BN was grown. The reduction of surface impurities and roughness reduces the nucleation density of the grown H-BN. A precursor of Ammonia Borane was thermally decomposed up stream to act as a feedstock.

An alternative method has been shown to produce H-BN with larger domain sizes. By creating an alloy of copper and nickel for use as a substrate for growth, single crystal domains with widths of 7500 μm have been produced,[77] as shown in Fig. 6. Copper substrates are electroplated in nickel. The plated copper is heated under a hydrogen atmosphere at 1050°C, where the nickel diffuses into the copper foil producing an alloy with a low nickel concentration. The nickel concentration was found to directly affect the rate of nucleation, increasing the concentration of nickel to 10–20% showed H-BN island sizes increase from ~ 10 μm from native copper to 70–90 μm. Increasing the concentration further to 30% was found to significantly reduce the nucleation density. For greater concentrations of nickel full suppression of nucleation was observed. The increase in nickel composition was also directly correlated to a reduction in the growth rate of H-BN islands.

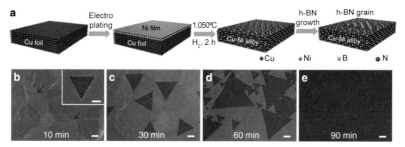

**Fig. 6.** An example of large single crystal growth of Hexagonal Boron Nitride on a Copper-Nickel alloy substrate. (a) An illustration demonstrating the formation of the Copper-Nickel alloy substrate used for H-BN growth. (b)–(e) SEM micrographs showing the growth of single crystal H-BN for increasing time until film coalescence is achieved in (e). The scale bars in (b)–(e) are 20 μm, and in inset in b is 2 μm. Reprinted with permission from Ref. 77. Copyright 2015 NPG group.

This is attributed to by the different growth mechanism of H-BN on copper and on nickel substrates. Copper has a surface mediated growth mechanism, conversely, for nickel there is no dominating mechanism either from surface mediated growth or the segregation and precipitation of B and N. The increased nickel concentration acts to suppress the H-BN nucleation on the copper alloy surface, however with excessive concentrations of nickel inhibit the action of the copper surface growth mechanism to halt any nucleation from occurring. By tailoring the composition of the substrate a balance between nucleation and growth rate, nucleation density was reduced to as little as 60 per mm$^2$ allowing for domains with widths of up to 7500 μm to be produced.[77]

# Transition Metal Dichalcogenide

Transition metal dichalcogenides are a large family of material the have a wide range of behaviours from superconducting and metallic, to semiconducting and insulating.[78] CVD processes for the manufacture of TMDC were first developed for molybdenum disulphide (MoS$_2$) and tungsten disulphide (WS$_2$), using MOCVD.[79] The decomposition of hexacarbonyl specie containing Mo or W, combined with vapourised sulphur, allowed for the growth of films on quartz, mica and LiF substrates. However, the produced films were thick (> 100 nm) and of polycrystalline nature, diluting the exotic properties required for next generation technologies. Producing large-area, 2D crystal TMDCs with enhanced fundamental properties are required to push the bounds of electronic devices, optics and ultimately technology.[80]

The primary focus of this section is on the CVD growth of the semiconducting TMDCs with direct band gaps such as MoS$_2$ and WS$_2$.

In the typical CVD formation of MoS$_2$ substrate preparation is key. An ultra-thin layer of molybdenum is evaporated onto an inert and compatible substrate.[81] Due to the high temperatures and thickness control to physically deposit refractory metals, electron beam evaporation is used. The pre-deposited substrate and sulphur powder are placed in a two zone tube furnace. The system is purged using high purity N$_2$ gas, which is flowed at 150–200 SCCM throughout the process also acting as a carrier gas. The substrate zone is gradually raised to 750°C over the course of 90 minutes. During this process, sulphur is heated upstream, just above the melting point, 113°C. Vaporized sulphur, carried by the nitrogen, passes over the substrate and reacts to form MoS$_2$ on the substrate surface. This process typically produces films between 1 and 5 nm thick. The ready reaction between the S and Mo leads to a high nucleation density and hence a polycrystalline film. The high density of defects at the boundaries of the polycrystalline films reduces the quality of the material. In order to produce a viable material for industrial use large single crystal domains are required.

## Growing Large area Single Crystal TMDC

Taking a different approach, by introducing the transition metal as a precursor that reacts with sulphur on the surface of the substrate has been shown to suppress the high nucleation, allowing WS$_2$ crystals as large as 370 μm to be grown.[82]

For the large crystal growth of MoS$_2$,[83] MoO$_2$ micro-platelets are precipitated on a SiO$_2$/Si substrate, by thermally evaporating molybdenum(III) oxide (MoO$_3$)

powder under a reducing sulphur atmosphere at a temperature between 650–850ºC. The resulting platelets are rhomboidal and act as the sites of $MoS_2$ island growth. The platelets are annealed using an upstream sulphur vapour, carried by Ar at a temperature of 850–950ºC. During the annealing process the $MoO_2$ platelets are reduced to form $MoS_2$ single crystals, with widths up to 10 µm. Electrical measurements of the resulting $MoS_2$ single crystals show a high on/off ratio of $10^8$ and a charge carrier mobility of > 200 cm$^2$V$^{-1}$S$^{-1}$.

Through process optimisation, this technique has been utilised to produce single $MoS_2$ crystals as large as 120 µm. Furthermore this process has a high yield and can produce 1–4 layers of $MoS_2$ with a high degree of precision.[84]

Further advances in the production of large $WS_2$ 2D crystals has also been demonstrated,[85] yielding single crystals with widths up to 50 µm. For this process sulphur is vaporized at 100ºC and carried to the growth zone using Ar gas. Tungsten(III) oxide ($WO_3$) powder and a sapphire ($Al_2O_3$) substrate are heated to 900ºC under an argon atmosphere at a pressure of 0.3 mBar. The $WO_3$ precipitates on the substrate and is reduced to $WS_2$. The similarly matched hexagonal lattice of the $Al_2O_3$ substrate coincides with that of the $WS_2$, aiding the single crystal formation. The addition of $H_2$ to the carrier gas causes the formation of triangular $WS_2$ crystals with widths up to 50 µm. However, the reported electrical quality of the grown $WS_2$ is worse than that of the smaller $MoS_2$ islands, with an on/off ratio of $10^2$ and a maximum observed charge carrier mobility of 0.46 cm$^2$V$^{-1}$S$^{-1}$.

Through the optimisation of growth conditions, larger domains of $WS_2$ have been achieved.[82] Using higher growth and vaporisation temperatures, it has been shown that single crystals as large a 370 µm have been grown directly on $SiO_2$/Si substrates. However, the electrical quality of the produced $WS_2$ was not assessed.

### Continuous Layers of Grown TMDC

Although there have been great advances in the growth size and quality of both $MoS_2$ and $WS_2$, these methods CVD growth form fragmented and discontinuous films. Until high quality and continuous films of $MoS_2$ and $WS_2$ can be produced there is a fundamental limitation on the industrial viability of these materials.

Furthermore, mechanically exfoliated flakes of $WS_2$ and $MoS_2$ have been shown to be susceptible to oxidation.[86] Manufacture of devices from a continuous film of $WS_2$ or $MoS_2$ require extra precautions, such as the addition of barrier layers, graphene or H-BN encapsulation, or performing sensitive manufacturing in an inert atmosphere. These factors add additional cost to manufacture processes and further reduce the viability of current day production methods.

## Conclusions

In this chapter, CVD production of graphene, H-BN and semiconducting TMDCs were discussed. Since the first implementation of CVD for the production of graphene, there has been rapid development on the growth of graphene and the other 2D materials.

The optimisation of graphene growth techniques have allowed for the development from large area low quality polycrystalline graphene films to high quality single crystalline domains that extend over many millimeters. Furthermore, research into cost effective graphene production is ever reducing the environmental and monetary cost of graphene production. With the current rate of development, graphene will soon surface as a viable replacement for current generation nano conductors and transparent conductors.

The growth of H-BN and TMDCs, although are still in their infancy have had rapid process in manufacturing high quality materials. The constant optimisation of CVD techniques driven by fundamental research and industries requirements has yielded leaps in film quality and production efficiencies.

In the coming years, expect high quality, large-area 2D crystals, each with unique and desirable properties, that will add to the ever expanding engineering toolbox of next-generation materials.

# References

1. X. Wang, Q. Weng, Y. Yang, Y. Bando and D. Golberg, *Chem. Soc. Rev.*, 2016, **45**, 4042–4073.
2. A. K. Geim and K. S. Novoselov, *Nature Materials*, 2007, **6**, 183–191.
3. H. Kim, K. Park, J. Hong and K. Kang, *Sci. Rep.*, 2014, **4**, 5278.
4. L. Yu, Y. Lim, J. H. Han, K. Kim, J. Y. Kim, S. Choi and K. Shi, *Synthetic Materials*, 2012, **162**, 710–714.
5. Y. Su, V. G. Kravets, S. L. Wong, J. Waters, A. K. Geim and R. R. Nair, *Nature Comm.*, 2014, **5**, 4843.
6. I. Khrapach, F. Withers, T. H. Bointon, D. K. Polyushkin, W. L. Barnes, S. Russo and M. F. Craciun, *Adv. Materials*, 2012, **24**, 2844–2849.
7. F. Withers, T. H. Bointon, M. F. Craciun and S. Russo, *ACS Nano*, 2013, **7**, 5052–5057.
8. T. H. Bointon, I. Khrapach, R. Yakimova, A. V. Shytov, M. F. Craciun and S. Russo, *Nano Lett.*, 2014, **14**, 1751–1755.
9. D. J. Wehenkel, T. H. Bointon, T. Booth, P. Bøggild, M. F. Craciun and S. Russo, *Sci. Rep.*, 2015, **5**, 7609.
10. T. H. Bointon, G. F. Jones, A. De Sanctis, R. Hill-Pearce, M. F. Craciun and Saverio Russo, *Sci. Rep.*, 2015, **5**, 16464.
11. J. Jung, A. M. DaSilva, A. H. MacDonald and S. Adam, *Nature Comm.*, 2015, **6**, 6308.
12. Y. Lin and J. W. Connell, *Nanoscale*, 2012, **4**, 6908–39.
13. N. Kostoglou, K. Polychronopoulou and C. Rebholz, *Vacuum*, 2015, **112**, 42–45.
14. C. R. Dean, F. Young, I. Meric, C. Lee, L. Wang, S. Sorgenfrei, K. Watanabe, T. Taniguchi, P. Kim, K. L. Shepard and J. Hone, *Nature Nanotechnology*, 2010, **5**, 722–726.
15. K. Watanabe, T. Taniguchi and H. Kanda, *Nature Materials*, 2004, **3**, 404–409.
16. G. L. Yu, R. V. Gorbachev, J. S. Tu, A. V. Kretinin, Y. Cao, R. Jalil, F. Withers, L. A. Ponomarenko, B. A. Piot, M. Potemski, D. C. Elias, X. Chen, K. Watanabe, T. Taniguchi, I. V. Grigorieva, K. S. Novoselov, V. I. Fal'ko, A. K. Geim and A. Mishchenko, *Nature Physics*, 2014, **10**, 525–529.
17. V. E. Calado, S. Goswami, G. Nanda, M. Diez, A. R. Akhmerov, K. Watanabe, T. Taniguchi, T. M. Klapwijk and L. M. K. Vandersypen, *Nature Nanotechnology*, 2015, **10**, 761–764.
18. S. Ahn, G. Kim, P. K. Nayak, S. I. Yoon, H. Lim, H. Shin and H. S. Shin, *ACS Nano*, 2016, **10**, 8973–8979.
19. V. Shautsova, A. M. Gilbertson, N. C. G. Black, S. A. Maier and L. F. Cohen, *Sci. Rep.*, 2016, **6**, 30210.
20. Q. H. Wang, K. Kalantar-Zadeh, A. Kis, J. N. Coleman and M. S. Strano, *Nature Nanotechnology*, 2012, **7**, 699–712.
21. K. S. Novoselov, D. Jiang, F. Schedin, T. J. Booth, V. V. Khotkevich, S. V. Morozov and A. K. Geim, *PNAS*, 2005, **102**, 10451–10453.
22. X. Li, W. Cai, J. An, S. Kim, J. Nah, D. Yang, R. Piner, A. Velamakanni, I. Jung, E. Tutuc, S. K. Banerjee, L. Colombo, R. S. Ruoff, *Science*, 2009, **324**, 1312–1314.

23. S. Bae, H. Kim, Y. Lee, X. Xu, J. Park, Y. Zheng, J. Balakrishnan, T. Lei, H. R. Kim, Y. I. Song, Y. Kim, K. S. Kim, B. Özyilmaz, J. Ahn, B. H. Hong and S. Iijima, *Nature Nanotechnology*, 2010, **5**, 574.
24. X. Li, Y. Zhu, W. Cai, M. Borysiak, B. Han†, D. Chen, R. D. Piner, L. Colombo and R. S. Ruoff, *Nano Lett.*, 2009, **9**, 4359–4363.
25. L. Baraton, Z. B. He, C. S. Lee, C. S. Cojocaru, M. Châtelet, J.-L. Maurice, Y. H. Lee and D. Pribat, *EPL*, 2011, **96**, 46003.
26. L. Huang, Q. H. Chang, G. L. Guo, Y. Liu, Y. Q. Xie, T. Wang, B. Ling and H.F. Yang, *Carbon*, 2012, **50**, 551–556.
27. A. Reina, S. Thiele, X. Jia, S. Bhaviripudi, M. S. Dresselhaus, J. A. Schaefer and J. Kong, *Nano Research*, 2009, **2**, 509–516.
28. J. W. Suk, A. Kitt, C. W. Magnuson, Y. Hao, S. Ahmed, J. An, A. K. Swan, B. B. Goldberg and R. S. Ruoff, *ACS Nano*, 2011, **5**, 6916–6924.
29. A. Ibrahima, S. Akhtara, M. Atiehb, R. Karnikc and T. Laouia, *Carbon*, 2015, **94**, 369–377.
30. K. L. Chavez and D. W. J. Hess, *J. Electrochem. Soc.*, 2001, **148**, 640.
31. Y. Lee, S. Bae, H. Jang, S. Jang, S. Zhu, S. H. Sim, Y. I. Song, B. H. Hong and J. Ahn, *Nano Lett.*, 2010, **10**, 490.
32. S. Bhaviripudi, X. Jia, M. S. Dresselhaus and J. Kong, *Nano Lett.*, 2010, **10**, 4128–4133.
33. H. Kim, C. Mattevi, M. R. Calvo, J. C. Oberg, L. Artiglia, S. Agnoli, C. F. Hirjibehedin, M. Chhowalla and E. Saiz, *ACS Nano*, 2012, **6**, 3614–3623.
34. X. Li, C. W. Magnuson, A. Venugopal, J. An, J. W. Suk, B. Han, M. Borysiak, W. Cai, A. Velamakanni, Y. Zhu, L. Fu, E. M. Vogel, E. Voelkl, L. Colombo and R. S. Ruoff, *Nano Lett.*, 2010, **10**, 4328–4334.
35. J. Hwang, C. Kuo, L. Chen and K. Chen, *Nanotechnology*, 2010, **21**, 465705.
36. G. Amato, G. Milano, U. Vignolo and E. Vittone, *Nano Research*, 2015, **8**, 3972–3981.
37. S. Kim, J. Kim, K. Kim, Y. Hwangbo, J. Yoon, E. Lee, J. Ryu, H. Lee, S. Cho and S. Lee, *Nanoscale*, 2014, **6**, 4728–4734.
38. H. Wang, G. Wang, P. Bao, S. Yang, W. Zhu, X. Xie and W. J. Zhang, *J. Am. Chem. Soc.*, 2012, **134**, 18476.
39. S. M. Kim, A. Hsu, Y. Lee, M. Dresselhaus, T. Palacios, K. K. Kim and J. Kong, *Nanotechnology*, 2013, **24**, 365602.
40. G. D. Kwon, E. Moyen, Y. J. Lee, Y. W. Kim, S. H. Baik and D. Pribat, *Materials Research Express*, 2017, **4**, 015604.
41. J. Čermák, T. Yamada, K. Ganzerová and B. Rezek, *Adv. Mater. Interfaces*, 2016, **3**, 1600166.
42. Z. Luo, Y. Lu, D. W. Singer, M. E. Berck, L. A. Somers, B. R. Goldsmith and A. T. C. Johnson, *Chem. Mater.*, 2011, **23**, 1441–1447.
43. D. Geng, B. Wu, Y. Guo, L. Huang, Y. Xue, J. Chen, G. Yu, L. Jiang, W. Hu and Y. Liu, *PNAS*, 2012, **109**, 21.
44. A. T. Murdock, A. Koos, T. B. Britton, L. Houben, T. Batten, T. Zhang, A. J. Wilkinson, R. E. Dunin-Borkowski, C. E. Lekka and N. Grobert, *ACS Nano*, 2013, **7**, 1351–1359.
45. T. H. R. Cunha, J. Ek-Weis, R. G. Lacerda and A. S. Ferlauto, *App. Phys. Lett.*, 2014, **105**, 4893696.
46. D. H. Jung, C. Kang, J. E. Nam, H. K. Jeong and J. S. Lee, *Sci. Rep.*, 2016, **6**, 21136.
47. X. Li, W. Cai, L. Colombo and R. S. Ruoff, *Nano Lett.*, 2009, **9**, 4268–4272.
48. X. Li, C. W. Magnuson, A. Venugopal, R. M. Tromp, J. B. Hannon, E. M. Vogel, L. Colombo and R. S. Ruoff, *J. Am. Chem. Soc.*, 2011, **133**, 2816–2819.
49. J. Li, X. Wang, X. Liu, Z. Jin, D. Wang and L. Wan, *J. Mater. Chem. C*, 2015, **3**, 3530–3535.
50. Z. Yan, Z. Lin, Z. Peng, Z. Sun, Y. Zhu, L. Li, C. Xiang, E. L. Samuel, C. Kittrell and J. M. Tour, *ACS Nano*, 2012, **6**, 9110–9117.
51. T. H. Bointon, M. D. Barnes, S. Russo and M. F. Craciun, *Adv. Materials*, 2015, **27**, 4200–4206.
52. Z. Tu, Z. Liu, Y. Li, F. Yang, L. Zhang, Z. Zhao, C. Xua, S. Wub, H. Liu, H. Yang and P. Richard, *Carbon*, 2014, **73**, 252–258.
53. C. Mattevi, H. Kim and M. Chhowalla, *J. Mater. Chem.*, 2011, **21**, 3324–3334.
54. I. Vlassiouk, M. Regmi, P. Fulvio, S. Dai, P. Datskos, G. Eres and S. Smirnov, *ACS Nano*, 2011, **5**, 6069–6076.
55. T. Terasawa and K. Saiki, *Carbon*, 2012, **50**, 869–874.
56. Q. Yu, L. A. Jauregui, W. Wu, R. Colby, J. Tian, Z. Su, H. Cao, Z. Liu, D. Pandey, D. Wei, T. F. Chung, P. Peng, N. P. Guisinger, E. A. Stach, J. Bao, S. Pei and Y. P. Chen, *Nature Materials*, 2011, **10**, 443–449.

57. Y. A. Wu, Y. Fan, S. Speller, G. L. Creeth, J. T. Sadowski, K. He, A. W. Robertson, C. S. Allen and J. H. Warner, *ACS Nano*, 2012, **6**, 5010–5017.
58. V. Miseikis, D. Convertino, N. Mishra, M. Gemmi, T. Mashoff, S. Heun, N. Haghighian, F. Bisio, M. Canepa and V. Piazza, *2D Materials*, 2015, **2**, 014006.
59. H. Zhou, W. J. Yu, L. Liu, R. Cheng, Y. Chen, X. Huang, Y. Liu, Y. Wang, Y. Huang and X. Duan, *Nature Comm.*, 2013, **4**, 2096.
60. T. Wu, X. Zhang, Q. Yuan, J. Xue, G. Lu, Z. Liu, H. Wang, H. Wang, F. Ding, Q. Yu, X. Xie and M. Jiang, *Nature Materials*, 2016, **15**, 43–47.
61. P. Sutter, J. Lahiri, P. Albrecht and E. Sutter, *ACS Nano*, 2011, **5**, 7303–7309.
62. J.-H. Park, J. C. Park, S. J. Yun, H. Kim, D. H. Luong, S. M. Kim, S. H. Choi, W. Yang, J. Kong, K. K. Kim and Y. H. Lee, *ACS Nano*, 2014, **8**, 8520–8528.
63. F. Orlando, R. Larciprete, P. Lacovig, I. Boscarato, A. Baraldi and S. Lizzit, *J. of Phys. Chem. C*, 2012, **116**, 157–164.
64. M. Morscher, M. Corso, T. Greber and J. Osterwalder, *Surface Science*, 2006, **600**, 3280–3284.
65. F. Müller and S. Grandthyll, *Surface Science*, 2013, **617**, 207–210.
66. T. Brugger, H. Ma, M. Iannuzzi, S. Berner, A. Winkler, J. Hutter, J. Oster-walder and T. Greber, *Angewandte Chemie - International Editions*, 2010, **49**, 6120–6124.
67. N. Guo, J. Wei, L. Fan, Y. Jia, D. Liang, H. Zhu, K. Wang and D. Wu, *Nanotechnology*, 2012, **23**, 415605.
68. A. L. Gibb, N. Alem, J. Chen, K. J. Erickson, J. Ciston, A. Gautam, M. Linck and A. Zettl, *J. Am. Chem. Soc.*, 2013, **135**, 6758–6761.
69. O. Cretu, Y. Lin and K. Suenaga, *Nano Lett.*, 2014, **14**, 1064–1068.
70. M. S. Bresnehan, M. J. Hollander, M. Wetherington, M. LaBella, K. A. Trumbull, R. Cavalero, D. W. Snyder and J. A. Robinson, *ACS Nano*, 2012, **6**, 5234–5241.
71. S. K. Jang, J. Youn, Y. J. Song and S. Lee, *Sci. Rep.*, 2016, **6**, 30449.
72. A. Ismach, H. Chou, D. A. Ferrer, Y. Wu, S. McDonnell, H. C. Floresca, A. Covacevich, C. Pope, R. Piner, M. J. Kim, R. M. Wallace, L. Colombo and R. S. Ruoff, *ACS Nano*, 2012, **6**, 6378–6385.
73. L. Wang, B. Wu, J. Chen, H. Liu, P. Hu and Y. Liu, *Adv. Mater.*, 2014, **26**, 1559–1564.
74. E. Yamaguchi, M. Minakata, *J. of Appl. Phys.*, 1984, **55**, 3098–3102.
75. Q. Wu, J. Park, S. Park, S. J. Jung, H. Suh, N. Park, W. Wongwiriyapan, S. Lee, Y. H. Lee and Y. J. Song, *Sci. Rep.*, 2015, **5**, 16159.
76. X. Song, J. Gao, Y. Nie, T. Gao, J. Sun, D. Ma, Q. Li, Y. Chen, C. Jin, A. Bachmatiuk, M. H. Rümmeli, F. Ding, Y. Zhang and Z. Liu, *Nano Res.*, 2015, **8**, 3164–3176.
77. G. Lu, T. Wu, Q. Yuan, H. Wang, H. Wang, F. Ding, X. Xie and M. Jiang, *Nature Comm.*, 2015, **6**, 6160.
78. Q. H. Wang, K. Kalantar-Zadeh, A. Kis, J. N. Coleman and M. S. Strano, *Nature Nanotechnology*, 2012, **7**, 699–712.
79. W. Hofmann, *J. Mater. Sci.*, 1988, **23**, 3981–3986.
80. J. Yu , J. Li , W. Zhang and H. Chang, *Chem. Sci.*, 2015, **6**, 6705–6716.
81. Y. Zhan, Z. Liu, S. Najmaei, P. M. Ajayan and J. Lou, *Small*, 2012, **8**, 966–971.
82. Y. Rong, Y. Fan, A. L. Koh, A. W. Robertson, K. He, S. Wang, H. Tan, R. Sinclair and J. H. Warner, *Nanoscale*, 2014, **6**, 12096–12103.
83. X. Wang, H. Feng, Y. Wu and L. Jiao, *J. Am. Chem. Soc.* 2013, **135**, 5304–5307.
84. A. M. van der Zande, P. Y. Huang, D. A. Chenet, T. C. Berkelbach, Y. You, G. H. Lee, T. F. Heinz, D. R. Reichman, D. A. Muller and J. C. Hone, *Nat. Mater.*, 2013, **12**, 554–56.
85. Y. Zhang, Y. Zhang, Q. Ji, J. Ju, H. Yuan, J. Shi, T. Gao, D. Ma, M. Liu, Y. Chen, X. Song, H. Y. Hwang, Y. Cui and Z. Liu, *ACS Nano*, 2013, **7**, 8963–8971.
86. F. Withers, T.H. Bointon, D. C. Hudson, M. F. Craciun and S. Russo, *Sci. Rep.*, 2014, **4**, 4967.

# CHAPTER 2

# Graphene and 2D Material in Biomedical Applications

*Reshma S. Cherian,[#] S. Syama[#] and P. V. Mohanan[\*]*

## Introduction

Elemental carbon is an interesting element in the periodic table as it is capable of forming many allotropes such as diamond, graphite, fullerene, carbon nanotube and graphene.[1] One atom thickness 2D graphene was first isolated from graphite by Andre Geim and Konstantin Novoselov form the University of Manchester in 2004.[2] Graphene is a $sp^2$ hybridized allotrope of carbon consisting of carbon atoms arranged in a honeycombed lattice network. Geim and Novoselov used a technique called micromechanical cleavage for obtaining 2D graphene sheets from 3D graphite. Geim and Novoselov were awarded the Nobel Prize in Physics (2010) for the ground breaking discovery and research conducted on graphene.

Graphene consists of numerous intriguing properties such as large surface area, superior mechanical strength, thermal and electrical conductivity and excellent optical transmittance. The unique properties of graphene make it a potential candidate for various electronic applications, transistor, optoelectronics and biomedical applications. Graphene and its derivatives are broadly classified into Graphene Family of nanomaterials (GFNs) based on their surface functionalization, purity and composition. Coming under this umbrella are monolayer graphene or pristine graphene, few layer graphene (FLG), graphene oxide (GO), reduced graphene oxide (rGO) and graphene quantum dots (GQD). These materials differ in its properties and have applications in various fields. Especially in biomedicine GFNs are used in drug/gene delivery,

Toxicology Division, Biomedical Technology Wing, Sree Chitra Tirunal Institute for Medical Sciences and Technology, Poojapura, Trivandrum 695 012, Kerala, India.
[\*] Corresponding author: mohanpv10@gmail.com
[#] Equal contribution

bioimaging, biosensing, photothermal therapy, tissue engineering and anti bacterial material. However as with other nanomaterials it is important to scrutinize the toxicity aspect of GFNs upon interaction with the biological system. This chapter briefly reviews the major methods of synthesis, biomedical applications of GFNs.

## Synthesis of GFNs

Over the years, numerous methods for synthesizing GFNs have been employed by the scientific community. Some of the synthesis methods are discussed below.

### *Monolayer graphene*

Monolayer graphene or pristine graphene is a mono-atomic layer of graphite. It has large surface area with a delocalized network of $\pi$ electrons. The electrons have high intrinsic mobility ($200,000$ cm$^2$ v$^{-1}$ s$^{-1}$) that contributes to its superior conductivity.[3] The increased mechanical properties of graphene are attributed to the high Young's modulus ($\sim 1.0$ TPa).[4] Pristine graphene also has good optical transmittance ($\sim 97.7\%$)[4] and thermal conductivity ($\sim 5000$ Wm$^{-1}$ K$^{-1}$).[5] The various application of monolayer graphene such as field effect transistor, sensors, transparent conductive films, clean energy devices and nanocomposites is reviewed extensively by Zhu et al.[6] Four primary synthesis methods exist for the synthesis of monolayer graphene or pristine graphene: Epitaxial growth, micromechanical exfoliation, liquid phase exfoliation and other methods.[7]

It is possible to grow monolayer graphene epitaxially on a hexagonal substrate like silicon carbide (SiC) and close packed metal substrates. Thermal decomposition of SiC consists of heating the substrate at very high temperatures (1273–1773 K) under ultra high vacuum (UHV). This results in the sublimation of Si leaving a carbon rich surface. Epitaxial growth occur in the either (0001) Si terminated or (000–1) carbon-terminated face of the SiC crystal. Growth on (0001) face gives graphene upto monolayers thickness and the reaction is self terminating. Growth on the (000 1) face gives rise to relatively thick samples (5–100 layers) and it is not self terminating.[8] An alternate route of epitaxial growth of graphene is chemical vapour deposition (CVD). Here carbon is supplied in the form of gas and metal is utilized as a catalyst and a substrate.[9] Substrates such as Cu and Ni are heated under low vacuum to increase the domain size. Subsequently, methane and hydrogen gas is flowed through the furnace. Hydrogen catalyzes the reaction between methane and the metal surface. This results in the deposition of carbon atoms from the methane gas on to the metal substrate.[10]

Micromechanical cleavage involves peeling of graphene from graphite using a Scotch tape. The method involves manually searching for single layer graphene among the various multi-layer-graphene. The presence of single layer graphene is identified by transferring it on to a suitable substrate like SiO$_2$/Si that provides ample.[11]

In liquid phase exfoliation (LPE) graphite is exfoliated using ultrasound from an ultrasonic water bath. This leads to cavitation induced shearing force on graphite thereby exfoliating it. Solvent plays a major role in this method as the energy required to exfoliate graphite needs to be balanced by the graphene-solvent interaction. Some of the solvents that meet this requirement are dimethylformamide (DMF), N-methyl-

2-pyrrolidinone (NMP), γ-butyrolactone and N,N-benzyl benzoate. Addition of small molecules such as sodium dodecyl sulphate (SDS), lignin molecules or porphyrins prevents the restacking of the exfoliated sheets.[12]

Other methods include substrate free gas phase synthesis of graphene platelets. In this method, graphene platelets were synthesized in a microwave plasma reactor. Argon gas and ethanol droplets were passed directly into the argon plasma inside the reactor. The ethanol droplets evaporate rapidly and dissociated in the plasma to form solid matter. The solid matter was cooled and collected in a nylon membrane filter. The collected graphene sheets were dispersed by sonication to form a black homogenous suspension.[13] Unzipping of carbon nanotubes (CNT) is another method for the synthesis of graphene. CNT can be opened longitudinally by intercalation of lithium (Li) and ammonia,[14] by plasma etching[15] or by chemical treatment.[16]

## Few layer graphene (FLG)

FLG consists of 2–10 layers of graphene sheets stacked together. It was originally a by-product of graphene synthesis. FLG shows no band gap and becomes increasingly metallic with an increase in layer number.[17] Functionalization of FLG to aid in dispersion is fairly facile. FLG has been used as a reinforcement agent for composite materials.[18]

Wang et al.,[19] synthesized FLG by electrochemically charging a negative graphite electrode and expanding it in an electrolyte containing Li salt and organic solvents under high current. Further sonication leads to the exfoliation into FLG. Park et al.,[20] utilized CVD on nickel (Ni) substrate for the synthesis of FLG. In another study FLG was synthesized using high pressure homogenization (HPH). During HPH a pressurized fluid is allowed to pass through a narrow gap valve and is subjected to three different mechanisms—impact effect, cavitation effect and high shear stress effect. Based on this HPH was utilized to exfoliate graphite to give FLG. Here graphite in a solvent (DMF) was passed through the narrow gap valve at a high pressure (100 MPa) for 10 min in 25°C to yield a stable dispersion of FLG in DMF.[21] Ball-milling of graphite through interaction with melamine (2,4,6-triamine-1,3,5-triazine) was employed to exfoliate to FLG in a nitrogen atmosphere. The resultant solution was dispersed in DMF.[22]

## Graphene oxide (GO)

GO is a highly oxidized form of graphene. It consists of oxygen functional groups such as carboxyl group at it edges and epoxy and hydroxyl group at the basal plane.[23] The widely accepted model for GO was put forth by Lerf and Klinowski.[24] It is a non-stoichiometric model where hydroxyl and epoxy groups are decorated around the carbon plane and the carboxyl group is present along the sheet edge. The presence of oxygen functional groups confer hydrophilicity to GO. Unmodified hydrophobic graphitic domains are also present in GO. Potential applications of GO includes sensing, microcircuits and in biomedicine as a drug/gene delivery nanocarrier, bioimaging, cancer therapy, etc.

B.C. Brodie, a British chemist, was the first to synthesize GO in 1859.[25] GO was synthesized by mixing potassium chlorate to a mixture of graphite in fuming nitric acid. The resultant material was oxidized and dispersible in water. L. Staudenmaier slightly modified Brodie's method by adding potassium chlorate in aliquots during the course of the reaction.[26] Sulfuric acid was also added to increase the acidity of the media. This was a more practical approach as the reaction was carried out in a single vessel. Modern chemistry utilizes the Hummer's method[27] for synthesis of GO and it is a modification of process developed by Staudenmaier. Here, graphite is mixed with potassium permanganate ($KMnO_4$), sodium nitrate ($NaNO_3$) and concentrated sulfuric acid. $KMnO_4$ and sulfuric acid react to generate an active oxidizing species: dimanganese heptoxide. The reaction is as follows:

$$KMnO_4 + 3H_2SO_4 \rightarrow K^+ + MnO_3^+ + H_3O^+ + 3HSO_4^-$$
$$MnO_3^+ + MnO_4^- \rightarrow Mn_2O_7$$

Modified Hummer's method by Kovtyukhova et al.,[28] utilized a pre-oxidized graphite powder. Here graphite powder was treated with sulfuric acid, potassium persulfate and phosphorous pentoxide to avoid incomplete oxidation of graphite. Recently Marcano et al.,[29] utilized an improved Hummer's method for synthesis of GO. Exclusion of Sodium nitrate ($NaNO_3$) avoids the evolution of toxic $NO_2/N_2O_4$ gases. Carrying out the reaction in 9:1 part sulfuric acid:nitric acid was found to increase the efficiency of the reaction. This method increased the fraction of oxidized carbon material in comparison to the traditional Hummer's method.

### Reduced graphene oxide (rGO)

rGO is obtained chiefly by the reduction of GO using various methods. The goal of reduction of GO is to obtain material similar to pristine graphene in property. GO is a poor conductor of electricity due to the presence of numerous oxygen functional groups. Reduction of GO partially restores the electrical properties of the graphene sheet. However, presence of numerous residual functional groups alters the structure and properties of the materials substantially.[30] Nonetheless reduction of GO is a hot topic in research as it is one of the favoured methods for the large scale, facile and inexpensive synthesis of graphene. rGO can be synthesized using various strategies such as thermal reduction, chemical reduction, biological reduction and other method.

Thermal reduction utilizes heat to reduce GO to rGO. This process is also called thermal annealing reduction. In this process GO is rapidly heated to 2000°C to exfoliate GO to rGO.[31] Normally thermal reduction is carried out by thermal irradiation, microwave irradiation[6] or photo-irradiation.[32] During this method CO and $CO_2$ gases evolve between the spaces of the GO sheets. This leads to decomposition of functional groups and an increase in pressure between the stacked layers, thereby separating it. However the procedure resulted in the formation of small, wrinkled graphene sheets as high temperature causes removal of carbon atoms from the plane.[33] Structural damage to rGO sheets were also noticed due to thermal reduction. Presence of defects affects the electrical properties of rGO as it introduces scatter centers and hinders the ballistic transport.

Chemical reduction uses various chemicals as reducing agent for synthesis of rGO from GO. Stankovich et al.[34] was the first to employ hydrazine hydrate for reduction of GO. Reduction by this method significantly decreases the oxygen moieties and produced electrically conductive rGO (99.6 S/cm). However large scale production of rGO using hydrazine hydrate is not favoured as it is highly explosive and carcinogenic. Sodium borohydride was utilized by Shin et al.[35] and showed similar reducing capacity as that of hydrazine hydrate. It was noticed that the sheet resistance was much lower than hydrazine hydrate reduced rGO. Sodium borohydride was found to be efficient in reducing C=O group but not epoxy and carboxylic groups. Moon et al.[36] reported HI as another strong reducing agent. HI was found to be a superior reducing agent when compared to hydrazine hydrate and sodium borohydride. Reduction of GO by HI produced rGO with improved tensile strength and electrical conductivity around 300 S/cm. The above reducing agents are highly toxic in nature. Presence of trace amounts of these chemicals due to improper washing may lead to adverse effects on exposure to the biological system.

Research is focused on finding an alternate green route for synthesis. L-Ascorbic acid or Vitamin C is an eco-friendly reducing agent that was employed by Fernandez-Merino.[37] L-Ascorbic acid is a naturally occurring antioxidant. It was found to be an ideal substitute for hydrazine hydrate as the reducing capacity was similar to it. Zhang et al.[38] found the conductivity of L-ascorbic acid reduced GO showed a conductivity of ~ 800 S/m. Moreover GO reduced by ascorbic acid forms a stable dispersion without adding any surfactants. L-ascorbic acid releases two protons to form dehydroascorbic acid that subsequently reacts with water molecules to form glucuronic acid and oxalic acid. Both glucuronic acid and oxalic acid can form bonds with residual oxygen moiety in rGO to prevent restacking and aggregation of rGO sheet.[38] Other reducing agents employed for reduction includes hydroquinone,[38] lithium aluminium hydride,[40] thiourea,[41] hydroxylamine,[42] etc.

Biological reducing agents such as microbes and plant extracts have been employed as a natural substitute for the reduction of GO. Salas et al.[43] used *Shewanella* genus microbes for the reduction of GO to rGO. The mechanism of reduction was attributed to ability of GO to act as a terminal electron acceptor during the bacterial respiration process thereby facilitating reduction. Akhavan and Ghaderi[44] highlight the use of *Escherichia coli* for the reduction of GO by utilizing mixed acid fermentation pathway in aerobic conditions. Other microbes utilized for reduction includes *Bacillus marisflavi*[45] and extremophiles such as *Halomonas maura* and *Halomonas eurihalina.*[46] Yin et al.[47] utilized green tea as reducing agent since it contained numerous pyrogallol and catechol moieties which aid in reduction of GO to rGO. *Rosa damascene* (rose) was utilized as a reducing agent as it contains numerous natural antioxidants such as flavonol glycosides and phenolic compounds.[48] Reduction using rose produced well dispersed rGO that was stable for one month. Other plant extracts such Asian red gingseng,[49] roots of *Salvadora persica* L.[50] cherry leaves[51] and roots of wild carrot[52] have also been explored as eco-friendly reducing agents.

Other method of synthesis includes utilization of irradiation of GO with microwaves to form rGO. Voiry et al.[53] synthesized high quality rGO after reduction by irradiation of GO. It was reported that initial reduction of GO slightly with thermal annealing prior to microwave exposure led to increased absorption of microwave. This led to rapid heating of GO prompting removal of oxygen functional groups and reordering the graphitic plane.

## Graphene quantum dots (GQDs)

GQDs are 0D materials containing single or few layer graphene with lateral dimension ranging between 2–20 nm. GQDs are a promising new class of fluorophores that retain the properties of graphene such a large surface area, high carrier transport mobility and superior thermal conductivity. Moreover it is well dispersed in aqueous solution due to the presence of carboxylic functional group at the edges.[54] However the properties of GQDs are distinct from graphene sheet due to quantum confinement and edge effect that confers photoluminescence and slow hot carrier relaxation properties to the material. In light of this GQDs have potential application in bio-imaging, drug/gene delivery, photovoltaics, light emitting diodes and sensors.

GQDs can be synthesized by bottom-up or top-down methods. Bottom-up synthesis of GQDs is performed mostly by condensation of aromatic compound in liquid phase, using unsubstituted hexa-peri-hexabenzocoronene as a precursor for synthesis[55] and CVD. Top-down method includes cutting of large sheets of graphene into GQDs of 2–20 nm. This is achieved by electrochemical oxidation,[56] hydrothermal process[57] and chemical exfoliation.[58]

## Applications of Graphene

### Drug delivery

GFNs are being projected as a promising nano-carrier for drug delivery applications (Fig. 1). The large surface area of GFNs facilitate high drug loading capacity unlike other nanoparticles.[59] The relative ease of covalent and non-covalent conjugation drugs on to GFNs makes it an attractive candidate for drug delivery applications. Moreover, the presence of delocalized π electrons promotes the π–π interaction with other π electron rich drugs such as doxorubicin (Dox).[60] Non covalent interactions also aid in loading of poorly soluble drugs. Drug delivery inside the cells is also assisted by the lipophilic nature of graphene that promotes the penetration of lipid bilayer membrane of the cell.[61]

GFNs have been utilized in drug delivery on various aspects such as cancer therapy, targeted delivery and controlled drug release. *In vivo* efficacy of PEGylated graphene oxide nanosheets (pGO) were demonstrated by Miao et al.[62] pGO was loaded with photosensitizer Ce6 and Dox. Passive tumor accumulation of Ce6/Dox/

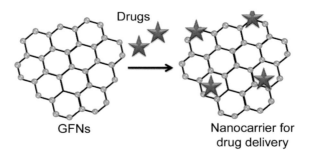

**Fig. 1.** GFNs as nanocarriers for drug delivery.

pGO was noticed in SCC7 tumor-bearing mice by enhanced permeation and retention (EPR) effect. Photodynamic therapy in conjunction with the anticancer activity of Dox reduced the tumor volume significantly in the mice, indicating the efficient eradication of the tumor. Enhanced accumulation of Dox in the nucleus of cancer cells were reported after conjugation with GQDs.[63] This indicated the increased efficacy of the conjugates in chemotherapeutic applications. Zhang et al.[64] reported the synthesis of folate conjugated nano graphene oxide (NGO) for targeted delivery of drugs to cancer cells. MCF-7 breast cancer cells line over expressing folate receptors were chosen as the *in vitro* test system in the study. Campothecin and Dox were loaded as anticancer drugs for destroying the cancer cells. The study demonstrated enhanced uptake of NGO conjugated with folate, camptothecin and Dox by MCF-7 cells lines with cytotoxic effects.

Glioma bearing rats were injected with PEGylated nanoscale graphene oxide (PEG-GO) conjugated with Transferrin (Tf) and Dox.[65] Tf is an iron transportation glycoprotein and Tf receptors are over expressed in gliomas. The lifespan of glioma bearing rats increased significantly with marked decrease in tumor volume. Controlled release of Dox was achieved by conjugating it with GO followed by encapsulation of GO-Dox in folate conjugated chitosan. A strong pH dependent release of Dox from the nanocarrier was noticed.[66] However numerous challenges such as toxicity, elimination and bioaccumulation of GFNs need to be scrutinized before commercialization for drug delivery applications.

## Gene delivery

Gene delivery is being pursued actively in the research field as an alternate to cure diseases that are difficult to treat by traditional methods. The major limitation of gene delivery is to formulate an efficient and safe gene delivery vector. GFNs have been proposed as an alternate nanocarrier for gene therapy applications (Fig. 2). Graphene reportedly binds strongly with the nucleobases of ssDNA by π–π interaction.[67] However it binds weakly with dsDNA as the nucleobases are hidden within the DNA double helix structure rendering it unavailable for binding. Moreover Lu et al.[68] showed that the oligonucleotides were protected from nuclease enzyme within the cells by means of steric hindrance.

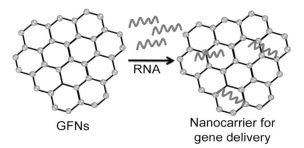

**Fig. 2.** GFNs as nanocarriers for gene delivery.

Feng et al.,[69] demonstrated the transfection efficiency of enhanced green fluorescence protein (EGFP) loaded on to GO-polyethyleneimine (PEI) nanocarrier in Hela cells. An improved transfection efficiency and lower cytotoxicity was noted in GO-PEI when compared with PEI alone. Similarly GO functionalized with chitosan aided in condensation of plasmid into stable complexes with improved transfection efficiency.[70] Kim et al.[71] employed GO-PEI nano-construct as a gene delivery vector and also as a bioimaging tool due to its intrinsic, tunable optical properties. In an interesting study by Feng et al.,[69] it was reported that laser irradiation increased and enhanced the delivery of small interfering RNA (siRNA) conjugated onto nano-GO-PEG-PEI construction. Increase in cell membrane permeability was noted when cells were exposed to a 0.5 W cm$^{-2}$ power density of laser irradiation. This aided in the silence of target RNA Polo-like kinase 1 by the siRNA in the cells. Paul et al.[72] formulated an injectable GO based hydrogel for sustainable and controlled release of vascular endothelial growth factor (VEGF) DNA to induce vasculogenesis and aid cardiac repair. PEI functionalized GO nanosheets loaded with DNA$_{VEGF}$ was incorporated in methacrylated gelatin (GelMA) and injected intramyocardially into rats with acute myocardial infarction. The transfected cardiomyocytes showed increased mitotic activity and an increase in myocardial capillary density in the infracted region due to VEGF release. This indicates a promising gene therapy application for ischemic heart disease.

## *Phototherapy*

Phototherapy is a rising approach to annihilate cancer cells by photo ablation. It is a non-invasive technique utilizing photosensitizer molecules that upon light exposure cause:

1) generation of heat (photothermal therapy/PTT) or
2) generation of reactive oxygen species (ROS) (photodynamic therapy/PDT). This is shown in Fig. 3.

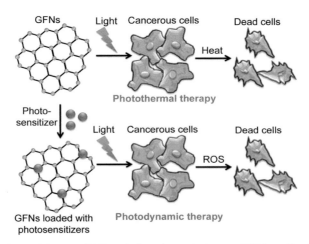

**Fig. 3.** Photothermal therapy (PTT) and photodynamic therapy (PDT) application of GFNs.

GFNs show excellent NIR absorption which aids in PTT. In PTT, laser exposure causes heat generation followed by photo ablation of the cells. In 2011, Robinson et al.,[73] demonstrated the ability of ultra small reduced graphene oxide (nano-rGO) to absorb at the NIR region. Nano-rGO tagged with RGD peptide sequence for targeted delivery showed efficient killing of U87MG cancer cells upon NIR irradiation. In the absence of irradiation these nanomaterial were cytocompatible. Yang et al.[74] utilized PTT for eradicating tumor in mice. 4T1 tumor bearing Balb/c were injected with PEGylated nanographene sheets (NGS). The strong NIR absorption of NGS was utilized for ablation of the tumor with evident decrease in tumor volume. Combining a Dox with PEGylated nanographene oxide (NGO-PEG-DOX) provided a synergistic effect of chemotherapy and PTT in eliminating cancer.[75]

PDT utilizes photosensitizer molecules that absorb light energy and transfer the energy to the oxygen present in tissue. This cause ROS generation and subsequently damages the cellular organelles and kills the cell. The large surface area of GFNs facilitates superior loading capacity of photosensitizer molecules for PDT. Huang et al.[76] formulated graphene oxide conjugated with folic acid and a photosensitizer molecule Ce6 (FA-GO-Ce6) for targeted PDT of cancer cells. MGC803 cancer cells were treated with FA-GO-Ce6 and photo irradiated using a laser. Upon laser irradiation Ce6 generated ROS and successfully killed the cancer cells by oxidative damage. Most of the photosensitizers presently used face the drawback of poor solubility, photostability and the inability to absorb light at the NIR region. Ge et al.[77] addressed the issue by using GQDs as photosensitizers for PDT. It was reported that GQDs showed the highest $^1O_2$ quantum yield and excellent photostability when compared to the conventional photosensitizers. Moreover GQDs exhibited broad absorption range between UV region and the visible region with a strong red emission.

## *Bioimaging*

Graphene oxide (GO) exhibits photoluminescence (PL) under wide range of wavelength from ultraviolet to near infrared (NIR). The intrinsic fluorescent property of graphene oxide (GO) was first studied by Dai's et al.[78] In their work, polyethylene glycol (PEG) functionalized GO conjugated with Rituxan antibody was developed that specifically target B cell lymphoma. These functionalized GO enable cellular imaging under NIR region with low background fluorescence. Similarly, GO nanoparticles conjugated with transferrin and PEG was synthesized for simultaneous targeting and imaging of cancer cells.[79] This surface functionalized GO nanoparticles exhibit excellent stability and strong two photon luminescence (TPL) for cancer imaging. The versatility and reliability of graphene based materials for bioimaging were further improved using multifunctional GO as contrast agent for multiple signaling technologies like photoacoustic and thermoacoustic tomography and magnetic resonance imaging. For example, fluorescein-o-methacrylate (FMA) labeled magnetic graphene were developed for *in vivo* imaging of zebrafish which showed excellent biocompatibility, co-localization and biodistribution from head to tail.[80] PEG functionalized reduced graphene oxide (rGO)-iron oxide nanoparticles hybrids were employed in multimodal imaging-guided photothermal therapy for cancer.[81] Besides this, GO can be used to deliver DNA/aptamer probes for *in vivo* imaging of biomolecules inside the cells.

The fluorescent property of graphene can also be used for the development of optical biosensor to detect various biomolecules such as ions, DNA, RNA and proteins. GO can protect the DNA to be delivered from hydrolysis by deoxyribonuclease (DNase). GO prevents the binding of DNase onto the surface by steric hindrance thereby protecting the DNA from cleavage. GO can be conjugated to different fluorescent probes for *in vivo* imaging. PEG functionalized nanographene sheets were labeled with Cy7 for *in vivo* fluorescence imaging. The efficient uptake of these nanographene sheets were noted in xenograft 4T1 murine breast cancer tumor, KB human epidermoid carcinoma tumor and U87MG human glioblastoma tumor mouse models.[82]

Zero dimensional graphene quantum dots (GQDs) have attracted tremendous attention in recent years and exhibits unique properties like quantum confinement and edge effects. Also it holds excellent stability, solubility in physiological conditions, low toxicity, tunable photoluminescence, ease of functionalization, up-conversion emission and resistant to photobleaching. GQDs were first synthesized by Pan et al. which are made of crystalline structure of single or few layered graphene.[83] GQDs possess strong optical absorption at 260–320 nm ($\pi$–$\pi$* transition of C=C bond) with shoulder peak around 270–390 nm (n–$\pi$* transition of C=O). PL emission of GQDs arises from the quantum confinement of conjugated $\pi$ electrons in sp$^2$ carbon network. GQDs offer several advantages in biomedical application compared to traditional semi-conductor quantum dots. Luminous emission of fluorescent graphene depends on size, defect and wavelength used. The fluorescent property of GQDs can be enhanced using doping or functionalization for cellular imaging (Fig. 4). Nitrogen-doped GQDs (N-GQDs) developed by Gong's group showed enhanced bright TPL fluorescent for cellular and deep-tissue imaging.[84] The small size of the GQDs brings minimal invasion on the cell membrane. Use of GQDs for cellular imaging drastically reduces the background noise since it has long fluorescent lifetime.

Fluorescent GQDs synthesized from chemical oxidation of graphite and supersonic isolation has lifetime of ~ 2 ns which is much longer than organic dyes. No obvious loss of fluorescence intensity was observed even after 30 min of excitation under 405 nm. Different types of cells such as HeLa, murine alveolar macrophage, human hepatic cancer cells, MCF-7 cells and stem cells including neurospheres cells, pancreas progenitor cells, cardiac progenitor cells and neural stem cells has been labeled using GQDs. GQDs synthesized from activated carbon using microwave purification and hydrothermal reaction was seems to be effectively taken up by the HeLa cells. The fluorescent images showed blue emitting GQDs inside HeLa cells under UV irradiation indicating the possible use of GQDs in optical imaging.[85] Similarly, GQDs obtained

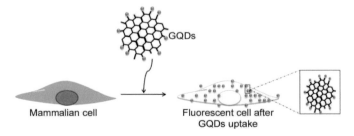

**Fig. 4.** Fluorescence emission of mammalian cells following GQDs uptake.

by hydrothermal cutting of GO, accumulated around the nucleus of HeLa cells without any nuclear localization signal and showed no decrease in PL intensity after continuous excitation. The presence of surface functional groups and the charge greatly influence the intracellular localization of GQDs. Poly l-lactide-Polyethylene glycol (PLA-PEG) grafted GQDs were synthesized for simultaneous imaging and therapy. The photoluminescent property exhibited by these GQDs was stable over a wide range of pH. In addition, this functionalized GQDs were seems to be effectively deliver miRNA probe for intracellular miRNA imaging and gene regulation. Fluorescein conjugated PEGylated GO was developed by Peng et al. which showed pH tunable fluorescent properties that can be used for intracellular imaging.[86] Here, PEG act as a bridge linking fluorescein with GO and also inhibit the fluorescence quenching property of GO. PEG increases stability, water dispersibility and cellular uptake of GQDs. For *in vivo* imaging, the fluorescent dye should be bright, biocompatible and emit fluorescence at longer wavelength. Up-conversion property of GQDs is desirable for *in vivo* application because of deep tissue penetration, bright fluorescence, resistant to photobleaching, molecular imaging with high spatial resolution, low background interference and low toxicity.

## Tissue engineering

Graphene holds a great promise in tissue engineering applications; it acts as an ideal scaffold for stem cells culture and maintenance. Maintenance of induced pluripotent stem cells (iPSCs) is a challenging task that hinders their potential application in regenerative medicines. It was stated that both graphene and GO helps in culture and differentiation of iPSCs. The different surface properties exhibited by graphene (hydrophobic) and GO (hydrophilic) activates different signal transduction pathway of iPSCs differentiation. Using graphene coated substrates iPSCs can be subcultured and expanded which evade the use of feeder cells for iPSCs maintenance. Cells grown on graphene showed similar attachment and proliferation capacity compared to cells grown on glass surfaces and maintained the cells in undifferentiated state. Cells grown on GO substrate showed increased adherence and faster proliferation and promote spontaneous differentiation towards ectodermal, mesodermal and endodermal lineages while graphene maintains the cells in undifferentiated states and suppresses endodermal lineage differentiation.[87] This discrepancy in the differentiation pattern might be due to the difference in surface groups on G and GO. iPSCs differentiation towards endodermal lineages can be achieved by culturing cells in 3D porous GO coated scaffolds. The development of graphene scaffold owns great promises for cell replacement therapy in future.

Graphene acts as a substrate for the attachment of various growth factors, proteins and secretory molecules that aid in cell-cell communication. Graphene can bind bone morphogenetic protein (BMP), trypsin, heparin, lysozyme and peptides via non-covalent interactions particularly $\pi$–$\pi$ interaction. Graphene promotes osteogenic differentiation of mesenchymal stem cells (Fig. 5) by binding dexamethasone and $\beta$-glycerophosphate. Even in the absence of growth factors such as BMP, graphene has been found to enhance osteogenic differentiation with increase calcium deposits in the extracellular matrix.[88] The ripples and wrinkles found on the graphene sheets

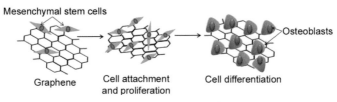

**Fig. 5.** Attachment, proliferation and differentiation of stem cells in graphene scaffold.

helps in protein adsorption, cell adhesion and differentiation. Graphene and GO can adsorb up to 8% and 25% of serum proteins via hydrophobic or electrostatic interactions. GO possess high affinity for insulin that helps in adipogenic differentiation whereas graphene denatures insulin due to π–π stacking. Pluronic coated GO-CaP nanocomposite was shown to possess synergistic osteoinduction potential on human mesenchymal stem cells and also increases the calcium deposits. Addition of GO into CaP, increases the material stiffness which causes mechanotransduction effect to regulate stem cells fate.[89] Similarly, calcium silicate/rGO composite was found to increase the proliferation and alkaline phosphatase (ALP) activity of human osteoblast cells which promises their potential application in bone tissue engineering.

Good electrical conductivity owned by graphene makes them an ideal platform for neural culture. Graphene/polyethylene terephthalate (PET) film was found to increases neural cell-cell coupling under electrical stimulation.[90] The cells also overexpressed F-actin promoting the formation of filopodia. Neural cells cultured on graphene substrate showed extended neurite growth, increased viability and enhanced expression of growth-associate protein-43 (GAP43), a factor for neurite sprouting and outgrowth. Fluorinated graphene induces the differentiation of MSCs towards neural lineages by forming extended cytoskeleton and nucleus elongation. Neuronal proteins such as Tuj1 and MAP2 were highly expressed in MSCs cultured on fluorinated graphene.[91]

### *Antibacterial property*

There are series of investigations reported on the antimicrobial property of graphene derivatives. Size, surface charge, surface chemistry, lipid solubility, density of the functional groups and conductivity are the key factors that control the antibacterial property of graphene materials. The edges of the graphene have significant roughness while the basal plane is smooth; the rough surface promotes antibacterial activity. It was also proposed that the orientation of graphene surface determines how it interacts with the bacteria. The high aspect ratio possessed by both graphene and GO either promotes or inhibit bacterial attachment. Among the graphene family materials, graphite (Gt) and graphite oxide (GtO) were less active against *E. coli* whereas GO and rGO retains antibacterial activity.[92] Time and dose dependent loss of viability was noticed in culture exposed to both GO and rGO. Owing to their high conductive nature, both rGO and Gt cause higher glutathione (GSH) oxidation compared to GtO and GO. Gram negative bacteria were least susceptible compared to Gram positive bacteria because of the presence of outer layer. For gram negative bacteria such as *E. coli* and *S. typhimurium*, the MIC (minimum inhibitory concentration) of graphene was found to be 1 µg/ml and 4–8 µg/ml for Gram positive bacteria *B. subtilis*.[93] Physical interaction of graphene with the bacterial cell membrane leads to irreversible membrane damage. The sharp edges of graphene penetrate the cell membrane mechanism known as insertion mode or penetration mode which causes membrane stress resulting in leakage of intracellular cytoplasmic contents. The strong interaction and better charge transfer between graphene and bacteria enhances their susceptibility. Since graphene is a good electron acceptor, it also induces cell death in bacteria by acting as an electron acceptor. The continuous removal of electrons from the bacterial cell membrane affects the respiratory chain causing depletion of adenosine triphosphate (ATP).

It was also stated that the larger sheets affects the growth of the bacteria compared to the smaller sheets. Because of the larger size these sheets can individually wrap bacteria thereby preventing their proliferation. A three step mechanism of cytotoxicity was proposed for antibacterial activity of graphene: first the cells deposit on the graphene based materials; second the sharp edges of the graphene nanosheets act as cutters on the cell membrane leading to membrane leakage and release of intracellular contents. Third, it also oxidizes cellular components such as membrane lipids, proteins, thiol containing GSH via ROS independent mechanism of oxidative stress. Graphene nanosheets induce increased ROS formation inside the cells which oxidizes cellular components such as lipid, protein and DNA. The fatty acid oxidation is a chain reaction that results in lipid peroxides formation which in turn causes disintegration of cell membrane. For example, dose and time dependent cell death was observed in *P. aeruginosa* exposed to GO and rGO. Both GO and rGO induced increased superoxide radical formation which in turn causes DNA fragmentation leading to cell death.[94] The below Fig. 6 shows the different mechanism of graphene toxicity towards bacteria.

Surface modification of graphene oxide with silver nanoparticles was employed to increase their antibacterial efficiency. It was found that the very low concentration (10 ppm) of the modified GO reduced the bacterial growth to 99.9%.[95] While most of the studies unravel the antibacterial efficacy of graphene materials some study in contrast to the above results noticed that GO supports the growth of *E. coli*. The oxygen containing surfaces of the basal plane provides wettability which in turn helps in stronger interaction of particle surface with the bacterial lipid membrane thereby helping bacterial adhesion and growth.[96]

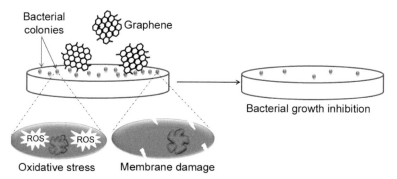

**Fig. 6.** Graphene induced oxidative stress and membrane damage inhibits bacterial growth resulting in less number of colonies.

## *Toxicity*

Both GO and rGO were known to induce dose dependent toxic responses such as cytotoxicity, DNA damage and oxidative stress in mammalian cells. However, the cellular responses and the consequences differs because of the distinct mechanism of toxicity provoked by GO and rGO. Increased ROS formation from hydrophilic GO exposure occurs through activation of nicotinamide adenine dinucleotide phosphate (NADPH) oxidase resulting in deregulation of antioxidant/DNA repair/apoptosis related genes. Transforming growth factor beta 1 (TGF-β1) mediated signaling pathway plays the central role in GO induced toxicological effect. TGF-β activates Smad proteins that regulate apoptosis by activating several pro apoptotic factors of Bcl2 family. rGO being hydrophobic was found to adsorbed at cell surface without internalization causes physical damage to the cell membrane with its sharp edges. rGO seems to activate toll like receptors-nuclear factor kappa beta (TLR4-NFκB) pathway. ROS generation induced by rGO showed poor gene regulation. Because of tendency to aggregates on cell surface, graphene can limits the availability of nutrients to the cells and also increases ROS generation.[97]

Graphene can enter into the cell either by penetrating the lipid membrane or via endocytic pathway and are distributed in the cytoplasm and absence in nucleus. The internalized GO was found to be mainly concentrated in lysosomes, mitochondria and endoplasm. The amount of intracellular uptake increases with dose and time period. Cells overcome the tensile strength due to graphene penetration by forming thickened intermediate filaments. Once inside the cell graphene influences various cellular functions or enclosed within vesicles to prevent interaction with the intracellular organelles. Graphene can affect the cellular adhesion by decreasing the expression of cell attachment proteins such as laminin, fibronectin, focal adhesion kinase (FAK).[98] Graphene induces apoptotic morphology like membrane vesicles formation, fragmentation and unclear cell boundary. Accumulation of graphene inside the intracellular organelles also interferes with the cellular metabolism. Dose and time dependent increase in ROS formation was observed when HepG2 cells were exposed to GO and graphene nanoplatelets. GO can act as an electron donor and participate in electron transport chain increasing mitochondrial respiration which in turn leads to increased ROS generation. It also causes mitochondrial depolarization by affecting the structural and functional integrity of mitochondria. Failure to overcome the oxidative stress or decreased activity of antioxidant enzymes guides the cell to undergo apoptosis. Several pro-apoptotic members such as *Bim*, *Bax* of Bcl2 family will be released from the depolarized mitochondria into the cytosol activating caspases.

In macrophages, graphene induces cell death by activating mitogen activated protein (MAP) kinase and TGF pathways. It also induces the release of several proinflammatory cytokines such as interleukins (IL-1α, IL-6, IL-10, tumor necrosis factor-α (TNF-α) and granulocyte macrophage colony stimulating factor (GM-CSF) and chemokines (macrophage inflammatory protein (MIP-1α, MIP-1β) by activating NFκB and TLR mediated pathway. Expression of mRNA encoding proinflammatory genes IL-1β, nitric oxide synthase (iNOS) and cyclooxygenase (COX-2) were reported to be increased upon graphene exposure. It also disrupts F actin filaments and induces membrane ruffling.[99] It disrupts podosomes, F-actin-rich integrin-based adhesions localized on macrophage protrusion which is responsible for cell adhesion. Graphene is also known to induce autophagy through TLR associated pathway. TLR 4 and TLR 9 are the two receptors that are significantly upregulated upon GO exposure in turn activates MyD88, TRAF-6 which promotes the formation of autophagosomes.[100] Bectin-1 and LC3-II are the markers of autophagy that are highly expressed in cells exposed to GO.

Graphene when enters into the body it comes in contact with the red blood corpuscles (RBCs) present in the circulatory system. The biocompatibility of nanoparticles with the blood cells is determined by hemolysis. Factors such as surface charge, oxygen content and particle size determines graphene haemocompatibility. Hemolytic activity exhibited by GO is due to the strong electrostatic interaction between its negatively charged oxygen groups and positively charged phosphotidylcholine on the surface of RBC membrane. Due to the presence of fewer amounts of oxygen groups, aggregated graphene sheets can induce haemagglutinin. It is suggested that GO can be surface passivated using chitosan that greatly reduce its hemolytic potential by masking the electrostatic interaction.[101] It is also reported that GO can induce platelet aggregation by enhancing the binding of fibrinogen to the integrins receptors on the cell surface. Adenosine diphosphate (ADP) andthromboxane A2 released by the activated platelets recruit nearby platelets causing aggregation.[102]

*In vitro* mutagenesis study showed GO can effectively interacts with genomic DNA and causes DNA damage in a concentration dependent manner resulting in base transition and base deletion. Also the expression pattern of the genes that plays key role in DNA damage control, cell cycle, metabolism and apoptosis seems to be affected. For example, GO exposure increases the expression of ATM and Rad51 at initial stages of DNA damage. Progression of cell cycle is arrested by inhibiting S phase and G0/G1 phase. In addition because of its planar structure, GO can intercalate between the bases and affect the flow of genetic information.

Toxicity studies using *in vivo* animal models showed that dose and surface modification are the two main factors that influence graphene biodistribution and toxicity. Animals administered with GO appeared physically inactive and death results due to the mechanical blockage of airways in lungs leading to suffocation, fibrosis and peribronchial inflammation. In addition, large number of inflammatory cells (Neutrophils, acidophiles, epithelioid granulomas and multinuclear giant cells) was seemed to be infiltered in lung alveolus interstitium which increases the thickness of the alveolar septa and cracks in alveoli resulting in leakage of proteins.[103] Excretion of GO occurs mainly through bile and very least amount is expelled out by kidney. GO can also increases the risk of cardiac arrest due to increase in plasma thrombin and anti-thrombin complexes. Surface functionalization of nanomaterials with polymers such as PEG increases its biocompatibility and blood circulation time. Graphene were seems to be accumulated inside the body even after a month following administration suggesting their long term retention in the body.

[125]I radiolabeled PEG functionalized GO and rGO were intragastrically given to female BALB/c mice, high level of radioactivity was observed in stomach and intestine after 4 h which decreases 1 day post feeding suggesting these materials are not effectively absorbed by the digestive system and are completely excreted after oral feeding. Unlike the oral feeding, intraperitoneal injection resulted in high

accumulation of graphene nanoparticles on liver and spleen.[104] The macrophages play a very important role in the uptake of functionalized graphene nanoparticles via reticuloendothelial system. GO is also shown to induce developmental toxicity in zebra fish embryo. It delays embryo hatching, induces morphological defects such as bent spine, tail malformations, etc.[105] GO functionalized with PEG/PLL (PEGylated poly L lysine) showed increased ROS production under normal physiological conditions in *Caenorhabiditis elegans*. GO induces the formation of more ·OH radical from $H_2O_2$. Release of cytC from the mitochondria and electron transfer between cytC/$H_2O_2$ affects the antioxidant defense system leading to toxicity in worms.

## Conclusion

GFNs are a promising class of nanomaterials with myriad application in electronics, storage devices, sensors and biomedicine. Novel and facile synthesis methods for GFNs production are being developed. The innumerous biomedical applications of GFNs have opened new horizons for treatment of cancer and other incapacitating diseases, tissue engineering, biosensing and bioimaging. However numerous challenges in scaling up have largely affected the commercialization of these nanomaterials. Due to the rapid growth in production and use of graphene in biomedicine, there is an increased risk of human exposure which is unavoidable. The risk posed by humans upon exposure remains unexplored. Human can be exposed to graphene during manufacturing (occupational risk) or in biomedical application (intentional exposure). It is difficult to outline appropriate exposure scenarios resulting in significant toxicity. Hence, these 2D nanomaterials must be subjected to regular risk assessment and toxicity investigation before consumption.

## Acknowledgement

The authors wish to express their thanks to the Director and Head, Biomedical Technology Wing, Sree Chitra Tirunal Institute for Medical Sciences and Technology, Thiruvananthapuram, Kerala, India for their support and for providing the infrastructure to carry out this work.

## References

1.  M. I. Katsnelson, *Mater. Today*, 2007, **10**, 20–27.
2.  K. S. Novoselov, A. K. Geim, S. V. Morozov, D. Jiang, Y. Zhang, S. V. Dubonos, I. V. Grigorieva and A. A. Firsov, *Science,* 2004, **306**, 666–9.
3.  K. I. Bolotin, K. J. Sikes, Z. Jiang, M. Klima, G. Fudenberg, J. Hone, P. Kim and H. L. Stormer, *Solid State Commun.*, 2008, **146**, 351.
4.  C. Lee, X. Wei, J. W. Kysar and J. Hone, *Science*, 2008, **321**, 385–388.
5.  A. A. Balandin, S. Ghosh, W. Z. Bao, I. Calizo, D. Teweldebrhan, F. Miao and C. N. Lau, *Nano Lett.*, 2008, **8**, 902–907.
6.  Y. Zhu, S. Murali, W. Cai, X. Li, J. W. Suk, J. R. Potts and R. S. Ruoff, *Adv. Mater.*, 2010, **22**, 3906–3924.
7.  Y. Zhu, S. Murali, M. D. Stoller, A. Velamakanni, R. D. Piner and R. S. Ruoff. *Carbon*, 2010, **48**, 2118–2122.

8.  X. Z. Yu, C. G. Hwang, C. M. Jozwiak, A. Köhl, A. K. Schmid and A. Lanzara, *J. Electron. Spectrosc. Relat. Phenom.*, 2010, **184**, 100–106.
9.  H. Tetlow, J. Posthuma de Boer, I. J. Forde, D. D. Vvedensky, J. Corauxc and L. Kantorovich, *Phys. Rep.*, 2014, **542**, 195–295.
10. K. S. Kim, Y. Zhao, H. Jang, S. Y. Lee, J. M. Kim, K. S. Kim, J. H. Ahn, P. Kim, J. Y. Choi and B. H. Hong, *Nature,* 2009, **457**, 706–710.
11. Q. Zheng and J.-K. Kim, *Graphene for Transparent Conductors.* pp. 29–94. *In*: Q. Zheng and J.-K. Kim (ed.). Springer, New York, 2015, vol. 23, ch. 2.
12. S. Haar, *Sci. Rep.*, 2015, **5**, 16684.
13. A. Dato, V. Radmilovic, Z. Lee, J. Phillips and M. Frenklach, *Nano Lett.*, 2008, **8**, 2012–2016.
14. A. G. Cano-Márquez, F. J. Rodríguez-Macías, J. Campos-Delgado, C. G. Espinosa-González, F. Trist´an-L´opez, D. Ram´ıre-Gonz´alez, D. A. Cullen, D. J. Smith, M. Terrones and Y. I. Vega-Cantú, *Nano Lett.,* 2009, **9**, 1527–1533.
15. L. Jiao, L. Zhang, X. Wang, G. Diankov and H. Dai, *Nature*, 2009, **458**, 877.
16. D. V. Kosynkin, A. L. Higginbotham, A. Sinitskii, J. R. Lomeda, A. Dimiev, B. K. Price and J. M. Tour, *Nature*, 2009, **458**, 872–876.
17. S. V. Morozov, K. S. Novoselov, F. Schedin, D. Jiang, A. A. Firsov and A. K. Geim, *Phys. Rev. B*, 2005, **72**, 201401.
18. V. C. Sanchez, A. Jachak, R. H. Hurt and A. B. Kane, *Chem. Res. Toxicol.*, 2012, **25**, 15–34.
19. J. Wang, K. K. Manga, Q. Bao and K. P. Loh, *J. Am. Chem. Soc.*, 2011, **133**, 8888–8891.
20. H. J. Park, J. Meyer, S. Roth and V. Skákalová, *Carbon*, 2010, **48**, 1088–1094.
21. J. Shang, F. Xue and E. Ding, *Chem. Commun.*, 2015, **51**, 15811–15814.
22. V. León, A. M. Rodriguez, P. Prieto, M. Prato and E.Vázquez, *ACS Nano*, 2014, **8**, 563–571.
23. D. R. Dreyer, S. Park, C. W. Bielawski and R. S. Ruoff, *Chem. Soc. Rev.*, 2010, **39**, 228–240.
24. A. Lerf, H. He, M. Forster and J. Klinowski. *J. Phys. Chem. B.*, 1998, **102**, 4477–4482.
25. B. C. Brodie, *Trans. R. Soc. London*, 1859, **149**, 249–259.
26. L. Staudenmaier, *Ber. Dtsch Chem. Ges.*, 1898, **31**, 1481–1487.
27. W. S. Hummers and R. E. Offeman, *J. Am. Chem. Soc.*, 1958, **80**, 1339.
28. N. I. Kovtyukhova, P. J. Ollivier, B. R. Martin, T. E. Mallouk, S. A. Chizhik, E. V. Buzaneva and A. D. Gorchinskiy, *Chem. Mater.*, 1999, **11**, 771–778.
29. D. C. Marcano, D. V. Kosynkin, J. M. Berlin, A. Sinitskii, Z. Sun, A. Slesarev, L. B. Alemany, W. Lu and J. M. Tour, *ACS Nano*, 2010, **4**, 4806–4814.
30. S. Pei and H.-M. Cheng, *Carbon*, 2012, **50**, 3210–3228.
31. M. J. McAllister, J.-L. Li, D. H. Adamson, H. C. Schniepp, A. A. Abdala, J. Liu, M. Herrera-Alonso, D. L. Milius, R. Car, R. K. Prud'homme and I. A. Aksay, *Chem. Mater.*, 2007, **19**, 4396–4404.
32. L. J. Cote, R. Cruz-Silva and J. Huang, *J. Am. Chem. Soc.,* 2009, **131**, 11027–11032.
33. H. C. Schniepp, J.-L. Li, M. J. McAllister, H. Sai, M. Herrera-Alonso, D. H. Adamson, R. K. Prud'homme, R. Car, D. A. Saville and I. A. Aksay, *J. Phys. Chem. B*, 2006, **110**, 8535–8539.
34. S. Stankovich, D. A. Dikin, R. D. Piner, K. A. Kohlhaas, A. Kleinhammes, Y. Jia, Y. Wuc, S. T. Nguyen and R. S. Ruoff, *Carbon*, 2007, **45**, 1558–1565.
35. H.-J. Shin, K. K. Kim, A. Benayad, S.-M. Yoon, H. K. Park, I.-S. Jung, M. H. Jin, H.-K. Jeong, J. M. Kim, J.-Y. Choi and Y. H. Lee, *Adv. Funct. Mater.,* 2009, **19**, 1987–1992.
36. I. K. Moon, J. Lee, R. S. Ruoff and H. Lee, *Nat. Commun.*, 2010, **1**, 73.
37. M. J. Fernandez-Merino, L. Guardia, J. I. Paredes, S. Villar-Rodil, P. Solis-Fernandez, A. Martinez-Alonso and J. M. D. Tascón, *J. Phys. Chem. C.,* 2010, **114**, 6426–6432.
38. J. Zhang, H. Yang, G. Shen, P. Cheng, J. Zhang and S. Guo, *Chem. Commun. (Camb).*, 2010, **46**, 1112–1114.
39. G. Wang, J. Yang, J. Park, X. Gou, B. Wang, H. Liu and J. Yao, *J. Phys. Chem. C*, 2008, **112**, 8192–8195.
40. A. Ambrosi, C. K. Chua, A. Bonanni and M. Pumera, *Chem. Mater.*, 2012, **24**, 2292–2298.
41. Y. Liu, Y. Li, Y. Yang, Y. Wen and M. Wang, *J. Nanosci. Nanotechnol.*, 2011, **11**, 10082–10086.
42. X. Zhou, J. Zhang, H. Wu, H. Yang, J. Zhang and S. Guo, *J. Phys. Chem. C*, 2011, **115**, 11957–11961.
43. E. C. Salas, Z. Sun, A. Luttge and J. M. Tour, *ACS Nano*, 2010, **4**, 4852–4856.
44. O. Akhavan and E. Ghaderi, *Carbon*, 2012, **50**, 1853–1860.
45. S. Gurunathan, J. W. Han, V. Eppakayala and J. H. Kim, *Colloids Surf. B Biointerfaces*, 2013, **102**, 772–777.

46. S. Raveendran, N. Chauhan, Y. Nakajima, H. Toshiaki, S. Kurosu, Y. Tanizawa, R. Tero, Y. Yoshida, T. Hanajiri, T. Maekawa, P. M. Ajayan, A. Sandhu and D. S. Kumar, *Part. Part. Syst. Char.*, 2013, **30**, 573–578.
47. Y. Wang, Z. Shi and J. Yin, *ACS Appl. Mater. Interfaces*, 2011, **3**, 1127–1133.
48. B. Haghighi and M. A. Tabrizi, *RSC Adv.*, 2013, **3**, 13365–13371.
49. O. Akhavan, E. Ghaderi, E. Abouei, S. Hatamie and E. Ghasemi, *Carbon*, 2014, **66**, 395–406.
50. M. Khan, A. H. Al-Marri, M. Khan, M. R. Shaik, N. Mohri, S. F. Adil, M. Kuniyil, H. Z. Alkhathlan, A. Al-Warthan, W. Tremel, M. N. Tahir and M. R. H. Siddiqui, *Nano Res. Lett.*, 2015, **10**, 1–9.
51. G. Lee and B. S. Kim, *Biotechnol. Prog.*, 2014, **30**: 463–469.
52. T. Kuila, S. Bose, P. Khanra, A. K. Mishra, N. H. Kim and J. H. Lee, *Carbon*, 2012, **50**, 914–921.
53. D. Voiry, J. Yang, J. Kupferberg, R. Fullon, C. Lee, H. Y. Jeong, H. S. Shin and M. Chhowalla, *Science*, 2016, **354**, 1413–1416.
54. Y. Zhu, G. Wang, H. Jiang, L. Chen and X. Zhang, *Chem. Commun.*, 2015, **51**, 948–951.
55. R. Liu, D. Wu, X. Feng and K. Mullen, *J. Am. Chem. Soc.*, 2011, **133**, 15221–15223.
56. M. Favaro, L. Ferrighi, G. Fazio, L. Colazzo, C. D. Valentin, C. Durante, F. Sedona, A. Gennaro, S. Agnoli and G. Granozzi, *ACS Catal.*, 2015, **5**, 129–144.
57. D. Qu, M. Zheng, L. Zhang, H. Zhao, Z. Xie, X. Jing, R. E. Haddad, H. Fan and Z. Sun, *Sci. Rep.*, 2014, **4**, 5294.
58. F. Liu, M.-H. Jang, H. D. Ha, J.-H. Kim, Y.-H. Cho and T. S. Seo, *Adv. Mater.*, 2013, **25**, 3657–3662.
59. J. Liu, L. Cui and D. Losic, *Acta Biomater.*, 2013, **9**, 9243–9257.
60. X. Yang, X. Zhang, Z. Liu, Y. Ma, Y. Huang and Y. Chen, *J. Phys. Chem. C*, 2008, **112**, 17554–17558.
61. K. S. Novoselov, V. I. Fal'ko, L. Colombo, P. R. Gellert, M. G. Schwab and K. Kim, *Nature*, 2012, **490**, 192–200.
62. W. Miao, G. Shim, S. Lee, S. Lee, Y. S. Choe and Y.-K. Oh, *Biomaterials*, 2013, **34**, 3402–3410.
63. C. Wang, C. Wu, X. Zhou, T. Han, X. Xin, J. Wu, J. Zhang and S. Guo, *Sci. Rep.*, 2013, **3**, 2852.
64. L. Zhang, J. Xia, Q. Zhao, L. Liu and Z. Zhang, *Small*, 2009, **6**, 537–544.
65. G. Liu, H. Shen, J. Mao, L. Zhang, Z. Jiang, T. Sun, Q. Lan and Z. Zhang, *ACS Appl. Mater. Interfaces*, 2013, **5**, 6909–6914.
66. D. Depan, J. Shah and R. D. K. Misra, *Mater. Sci. Eng. C.*, 2011, **31**, 1305–1312.
67. H. Kim, R. Namgung, K. Singha, I. K. Oh and W. J. Kim, *Bioconjug. Chem.*, 2011, **22**, 2558–2567.
68. C. H. Lu, C. L. Zhu, J. Li, J. J. Liu, X. Chen and H. H. Yang, *Chem. Commun. (Camb)*, 2010, **46**, 3116–3118.
69. L. Feng, X. Yang, X. Shi, X. Tan, R. Peng, J. Wang and Z. Liu, *Small*, 2013, **9**, 1989–1997.
70. H. Bao, Y. Pan, Y. Ping, N. G. Sahoo, T. Wu, L. Li, J. Li and L. H. Gan, *Small*, 2011, **7**, 1569–1578.
71. H. Kim, R. Namgung, K. Singha, I.-K. Oh and W. J. Kim, *Bioconjug. Chem.*, 2011, **22**, 2558–2567.
72. A. Paul, A. Hasan, H. A. Kindi, A. K. Gaharwar, V. T. S. Rao, M. Nikkhah, S. R. Shin, D. Krafft, M. R. Dokmeci, D. Shum-Tim and A. Khademhosseini, *ACS Nano*, 2014, **8**, 8050–8062.
73. J. T. Robinson, S. M. Tabakman, Y. Liang, H. Wang, H. S. Casalongue, D. Vinh and H. Dai, *J. Am. Chem. Soc.*, 2011, **133**, 6825–6831.
74. K. Yang, S. Zhang, G. Zhang, X. Sun, S.-T. Lee and Z. Liu, *Nano Lett.*, 2010, **10**, 3318–23.
75. W. Zhang, Z. Guo, D. Huang, Z. Liu, X. Guo and H. Zhong. *Biomaterials*, 2011, **32**, 8555–8561.
76. P. Huang, C. Xu, J. Lin, C. Wang, X. Wang, C. Zhang, X. Zhou, S. Guo and D. Cui. *Theranostics*, 2011, **1**, 240–250.
77. J. Ge, M. Lan, B. Zhou, W. Liu, L. Guo, H. Wang, Q. Jia, G. Niu, X. Huang, H. Zhou, X. Meng, P. Wang, C.-S. Lee, W. Zhang and X. Han. *Nat. Commun.*, 2014, **5**, 4596.
78. X. Sun, Z. Liu, K. Welsher, J. Robinson, A. Goodwin, S. Zaric and H. Dai, *Nano Res.*, 2008, **1**, 203–212.
79. J. L. Li, H. C. Bao, X. L. Hou, L. Sun, X. G. Wang and M. Gu, *Angew. Chem., Int. Ed.*, 2012, **51**, 1830–1834.
80. G. Gollavelli and Y. C. Ling, *Biomaterials*, 2012, **33**, 2532–2545.
81. K. Yang, L. Hu, X. Ma, S. Ye, L. Cheng, X. Shi, C. Li, Y. Li and Z. Liu, *Adv. Mater.*, 2012, **24**, 1868–1872.
82. K. Yang, S. Zhang, G. Zhang, X. Sun, S. T. Lee and Z. Liu, *Nano Lett.*, 2010, **9**, 3318–3323.
83. D. Pan, J. Zhang, Z. Li and M. Wu, *Adv. Mater.*, 2010, **22**, 734–738.
84. Q. Liu, B. Guo, Z. Rao, B. Zhang and J. R. Gong, *Nano Lett.*, 2013, **13**, 2436–2441.
85. Y. Huang, C. Bai, K. Cao, Y. Tian, Y. Luo, C. Xia, S. Ding, Y. Jin, L. Ma and S. Li, *RSC Adv.*, 2014, **4**, 43160–43165.

86.  C. Peng, W. Hu, Y. Zhou, C. Fan and Q. Huang, *Small*, 2010, **6**, 1686–1692.
87.  G. Y. Chen, D. W. P. Pang, S. M. Hwang, H. Y. Tuan and Y. C. Hu, *Biomaterials*, 2012, **33**, 418–427.
88.  T. R. Nayak, H. Andersen, V. S. Makam, C. Khaw, S. Bae, X. Xu, P. L. R. Ee, J. H. Ahn, B. H. Hong, G. Pastorin and B. Ozyilmaz, *ACS Nano*, 2011, **5**, 4670–4678.
89.  R. Tatavarty, H. Ding, G. Lu, R. J. Taylor and X. Bi, *Chem. Commun.*, 2014, **50**, 8484–8487.
90.  C. Heo, J. Yoo, S. Lee, A. Jo, S. Jung, H. Yoo, Y. H. Lee and M. Suh, *Biomaterials*, 2011, **32**, 19–27.
91.  S. Ryu and B. Kim, *Tissue Engineering and Regenerative Medicine*, 2013, **10**, 39–46.
92.  S. Liu, T. Zeng, M. Hofmann, E. Burcombe, J. Wei, R. Jiang, J. Kong and Y. Chen, *ACS Nano*, 2011, **5**, 6971–6980.
93.  K. Krishnamoorthy, M. Veerapandian, L. Zhang, K. Yun and S. Kim, *J. Phys. Chem. C.*, 2012, **116**, 17280–17287.
94.  S. Gurunathan, J. W. Han, A. A. Dayem, V. Eppakayala and J. Kim, *Int. J. Nanomedicine*, 2012, **7**, 5901–5914.
95.  D. Zhang, X. Liu and X. Wang, *J. Inorg. Biochem.*, 2011, **105**, 1181–1186.
96.  O. N. Ruiz, K. A. S. Fernando, B. Wang, N. A. Brown, P. G. Luo, N. D. McNamara, M. Vangsness, Y. P. Sun and C. E. Bunker, *ACS Nano*, 2011, **5**, 8100–8107.
97.  N. Chatterjee, H. Eom and J. Choi, *Biomaterials*, 2014, **35**, 1109–1127.
98.  S. Syama and P. V. Mohanan, *Int. J. Biol. Macromol.*, 2016, **86**, 546–555.
99.  H. Zhou, K. Zhao, W. Li, N. Yang, Y. Liu, C. Chen and T. Wei, *Biomaterials*, 2012, **33**, 6933–6942.
100.  G. Chen, H. Yang, C. Lu, Y. Chao, S. Hwang, C. Chen, K. Lo, L. Sung, W. Luo, H. Tuan and Y. Hu, *Biomaterials*, 2012, **33**, 6559–6569.
101.  K. Liao, Y. Lin, C. Macosko and C. Haynes, *ACS Appl. Mater. Interfaces*, 2011, **3**, 2607–2615.
102.  S. Singh, M. Singh, M. Nayak, S. Kumari, S. Shrivastava, J. Grácio and D. Dash, *ACS Nano*, 2011, **5**, 4987–4996.
103.  K. Wang, J. Ruan, H. Song, J. Zhang, Y. Wo, S. Guo and D. Cui, *Nanoscale Res. Lett.*, 2010, **6**, 8.
104.  K. Yang, J. Wan, S. Zhang, Y. Zhang, S. Lee and Z. Liu, *ACS Nano*, 2011, **5**, 516–522.
105.  L. Chen, P. Hu, L. Zhang, S. Huang, L. Luo and C. Huang, *Sci. China Chem.*, 2012, **55**, 2209–2216.

# CHAPTER 3

# Overview of Biological Interactions and Responses to 2D-nanomaterials

*Annette von dem Bussche*[1,*] and *Martin Winkler*[2]

Two dimensional nanomaterials constitute a large and diverse family of materials, characterized by a distinct planar morphology, unique physicochemical properties, significant electronic and thermal conductivity as well as interesting photoactive and catalytic features.[1-4]

The taxonomy of the described classes of 2D-nanomaterials is outlined in Table 1.

Synthesis and characterization of advanced 2D-materials are described in Chapter 1 and have been reviewed previously.[1,3,5,6]

This chapter will focus on the biological interactions and responses to graphene, layered transition metal dichalcogenides and transition metal dioxides as examples. These specific examples provide a mechanistic framework for understanding the physical and chemical properties responsible for the toxicity of 2D (two dimensional) nanomaterials.

Possible routes of exposure and potentially adverse health impacts will be discussed; followed by chemical transformation of 2D-nanomaterials . Physicochemical properties related to toxicity, the mechanism of toxicity and implications for a safer design will conclude this chapter.

[1] Pathology and Laboratory Medicine, Brown University, 70 Ship Street, Providence, RI, 02903, USA.
[2] AG Photobiotechnologie, Lehrstuhl Biochemie der Pflanzen, Ruhr-Universität Bochum, Universitätsstrasse 150, D-44801 Bochum, Germany.
Email: martin.winkler-2@rub.de
* Corresponding author: Annette_von_dem_Bussche@brown.edu

**Table 1.** Existing 2D nanomaterials and examples.

| 2D-material-family | Examples |
|---|---|
| Graphene | pristine graphene |
| | graphene oxide (GO) |
| | reduced GO (rGO) |
| Graphitic C3N4 | |
| Transition metal dichalcogenides (TMDs) | $MoS_2$, $WS_2$, $WeS_2$, $VS_2$, $VSe_2$, $VTe_2$, $TiS_2$, $ZrS_2$, $WS_2$, $GeS_2$, $HFS_2$, $ReS_2$, $PtS_2$, $SnSe_2$, $TaS_2$, $MoSe_2$, $WSe_2$, $TESe_2$, WSes, TcSe2, $TiTe_3$, $ZrTe_3$, $VTe_2$, $NbTe_2$, $TaTe_2$, $MoTe_2$, $WTe_2$, $CoTe_2$, $RhTe_2$, $IrTe_2$, $NiTe_2$, $PdTe_2$, $PtTe_2$, $CuSbTe_2$, $CuBiS_2$, $PdS_2$, ReSSe, $SbOS_2$, $TeS_2$, $SiS_2$, PdTe, TaSeTe, $CuTe_2$, SiTe, $CuSbS_2$, Pds, $Sb_2Os$ |
| Transition metal oxides (TMOs) | $MoO_3$, $MnO_2$, PbO, SnO, $NB_2O_3$, $V_2O_3$ |
| Hexagonal boron nitride | hBN |
| 2D clay materials | Bentonite, Egyptan Blue, Cariopilite, Greenalite, Amesite, Chlorite, Nepourite |
| Ultrathin black phosphorus | |

## Routes of Exposure and Potential Adverse Health Impacts

Two dimensional materials have several properties that are attractive for biomedical applications, including biosensing, drug delivery, bioimaging,[1–4,7,8] and tissue engineering.[9–14] Their large surface area and ease of chemical functionalization and mechanical strength render these emerging materials attractive for biomedical applications.[7,8,15–18] Direct implantation of biosensors or bioengineered tissues based on 2D-nanomaterials represents a feasible and promising approach for biomedical therapies, although the potential risk of large smooth 2D surfaces to induce foreign body carcinogenesis has not been investigated.[19] The main routes of human exposure to engineered nanomaterials are inhalation, ingestion, dermal contact and direct injection (Fig. 1A). In this part, common uptake routes are described, which have been mostly studied with zero and one dimensional (1D) particles. Nevertheless, if the particles exhibit similar physico-chemical properties, similar uptake mechanisms can be expected for 2D-nanomaterials . In the following paragraph uptake routes will be introduced as described for graphene that will provide information about the biodistribution of 2D-nanomaterials .

Inhalation is the most common type of exposure for workers involved in the synthesis, processing and fabrication of 2D-nanomaterials . During synthesis, graphene nanoparticles can be easily inhaled as dry powder or liquid aerosols.[19–22]

Following inhalation, 2D particles with lateral dimensions between 0.5 and 25 μm can deposit throughout the human respiratory tract, including nasal, tracheobronchial and alveolar region.[22] In the conducting airways, nanomaterials covered with mucus are moved by the cilia of the bronchial epithelial cells away from the lung into the pharynx; this is called the mucociliary escalator. From there they can be expelled either by coughing or swallowing into the gastrointestinal tract. If they escape the mucociliary escalator, their deposition in the respiratory tract mainly depends on their

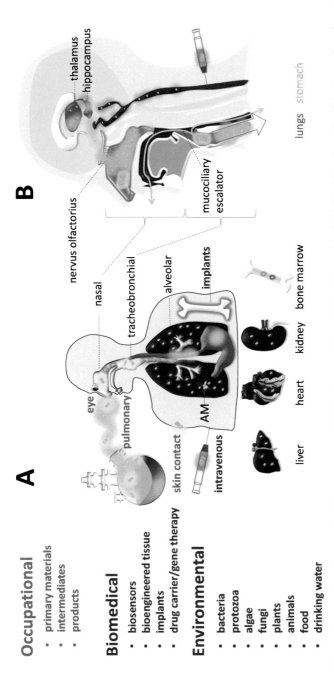

**Fig. 1.** Possible exposition and uptake routes for 2D nanomaterials. (A) Particles can interact with the human body by inhalation, ingestion, injection or dermal contact. Possible sources can be occupational by (e.g., inhalation, eye and skin contact in case of workers exposed to nanomaterials during the production process), environmental (uptake via food chain, or water), or biomedical (e.g., injection of drug carrier or from implants). After uptake, nanoparticles can accumulate in liver, heart, kidney or bone marrow or secondarily enter the brain (thalamus, hippocampus) via the nervus olfactorius (B). AM: alveolar macrophages. Green particles: pulmonary uptake; blue particles: injected rGO.

size, shape and density. These properties determine the aerodynamic diameter, an index that defines the respirability of particles. Particles with an aerodynamic diameter between 1 and 5 μm can be found in the conducting airways, whereas smaller particles < 1 μm are more likely to accumulate in the alveolar area[23,24] where they can be taken up (phagocytosis) by alveolar macrophages. Within macrophages the particles are sequestered in lysosomes that contain hydrolytic enzymes.[25] If the particles exceed the size of the macrophage and cannot be ingested, additional macrophages are recruited from the bloodstream, inducing a persistent inflammatory response, cell damage and collagen deposition (fibrosis) in the alveoli, which impairs gas exchange.[19]

For ultrafine particles, translocation across the lung epithelial barrier into blood vessels and lymphatics has been described.[24] Passing through the lymphatic system, nanoparticles can be distributed through the blood flow into different organs such as liver, kidney, heart and bone marrow. Here, based on their physicochemical properties, they are either metabolized, excreted or retained.[26] When intravenously injected, GO (1 nm thickness/10 to 800 nm lateral dimension) accumulates in the lung and, to a lower extent, within the reticulo-endothelial system of mice. At a dosage of 1 mg/kg a long blood circulation time but nevertheless good biocompatibility with red blood cells (erythrocytes) was determined. However, when exposed to a dosage of 10 mg/kg after 14 days, inflammation, pulmonary edema and granulomas occurred due to aggregation and prolonged retention.[27] Biodistribution and toxicity of GO were further monitored in C57/BL6 mice upon intratracheal instillation of a radioactive [125]I-labeled GO-derivative (size range 10 to 800 nm in lateral dimension and 1nm thickness). At a dosage of 10 mg/kg transfer into the blood circulation, liver and small and large intestines was described. GO clearance from the lungs took between 24 hours to 3 month after instillation. After penetrating the blood stream via the alveolar-capillary interface, small GO sheets were eliminated by the kidney. As alveolar macrophages accumulate GO in the bronchial lumen, mucociliary clearance from the body via the sputum has been suggested. A discoloration of the lung surface however indicates incomplete clearance.[28] Dextran-coated GO (lateral size distribution ~ 50 to 100 nm; thickness 2.8 nm) mainly accumulated within the spleen and liver after intravenous injection into mice with clearance within a week and no noticeable toxicity.[29] The thickness of graphene oxide appeared to play an important role in tissue distribution as well. Injection of 50 μg functionalized graphene (fGO) with two different thicknesses (thickness: ~ 6 nm and ~ 20 nm/lateral dimension 800 nm) into the vein tail of C57/BL6 mice revealed that the thinner fGO is mostly eliminated via renal route after 24 hours, whereas the thicker fGO accumulates to a greater extent within the liver and spleen.[30]

Inhaled nanoparticles can also be translocated via the *nervus olfactorius,* located within the nasal region, into the brain[31] (Fig. 1B). In case of reduced graphene oxide (rGO) a transfer across the blood brain barrier has been demonstrated.[17,31,32] When systemically administered to rats, rGO mainly accumulates in the thalamus and hippocampus. Measurements in thalamus and hippocampus indicate a strong increase in rGO content during the first three hours after application; rGO ($342 \pm 23.5$ nm) was able to enter the brain (Fig. 1B). As demonstrated by three separate experimental data sets rGO administration transiently decreases the paracellular tightness of the blood-brain barrier BBB, rendering it accessible for nanomaterials, chemicals and drugs.[32]

The skin as largest organ of the body regulates and maintains physiological conditions and acts as a barrier to protect against environmental threats (e.g., microorganisms and chemicals). Interference of nanoparticles with the function of the skin depends on the capability of the particle to penetrate the skin, which in turn is influenced by the size and physico-chemical properties of the material.[26]

The skin contains three major layers, comprising epidermis, dermis and subcutis.[33] Although the uptake and penetration of nanomaterials via the skin upon dermal exposure is still controversial, it should be pointed out that a 72-hour exposure of human skin cells (HaCaT keratinocytes) to graphene or GO showed no significant toxicity at concentrations below 30 μg/ml.

However, at high concentrations (> 30 μg/mL for FLG and > 1 μg/mL for GOs, respectively) and after an exposure time of three days, significant cellular damage of skin keratinocytes was determined for few-layer graphene (FLG) and GOs, with variable potencies depending on the oxidation state of the graphene-based material. However, this study has been performed using isolated keratinocytes and did not take into account the barrier properties of intact skin.[34]

The interactions with and biological responses to 2D-nanomaterials that determine their toxicity are caused by material specific properties that again are defined by their physical and chemical features. In general, we can differentiate between mechanical, chemical and electronic interactions.

## Chemical Transformation of 2D-nanomaterials

When immersed in environmental liquids or biofluids, nanomaterials are subjected to non-equilibrium conditions which promote phase-material transformations such as hydrolysis,[35] oxidation,[36] sulfidation[37] or chalcogenide exchange[38,39] reactions leading to changes in surface-chemistry and structure, degradation and dissolution. Especially in biological systems, ligand exchange reactions[40] and the adsorption of ions, small molecules or entire proteins[41] have been described and are expected to be most prominent for mono- and few-layer nanomaterials due to their high surface to volume ratios. $WS_2$ nanosheets for instance, can be functionalized with bovine serum albumin via physical adsorption, based on nonpolar amino acid residues.[42]

However, in most studies physiologically-relevant fluid phases have not been used and thus the relevance of these findings for biological interactions remains to be clarified. Individual 2D materials including metal chalcogenides, metal oxides and hydroxides show different levels of susceptibility for biological dissolution (Fig. 2A).

The level of dissolution stability is a double-edged sword. Non-biopersistent nanoparticles, which readily dissolve under physiological condition, do not cause long-latency pathogenic effects such as lung fibrosis or cancer, while highly concentrated dissolution products on the other hand, can cause acute toxic effects. While silicic acid and the molybdate anion have been shown to be low toxicity species, the release of many metal ions or metal chalcogenides such as $Ni^{2+}$, $Cu^{2+}$, $Ag^+$, NiO, ZnO or CdSe as a consequence of hydrolysis or solid-phase redox-reactions with oxidizing or reducing agents, requires particular attention in terms of toxicity. Many metal oxides like $MoO_3$ $WO_3$[43,44] hydroxides (e.g., $Co(OH)_2$) and especially their 2D materials (LDHs) exhibit high solubilities[45] and readily dissolve at pH7 and independent from

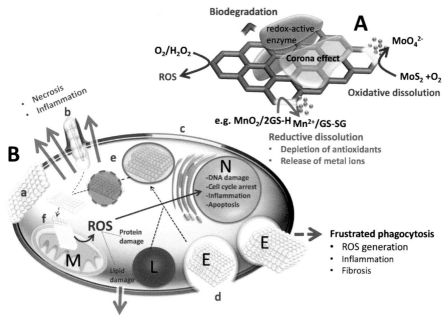

**Fig. 2.** Chemical and mechanical interactions of 2D-nanomaterials with and within cells. (A) Possible interactions between 2D nanomaterials and biofluids. Based on the physico-chemical properties of the particle and the specific local environment, nanoparticles can biodegrade due to the activity of redox active enzymes or undergo enzyme independent oxidative or reductive dissolution over time. Nanoparticles exposed to biomolecules such as proteins, can be subjected to a 'corona effect' which describes the adsorbtion of biomolecules to the nanomaterial surface and the accompanying changes in it's physico-chemical properties. (B) Cellular interactions and uptake of nanoparticles with possible cytotoxic consequences. Particles can be attached (a) to or integrated into the cell membrane (b–c) or internalized via endocytosis or phagocytosis (d–e). The uptake of particles can result in the generation of reactive oxygen species (ROS), as a consequence of mitochondrial damage (f), inducing the oxidative degradation of proteins, lipids and DNA resulting in cell cycle arrest and apoptosis. Frustrated phagocytosis occurs, if the particle size exceeds the size of the macrophage. N: nucleus; L: lysosome; E: endosome; M: mitochondrion.

redox active agents. LDH dissolution has been shown to be a rather complex process and was studied extensively under the aspect of mineral weathering.[46] However, for several metals such as Zn or Ni even a low solubility in the μM range can produce toxic ion concentrations.

Layered TMDs and metal phosphorus trichalcogenides can further undergo cationic substitution-intercalation reactions, upon which cations from the environment replace released toxic metal ions.[47,48,49,50] On the other hand layered nanomaterials can be employed to incorporate unwanted guest molecules for their removal into the interlayer between two nanosheets. In this way LDHs have been used to adsorb and remove anionic contaminants like fluoride, phosphate or even herbicides[51,52,53] and titanate nanosheets or vanadate based layered materials and clays like montmorillonite or vermiculite and have been successfully applied to remove radioactive or heavy metal ions.[54–58] Also, the negative charges and vacant sites in $MnO_2$ nanosheets lead to the adsorption of metal ions like $Co^{2+}$, $Zn^{2+}$ or $U^{6+}$ which can promote the reductive dissolution of $MnO_2$.

The sensitivity for redox dependent dissolution strongly depends on the thermodynamic oxidation potential of the nanomaterial. When falling below the corresponding value of the oxygen/water redox couple ($O_2/H_2O$), as in case of transition metal dichalcogenides (TMDs) oxidation induced dissolution (MX $\longleftrightarrow$ $M^{z+}$ + X + ze$^-$) is thermodynamically favored (Fig. 2A).

However, usually a passivating oxide layer prevents oxidation.[59] Only in case of ultrathin 2D materials could complete oxidative dissolution be expected to occur.

While exhibiting a comparatively low solubility, most metal sulfides are prone to environmental oxidation, leading to soluble products. Few-layered $MoS_2$ undergoes dissolution over the period of days in the metallic 1T phase when exposed to biological or environmental fluids[60] and is prone to photo-oxidative degradation, yielding soluble $MoO_x$ species.[61] Metal sulfides like $MoS_2$ or $WS_2$,[62] are also subjected to bioleaching processes, which lead to their oxidative conversion to sulfuric and molybdic acid[63] that can be induced through direct contact with enzymes from bacteria of genera like Thiobacillus or Leptospirillum[64] or indirectly via redox interaction with ferric iron, produced by those prokaryotes. Also for photoactive metal phosphorus trichalcogenides light-induced degradation or transformation can be assumed.

To undergo reductive dissolution (MX + ze$^-$ $\longleftrightarrow$ M + $X^{z-}$), nanomaterials need to have a reduction potential within or higher than the cellular redox potential range, largely comprising values between –0.4 and +0.4V with glutathione as main reducing agent (–0.24V) (Fig. 2A). Accordingly, reductive dissolution under physiological conditions is thermodynamically favored in the cases of $WO_3$, $MoO_3$ and $MnO_2$. Reductive dissolution of metal oxides ranks among the most relevant transformations in the geochemical cycling of electrons.[46] During reductive dissolution of $MoO_3$ in organic soil extracts containing humic acids Mo has been shown to change its redox state from VI to V or III.[65]

More than 30 different manganese oxides exist with different levels of environmental stability while the most common derivative $MnO_2$ under presence of reductants such as reduced metal ions or natural organic matter is pH-dependently reduced or photo-converted to soluble $Mn^{2+}$ which is quite stable due to its slow oxidation kinetics.[66] The redox-responsive behavior of nanomaterials is strongly dependent on their surface to volume ratio and especially in case of 2D materials often lacks experimental verification.

2D-nanosheets generated by physical exfoliation usually show low reactivity for most chemical processes at the saturated basal surfaces. Susceptible sites are the unsaturated edge planes and basal plane positions with surface material defects as demonstrated for $MoS_2$ flakes.[67] Alternative production processes such as chemical exfoliation based on lithium intercalation increase the number of basal plane defects but their easy functionalization with thiol terminated ligands can be exploited to enhance material stability and biocompatibility.[68,69,70,71]

In contrast to many chalcogenides or oxides, ultrathin exfoliated boron nitride (BN) films are significantly resistant against even extreme environmental conditions.[72,73]

## Physicochemical Properties related to Toxicity

Nanomaterials can be highly diverse with a broad range of physicochemical properties. The parameters defining these properties (Table 2) need to be addressed since they play an important role in biocompatibility.

**Table 2.** Parameters defining physicochemical properties of nanomaterials.

| Morphology | Chemical composition | Surface properties |
|---|---|---|
| Aspect ratio | Redox potential | Hydrophobicity |
| Shape, Size Thickness Density | Dissolution Ion release | Surface charge Chemical Functionalization |
| Stiffness | Crystallinity | |
| Aggregation | Biopersistence | Coating, Corona |

In regard to surface chemistry, the hydrophobicity has an indirect impact on cell viability due to the adsorption of molecules (amino acids and proteins) in physiological solutions. Depending on their level of hydrophobicity, biomolecules in physiological solutions such as amino acids or even proteins adsorb to graphene sheets thus, limiting their availability within the cell.[74]

The hydrophobicity of nanomaterials, e.g., graphene, can be modified by modulating the surface chemistry (functionalization). This in turn consequently leads to an alteration of the original properties of the nanomaterial. Functionalization can stabilize 2D-nanomaterials , inhibit aggregation tendencies and increase or decrease their potential to adsorb proteins and molecules onto their surface.[75–79]

Cell adhesion and particle internalization is also influenced by the modulation of surface chemistry. Reduced graphene oxide for example which exhibits a higher level of hydrophobicity has been shown to attach at the membrane of liver cells (Fig. 2B.a), whereas the more hydrophilic graphene oxide is internalized (Fig. 2B.b–f); the exposure to either GO or reduced GO resulted likewise in ROS generation, but each based on a different molecular mechanism.[80,81]

Also in terms of pathological responses to mechanical interactions, high aspect ratio materials with 1D and 2D geometries require particular attention.[22,82] Again, most of the available data have been gained from studies with graphene-based materials, but can be likewise referred to other types of materials with corresponding physical features.

The cytotoxic effects of graphene materials comprise abrasive or disruptive plasma membrane interactions[82–88] and disturbance or inactivation of respiratory mitochondrial function[87–90] causing oxidative stress responses.[87,89,91,92] This leads to the damage of proteins[93] and DNA/RNA,[94] and finally triggers necrotic or apoptotic cell death.[95] 2D materials interact differently with bio-membranes depending on their shape, surface features and bending stiffness. Carbon nanotubes for example have been described to behave as nanodarts that can penetrate the cell membrane or vesicles directly. Alternatively they may interact with DNA in the nucleus and with mitochondria generating reactive oxygen species.[96,97] To clarify the interaction of graphene with the cytoplasmic membrane[82] *in silico* studies were performed (coarse grained-molecular dynamics), employing all-atom steered molecular dynamics simulations, revealing that the thin edges of graphene sheets are able to penetrate the membrane due to a low local energy barrier resulting in cell injury.

Pristine graphene and GO sheets pierce into membranes and thereby extract lipid molecules which increasingly weakens the integrity of the membrane[83] (Fig. 2B.b). Combined cell studies and computer simulation suggest that stiffness and hydrophobicity of carbon-based nanomaterials are critical determinants of lysosomal membrane integrity. Lipid extraction and increased lysosomal permeability (Fig. 2B.e) can cause release of cell death inducing enzymes (pro-apoptotic factors).[88,98,99] Although one-dimensional carbon nanotubes differ in shape from graphene the biophysical properties associated with sharp corners or edges of two dimensional nanosheets predict that corners of graphene can interact with cell membranes and enter the cell.[82] Depending on what percentage of edge carbons are oxidized and carry hydrophilic oxygen groups, 2D sheets can be either positioned in plane between the two membrane layers ($\leq$ 5%) or cut across like transmembrane proteins ($\geq$ 10%)[100] (Fig. 2B.b). The lateral size dimension determines the possibility for this reorientation. For nano-scaled sheets the energy barrier is small enough to switch from a perpendicular orientation to an inter membrane position (Fig. 2B.c) while this is not the case for microscale sheets.[82,101]

Macrophages, as a first line of defense and important effector cells of the immune system, can actively take up nanomaterials (e.g., 2D-nanomaterials ) by phagocytosis (Fig. 2B.d).

Upon their phagocytosis, nanomaterials are either cleared by the reticuloendothelial system or retained within macrophages for biodegradation (Fig. 2).[102–106] Many 2D-nanomaterials are internalized and accumulate in lysosomes. However, features like size, shape, flexibility and hydrophobicity also influence the uptake mechanism of 2D-nanomaterials into cells with impact on cell viability. Lateral dimensions play an important role in uptake. Single layer GO nano-sheets with a lateral dimension of 2 μm wrinkle and thus can be folded into lysosomes of macrophages while smaller ones 350 nm in lateral dimensions can be internalized intact via filopodia demonstrating how sheet flexibility can determine the internalization pathway.[91] Few-layered graphene sheets with a lateral size of up to 5 μm are readily internalized into human THP-1 macrophages while macrophages exposed to 25 μm layered sheets only adhere and spread over their surface.[19]

Internalization into non-phagocytic cells occurs by three different endocytotic pathways: Macropinocytosis, clathrin-mediated, or caveolin-mediated. Uptake routes mainly depend on the physico-chemical properties of the particles. The actin-driven mechanism of macropinocytosis enables the internalization of micro-sized particles[107,108] and is used by human hepatoma cells (HepG2), murine macrophages (RAW264.7) and osteoblasts to take up GO nano-sheets.[109] Smaller protein coated GO nanoparticles (120–150 nm) enter HepG2 cells or RAW-264.7 macrophages primarily via clathrin-mediated endocytosis which is initiated following adsorption or binding to receptors[110] while an increase of size leads to a change in the uptake mechanism.[111] Caveolin-mediated endocytosis only allows for the internalization of particles between 50–80 nm and mainly occurs at the basolateral side of endothelial cells.[103,110] After internalization nanoparticles reach the acidic lysosomes following endo-lysosomal fusion; the endosome content is then exposed to digestive enzymes (glycosidases, proteases and sulfatases) leading to the degradation of particles and macromolecules at low pH. 2D-nanomaterials , e.g., $MoS_2$ can degrade over time, leading to the release of ions.[112] At high concentrations the degradation products of pristine $MoS_2$ has revealed an increased level of toxicity in cervix carcinoma cells (HeLa) and murine macrophages (RAW264.7) compared to functionalized $MoS_2$, which is more resistant to degradation by $H_2O_2$.[71] Depending on the released ions, the degrading 2D-nanomaterial can induce toxic effects such as ROS generation and activation of cell death-inducing enzymes (Fig. 2B.f).

Other materials reside, as they are less affected by the low pH and cannot be digested by lysosomal enzymes.[113] Graphene and its derivatives in general are considered to be biopersistent. However, several features can promote biodegradation including the number of layers, lateral dimension, tendency for aggregation, chemical modification, structural defects, degree of oxidation and surface functionalization. Especially in case of thin layered 2D materials, sensitivity towards peroxidases has been documented *in vitro* (Fig. 2A). In the presence of $H_2O_2$ horse radish peroxidase (HRP)[114] is able to oxidize and perforate GO flakes and myeloperoxidase, found in neutrophils, (and to a lesser extent in monocytes and macrophages) likewise enables the oxidative degradation of short CNTs[115] and highly dispersed GO while no degradation was determined in case of aggregated GO[116] or chemically reduced graphene oxide (rGO). In case of graphene oxide, the heme active site of HRP appears to be positioned in closer proximity to the surface compared to rGO, thus enhancing basal plane oxidation.[114] Raman spectroscopy enables monitoring of the time-dependent accumulation and degradation of carboxylated graphene following injection into the tail-vein of mice.[102] Graphene first aggregates and accumulates in lung, liver, kidney and spleen following degradation of the material over a period of three month. Biodegradation is prominent in resident tissue macrophages and the gene expression pattern of pro-inflammatory cytokines indicates a phagocytic immune response. *In vitro* tests with the same nanomaterial showed graphene degradation in lysosomes of murine macrophages within seven days.

Many nanomaterials are known to be biopersistent which is an important factor to determine toxicological effects within exposed cells. Due to lateral dimensions and hydrophobicity, carbon-based nanomaterials can interfere with lysosomal integrity and through lipid extraction affect lysosome permeability, resulting in the release of cell-death inducing enzymes (pro-apoptotic factors) (Fig. 2B.e–f).[98,99] If the particle volume of the 2D-nanomaterial exceeds the macrophage volume, the material can only be taken up partly (frustrated phagocytosis) (Fig. 2B), this in turn leads to the release of inflammatory cytokines and reactive oxygen species (ROS). Both can damage neighboring cells, e.g., epithelial cells in lung, causing impaired function and collagen deposition (fibrosis). Ultimately if the particles are genotoxic cell transformation (cancer) may develop. Frustrated phagocytosis resulting in ROS generation and cytokine release can induce the formation of granulomas which is well documented in the case of carbon nanotubes.[117,118]

Surface functionalization can alter biopersistence of nanomaterials. Toxicity of graphene oxide can be mitigated after protein coating (corona effect) (Fig. 2A) with bovine serum albumin (BSA) due to reduced cell uptake.[119] Molecular dynamics simulation revealed that the adsorbed BSA weakened the interaction between graphene oxide and the phospholipids of the cellular membrane. GO functionalization with polyethylene glycol (PEG) showed reduced cytotoxicity compared to pristine GO. Pristine GO permits the adsorption of serum proteins and activates the complement cascade leading to generation of anaphylatoxins, which are involved in local inflammatory responses. PEG functionalization significantly restricts protein binding and complement activation.[120] However, at the same time PEG functionalization prolongs circulation time and promotes bioaccumulation and resistance against biodegradation. To overcome this problem, disulfide bridges were used as cross-linkers between PEG and GO. In this way PEG provided higher biocompatibility to the GO-PEG hybrid but PEG is reductively uncoupled upon particle internalization, thus rendering GO susceptible to lysosomal degradation.[121,122] Although these results underline the importance of the corona effect, the composition of the proteins which participate, e.g., during inhalation and ingestion, differs from those used in cell culture experiments. Several studies suggest that deviations in size, surface functionalization and hydrophobicity lead to various corona profiles[123,124] resulting in different uptake patterns into cells with possible toxicological impact.

Often biomolecules interact and chemically react with the enlarged surfaces of internalized 2D nanoparticles, which as biopersistent inorganic catalysts can attain remarkable turnover numbers. Many semiconductor materials like $MoS_2$, $MnO_2$ and $TiO_2$ are well examined inorganic catalysts. However, very little is known about the bio-reactivity of $MoS_2$, or $MnO_2$. For $MoS_2$, $MnO_2$ nanosheets, graphene, and single-walled carbon nanotubes (SWCNTs), biological redox-activity can be assumed according to the overlaps between their conduction band edge energies and the cellular redox potential. For graphene and SWCNTs biological redox activity has indeed been reported.[125,126] Interactions with dissolved $O_2$ or $H_2O_2$ molecules can yield reactive oxygen species (ROS) (Fig. 2.A), which attack nucleic acids, proteins or lipids and cause oxidative stress (Fig. 2.B.f) if the ROS production cannot be scavenged by antioxidant defenses. Furthermore, redox-active nanomaterials can oxidize and deplete glutathione (Fig. 2.A), an essential cellular antioxidant which results in additional comprise to the oxidative stress response,[127] although the relevance of this *in vitro* assay has not been verified in intact cells or tissues.

Graphene based nanomaterials are further known for their anti-microbial,[83,84,88,125,128–131] antiviral[132,133] and antifungal activity[129,134,135] which is advantageous for biomedical applications. Particle dimensions here play a decisive role. GO sheets with an average area of 13 $\mu m^2$ have a bacteriostatic effect as they are able to fully cover and isolate the bacteria, thus, inhibiting their proliferation[125]. In case of yeast cells, the coverage with 2D materials in fact rather has a protecting influence and enhances cell viability by restricting the high level communication with the environment.[136,137,138] GO and sulfonated rGO have also been shown to prevent infections with herpes simplex virus type 1, pseudorabies virus or porcine epidemic diarrhea virus by inhibiting the initial cell attachment step. The antimicrobial effect of 2D materials can be partially explained by physical interactions between sharp edges of 2D sheets and microbial membranes, which damage membrane integrity and transmembrane potential, resulting in a loss of cell contents and oxidative stress. However, antimicrobial effects like these have their downsides when these materials are released into the environment. 2D-nanomaterials not only affect the viability of human pathogens but also other microorganisms including protozoans, fungi and algae (Fig. 1). The growth of microalgae like *Chlorella vulgaris* and *Raphidocelis subcapitata* is severely suppressed when algal cells are exposed to high concentrations of GO.[139,140] In addition to a strong decrease in photosynthetic activity due to the shading effect of the attaching nanomaterial, enhanced oxidative stress and membrane damage has been reported when the GO concentration exceeds 10 µg ml$^{-1}$. For *C. variabilis* GO sheets have been identified within the cells. Here they damage organelles and cause a massive production of harmful reactive oxygen species.[141] Comparable ecotoxicological effects were also demonstrated for protozoan species like *Eugenal gracilis*.[142]

## Mechanisms of Toxicity

### *In vitro and in vivo models*

Many nanomaterials have been shown to cause cytotoxic effects. *In vitro* cell culture systems and *in vivo* animal models have been exploited to understand the underlying mechanisms of their toxicity. Although *in vitro* experiments are generally more reproducible and less expensive than whole animal studies, limitations have to be considered. Not all cell types and cell lines exhibit a toxic response to these nanoparticles, which may be related to the fact that most cell lines are transformed. Transformed or immortalized cell lines often are less differentiated than primary cells and may have altered important signal transduction pathways. Additionally, *in vitro* studies don't always reflect the results obtained by *in vivo* experiments due to the challenges in recapitulating tissue structure and in dosimetry using exposure relevant quantities of nanoparticles that individuals might be exposed to over a lifetime. In this regard, e.g., airway modeling studies are suitable to understand the deposition and dose of different sizes of graphene particles[20] or other two dimensional nanomaterials in the human respiratory tract. *In vivo* mouse models assist to confirm *in vitro* results and to gain a better understanding of the intrinsic toxicity of the particle and a better estimation of possible adverse health impacts. Needless to mention, that the data obtained from mouse models do not always reflect the impact on human health over a lifetime of exposure.

Recently, three dimensional cell culture models have evolved as a suitable alternative to testings in animals and are being increasingly applied to study nanotoxicological effects.[143–145] Ranging from self-assembled microtissues to scaffold-based systems, these toxicity testing platforms have been developed for many different cell types and often display cellular morphology and physiology (cell shape, cellular function and cellular response) that differs from monolayer culture.[146–148] Compared to cells cultured on flat substrates, 3D microtissues may show increased differentiation, more tissue-like organization, and can be cultured for extended periods of time.[149–151] This extended viability can allow nanoparticles to be tested over a longer time period and the dosage of these particles can be reduced to physiologically relevant concentrations. Advanced cell culture methods can also be integrated with microfluidic systems to treat cells in a dynamic way analogous to exposure from the bloodstream.[147,148] It should also be noted that particles applied to assembled microtissues, especially micro-sized particles, may be unable to reach the cells at the microtissue center, leading to increased exposure at the periphery. Cells can be exposed during assembly for more even uptake of particles, but this can interfere with microtissue formation and may not reflect exposure *in vivo.*

## Caveats in cell based assays

To evaluate possible toxic responses of cells to a wide range of 2D-nanomaterials, a broad array of cell based assays are available to define cytotoxicity, cytokine release, necrosis, membrane integrity, changes in metabolic activity, etc.[16] One of the major caveats concerning the toxicity of nanomaterials results from contamination. Hydrophobic 2D materials with high surface to volume ratios can adsorb endotoxins which originate from the outer membrane of gram-negative bacteria. Contamination can occur during the production process. Air, distilled water, plastic containers and glassware are possible additional sources of contamination with endotoxins and due to their high thermal stability it is difficult to remove or inactivate them via heat-induced denaturation.[152] The interpretation of toxicological data gained from contaminated 2D-nanomaterials often results in false or exaggerated conclusions.

False results can also be the consequence of choosing the wrong *in vitro* assay. Many cell based *in vitro* tests are fluorescence or colorimetric assays. Due to its high specific surface area and hydrophobicity graphene tends to adsorb biomolecules and synthetic dyes. Furthermore, hydrophilic GO is photoactive and thus by itself exhibits both, light absorbance and emission resulting in spectroscopic artifacts during fluorescence or absorbance based cell assays.[74] It is therefore crucial to verify toxicological data using several different assays and to confirm that the materials are not interfering with the *in vitro* assays. To confirm the data obtained by any absorbance or fluorescence-based assay, microscopic and histological studies should be used to identify possible morphological changes of organelles, cells (*in vitro*) or morphological alterations in tissues of lung, kidney or spleen (*in vivo*) of test animals.

In the first part, possible uptake routes and physico-chemical properties related to toxicity were discussed. Here we will review possible toxic mechanisms induced by two dimensional nanomaterials. Most of the toxicity mechanisms discussed here are based on the example of members of the graphene family, since only a small number of studies exist which investigate the toxicity of other 2D-nanomaterials .

## Toxicity mechanisms induced by carbon based 2D-nanomaterials

### Graphene

Graphene is either produced bottom up through chemical vapor deposition (CVD) or top down via exfoliation, it consists of a sheet-like hexagonal configuration of sp$^2$ hybridized carbon atoms which can be turned into different allotropes such as marble-shaped 0D fullerenes, 1D-nanotubes or stacked to three-dimensional graphite.[16] However, due to the extraordinary high surface to volume ratio (2,600 m$^2$/g),[16] graphene is increasingly used in its original 2D flat sheet nanostructure that can be easily functionalized covalently or non-covalently to expand the spectrum of chemical features and support the immobilization of biomolecules like peptides or medical agents. Due to its 2D nanostructure graphene, provides a higher photo-thermal sensitivity and thus, is more effective compared to, e.g., carbon nanotubes.[153,154]

The lethal dose (LD50) of graphite has been reported to be 2 g per kg.[155] Accordingly, the level of toxicity of graphene has been estimated to be very low. However, the low solubility, high tendency for aggregation and long retention and circulation times of graphene[27] give rise to the concern that it might have long-term adverse health impacts. The extent of toxic responses relies on different physical (particle dimensions and geometry, surface to volume ratio and functionalization density) and chemical (chemistry of surface functionalization and purity) parameters, as well as dose.

## Graphene oxide

Graphene oxide as the most important member of the graphene family provides enhanced mechanical, optical, thermal and electro-conductive features and thus is exploited in different nanoelectronic, semiconductor-based and nanomedical applications ranging from tissue regeneration and biosensing/imaging over drug or gene-delivery agents to anti-bacterial/-viral and targeted cancer treatment thereby, e.g., making use of its photochemical activity in photo-thermal therapy.[17,18,156]

Chemical exfoliation of graphite employing strong oxidants is used for the production of graphene oxide (GO).[5,6] Due to its pH-dependent negative surface charge and colloid stability[157] the dispersion efficiency of graphene oxide is higher than graphene and allows for further functionalization and modification due to the electrostatic or covalent interaction of hydrophilic or hydrophobic foreign molecules.[5,121] It is noteworthy that the complexity of naturally occurring graphite with defects in the π-surface structure under the variety of reaction conditions in different laboratories results in a significant diversity of graphene oxide (GO) with different oxidation degrees and deviating layer numbers. This in turn has a great impact on the surface chemistry and the degree of possible functionalization and can lead to a different outcome in toxicological studies of different laboratories.[5,121,158]

## In vitro toxicity of graphene

After inhalation, nanomaterials can reach the alveolar area resulting in interaction with peripheral immune cells and from there translocate to different organs (Fig. 1A). Active uptake of nanomaterials into macrophages mainly occurs via receptor-mediated phagocytosis and is initiated upon particle adhesion to the cell membrane (Fig. 2B). Phagocytosis is promoted by protein adsorption at the particle surface which occurs in the lung lining fluid and during transport via the circulatory system (Fig. 2A).[107,108]

Depending on size, shape, oxidation state and functionalization, graphene based nanomaterials can impact immune cells individually, resulting in a different activation pattern of these cells.[159] Several studies suggest that graphene oxidation decreases toxicity in macrophages in comparison to pristine graphene.[160] Pristine graphene has been shown to activate oxidative stress pathways and to stimulate the secretion of cytokines and chemokines from macrophages.[161] The ROS triggered mitochondrial pathway of murine macrophages (RAW264.7) is up-regulated upon exposure to pristine graphene in a dose and time dependent manner. Due to the loss of mitochondrial membrane potential, accumulation of ROS (Fig. 2B.f) results in the activation of MAPK and TGF-β induced pathways. This consequently activates pro-apoptotic factors.[89]

Graphene oxide, on the other hand, induces autophagy and cell death in murine macrophages (RAW264.7) in a dose dependent manner. Toll-like receptor mediated inflammatory response and autophagy has been triggered in murine macrophages after exposure to GO sheets with two different lateral dimensions (2.4 μm and 350 nm). Autophagy induction by graphene oxide has been further demonstrated for cancer cells of liver (SNU-449 and mahlavu) and lung (A549) and human embryonic kidney cells (HEK293).[162]

An increase of oxygenated functional groups has been shown to increase ROS generation and to decrease viability of GO treated HUVEC cells (human umbilical vein endothelial cells). Exposure of HUVEC cells to 0.4 and 0.8 μm GOs or their derivative rGO at a concentration of 10 μg/ml leads to ~ 15% higher toxicity in GO exposed cells.[163]

*In vitro* studies in which lung epithelial cells (A549 cells) are exposed to 100 μg/ml purified graphene oxide with lateral dimensions below 500 nm show no toxicity. These results were further confirmed *in vivo* when an intraperitoneal injection of 50 μg/ml purified graphene into C57Bl/6 mice failed to induce inflammation or granulomas.[164]

### *In vivo toxicity of graphene*

For investigating the pulmonary toxicity of graphene, toxicological *in vivo* studies have been expanded to mice. Graphene nanoplatelets with an aerodynamic diameter of 3 μm induced an inflammatory response after pharyngeal aspiration or pleural injection into BALB/C mice.[165] The results strengthened the hypothesis that shape and size of nanoparticles can promote frustrated phagocytosis in macrophages, resulting in ROS generation and inflammatory response (Fig. 2B). However, six weeks after exposure to pristine graphene with different lateral dimensions (1, 5 and 20 μm) no signs of fibrosis were detectable, suggesting resolution over time.[166,176,168]

Aggregation and functionalization of graphene was shown to impact cell toxicity in mice. When comparing the effects of injecting 50 μg graphene dispersed in 2% surfactant containing pluronic acid (F108), aggregated graphene or graphene oxide into the lungs of mice, graphene oxide showed the highest level of toxicity, suggesting that functional oxide-groups of GO promote pulmonary toxicity. Aggregated graphene was retained in the small airways of the lung, triggering a mild fibrosis, whereas well dispersed graphene does not cause any fibrosis even 21 days after administration. The results suggest that dispersed graphene is more readily cleared by macrophages and is more biocompatible.[169]

The use of different coatings for surface modifications on the same nanomaterial results in conflicting toxicological responses *in vitro* and *in vivo*. Graphene dispersed in bovine serum albumin leads to an inflammatory response in human macrophages (THP-1) and lung epithelial cells (BEAS-2B) resulting in an inflammatory response (NLRP3 inflammasome activation) and causes fibrosis in mice, whereas graphene did not reveal any cytotoxicity *in vitro* and did not lead to fibrosis in mice (*in vivo*) when dispersed with the surfactant pluronic acid (F108).[170]

An explanation for the variations in graphene cytotoxicity reported from different laboratories might be based on the techniques of production and synthesis of graphene and its derivatives. The synthesis of graphene oxide, leads to a large variance of products with different oxidation degrees and according variations in physicochemical properties whose fractions can turn out differently even by slight deviations in the production process. The different results on the cytotoxicity of similar nanomaterials within the wide range of studies further imply that beside the intrinsic features of a 2D-nanomaterial (lateral dimension, degree of oxidation, aggregation in physiological solutions) its purity or possible contamination during synthesis, the examined cell type as well as the uptake mechanism can determine the toxicological outcome of the studies.

### Toxic mechanism of transition metals dichalcogenides

Transition metal dichalcogenides (TMDs) are one of the best studied families of 2D materials. TMDs are layered compounds of chalcogens (Te, S, Se, etc.) and transition metals (Mo, W, V, etc.), whereas the transition metal layer is positioned between two layers of chalcogens.[171] In the last couple of years several excellent reviews have been published about the synthesis of this nanomaterial via exfoliation of bulk material or bottom up synthesis.[3,4,172,173] Due to their unique optical properties TMDs have been evaluated for their possible applicability as sensors in biomedical approaches or in photothermal therapy.[8,174] There is only a limited number of toxicological studies available, therefore a complete understanding of possible toxic mechanisms is yet not attainable for this class. Many new 2D-nanomaterials which contain metal constituents can dissolute and degrade over time resulting in distinctive reactions for individual elements or dissolute species. There are extensive reviews about metal toxicity, which allow some predictions about possible[175,176] toxicity effects in biological systems if the route of uptake and location of particle accumulation is known. In the following paragraph the toxicological studies performed so far will be summarized.

A comparison of toxicity between $MoS_2$, $WS_2$ and $WeS_2$ suggests that the chalcogenides $MoS_2$ and $WS_2$ are not toxic in lung epithelial cells (A549) up to a concentration of 200 µg/ml, whereas $WSe_2$ causes 50% toxicity at concentrations of 100 µg/ml in a dose-dependent manner.[177] This result suggests that the chalcogen selenide is responsible for the toxic effect. A comparison between dichalcogenides of transition metal groups V ($VS_2$, $VSe_2$, $VTe_2$) and VI ($MoS_2$, $WS_2$ and $WSe_2$) reveals a stronger toxicity effect for chalcogenides of group V in lung epithelial cells (A549). In both studies metal-selenide compounds exhibited a more severe toxicity level than the corresponding metal sulfides.[171] This is not unanticipated as $Se^{2-}$ has been shown to have toxic effects when exceeding a concentration of 5 μM in human fibroblasts[178] as it can substitute sulfide constituents in biomolecules and bioinorganic redox-complexes such as FeS clusters yielding, e.g., redox enzyme-derivatives with deviating functionalities.[179] This study underlines the importance to investigate the dissolution of transition metals and their potential toxic effects in cells. Although the exact cytotoxic mechanisms are not fully understood yet, the high surface area to volume ratio of 2D materials clearly enhances the risk of releasing potentially toxic constituents such as heavy metal ions.

### Toxicity mechanism of MoS₂ in eukaryotic cells

One of the most studied TMD is molybdenum disulfide that is extensively used as an industrial lubricant. Due to many unique electronic, optical and mechanical features and its simple production through chemical exfoliation, 2D $MoS_2$ appears to be a promising candidate for electronic devices, but has also received attention in biomedical applications.

MoS₂ has shown only low toxicity and studies have been performed to evaluate its toxicity after exfoliation. When taken up by lung epithelial cells (BEAS-2B) or human macrophages (THP-1), exfoliated $MoS_2$ caused no severe toxic effects.[180] However, at concentrations of 50 μg/ml degraded pristine $MoS_2$ in comparison to functionalized $MoS_2$, revealed a significant level of toxicity in cervix carcinoma cells (HeLa) and murine macrophages (RAW264.7), suggesting that the degradation products can cause toxicity at concentrations above 50 μg/ml.[71]

### Toxic mechanism of transition metal oxides

The reactivity of two dimensional transition metal oxides and the limited dissolution rates of compounds out of this group determine their unique properties. The few available studies suggest that some members of this group can be used as therapeutics in low doses, as carriers for drugs and as antibacterial agents. Due to their high surface area, transition metal oxide nanoparticles also have a significant relevance for storage applications. However, there are only few reports about two dimensional transition metal oxides and their potential impact on human health and environmental implications available.

Molybdenum trioxide ($MoO_3$) nanoplates have been shown to disrupt the membrane of four different bacterial strains.[181] It has also been reported to activate caspase in the breast cancer cell line MCF-7, but not in immortalized human keratinocytes (HaCaT cells).[182] These studies suggest a potential use as antibacterial agent and possible biomedical application for breast cancer treatment.

Manganese oxides ($MnO_2$) have received some attention due to their relatively low toxicity.[183] If applied at very low concentrations, manganese oxide nanoplates showed an enhanced contrast of magnetic resonance images. At higher concentrations a significant decrease of cell viability in breast cancer cells (MCF-7) was observed, just as described for molybdenum trioxide ($MoO_3$). It might therefore be a promising contrast agent but needs to be monitored for its possible cytotoxic effects at higher doses to avoid cell damage.[184]

For a better understanding of their potential toxicity it is crucial to compare $MoO_3$ and $MnO_2$ to other transition metal oxides, to determine their ion release and to define further dissolution mechanisms.

## Implications for Safe Designs and Applications of 2D-nanomaterials

During the last 10 years, two dimensional nanomaterials have received great attention due to their promising unique physicochemical properties. Due to their high surface area, there is a wide variety of applications for these materials in various engineering and biomedical disciplines. Variations in lateral dimensions, surface chemistry and functionalization alter their physico-chemical properties and allow these nanomaterials to be used as drug carriers, bioimaging compounds, agents for cancer photothermal therapy, biosensors and nanostructural scaffolds for tissue regeneration. Because of their numerous applications, potential adverse environmental and human health impacts of 2D-nanomaterials need to be carefully assessed. Processing and synthesis methods for these nanomaterials including liquid phase growth, liquid phase exfoliation, vapor phase growth or dry exfoliation are of great concern due to the corresponding occupational risks for exposed production workers. The synthesis and production of these nanomaterials by dry exfoliation and the drying process of wet exfoliation can lead to the exposure by inhalation.[98] Another potential adverse health impact might result from the use of 2D-nanomaterials as drug carriers. Therefore, determination of their circulation times, distribution, biopersistence, biodurability and dissolution rates in biological systems are important parameters to access.[185] *In vitro* and *in vivo* assays are commonly used to assess biodurability. Inhalation exposure studies in which a wide variety of particles are monitored in mice over a period of three months are needed to assess particle clearance. Acellular assays are commonly used to provide information about the dissolution rate of nanomaterials and ion release under different biological and environmental conditions.[185] Macrophages are commonly used cells to assess impacts of the lateral dimension of particles although these cells are not in their native environment and therefore only partially reflect *in vivo* conditions. However, these assays allow for rapid assessment of acute toxicity that may predict possible long-term health impacts of particles and thus provide valuable information for safer drug design or production processes.

# Conclusion

This chapter summarizes the current state of knowledge about the human health and environmental impacts of two dimensional nanomaterials. The influence of individual physico-chemical properties on the outcome of *in vitro* and *in vivo* studies on nanomaterial toxicity were summarized and discussed. Taken together, the potential hazard of these nanomaterials can be estimated as a function of different extrinsic and intrinsic parameters such as dose and frequency of exposure, chemical composition, particle dimensions, surface chemistry and functionalization, tendencies to undergo the corona effect and aggregation. These parameters define material distribution, circulation time, cell internalization processes, dissolution, biopersistence and clearance, which determine the level of risk following short term and long term exposures.

Post-exposure transformation processes should also be considered when assessing nanoparticle toxicity including material phase transformations, redox-interactions with biomolecules as well as physico-mechanical interactions during the internalization process which promote or slow down surface oxidation/-reduction, degradation and dissolution. Based on the sheer number of newly designed 2D-nanomaterials it appears to be a mammoth task to gain detailed insights into the individual toxic implications of each material in each group. Instead, it might be more promising to deepen our understanding about the common principles that determine intrinsic toxicity, biopersistence and biodurability of nanoparticles in order to predict and prevent potential adverse impacts of newly designed materials.

The authors of this chapter are grateful for the invaluable advice provided by Professor Agnes Kane (Brown University), Professor Robert Hurt (Brown University) and Professor Alberto Bianco (University of Strasbourg).

# References

1. S. Z. Butler, S. M. Hollen, L. Cao, Y. Cui, J. A. Gupta, H. R. Gutierrez, T. F. Heinz, S. S. Hong, J. Huang, A. F. Ismach, E. Johnston-Halperin, M. Kuno, V. V. Plashnitsa, R. D. Robinson, R. S. Ruoff, S. Salahuddin, J. Shan, L. Shi, M. G. Spencer, M. Terrones, W. Windl and J. E. Goldberger, *ACS Nano*, 2013, **7**, 2898–2926.
2. V. Nicolosi, M. Chhowalla, M. G. Kanatzidis, M. S. Strano and J. N. Coleman, *Science*, 2013, **340**, 1226419.
3. M. Chhowalla, H. S. Shin, G. Eda, L. J. Li, K. P. Loh and H. Zhang, *Nature Chem.*, 2013, **5**, 263–275.
4. Q. H. Wang, K. Kalantar-Zadeh, A. Kis, J. N. Coleman and M. S. Strano, *Nature Nanotechnol.*, 2012, **7**, 699–712.
5. D. R. Dreyer, S. Park, C. W. Bielawski and R. S. Ruoff, *Chem. Soc. Rev.*, 2010, **39**, 228–240.
6. H. B. Zhang, C. Chen, J. W. Wang, Y. Yang, Z. H. Lu, Q. Yan and W. G. Zheng, *J. Nanosci. Nanotechnol.*, 2011, **11**, 10868–10870.
7. P. You, Y. Yang, M. Wang, X. Huang and X. Huang, *Current Pharmaceutical Design*, 2015, **21**, 3215–3222.
8. Kurapati, R., K. Kostarelos, M. Prato and A. Bianco, *Advanced Materials*, 2016, **28**, 6052–6074.
9. A. Servant, V. Leon, D. Jasim, L. Methven, P. Limousin, E. V. Fernandez-Pacheco, M. Prato and K. Kostarelos, *Adv. Health Mater*, 2014, **3**, 1334–1343.
10. A. Sahu, W. I. Choi and G. Tae, *Chemical Communications*, 2012, **48**, 5820–5822.
11. C. Heo, J. Yoo, S. Lee, A. Jo, S. Jung, H. Yoo, Y. H. Lee and M. Suh, *Biomaterials*, 2011, **32**, 19–27.
12. S. Agarwal, X. Zhou, F. Ye, Q. He, G. C. Chen, J. Soo, F. Boey, H. Zhang and P. Chen, *Langmuir*, 2010, **26**, 2244–2247.
13. S. Park, N. Mohanty, J. W. Suk, A. Nagaraja, J. An, R. D. Piner, W. Cal, D. R. Dreyer, V. Berry and R. S. Ruoff, *Adv. Mater.*, 2010, **22**, 1736–1740.
14. G. Lalwani, A. M. Henslee, B. Farshid, L. Lin, F. K. Kasper, Y. X. Qin, A. G. Mikos and B. Sitharaman, *Biomacromolecules*, 2013, **14**, 900–909.
15. D. Chimene, D. L. Alge and A. K. Gaharwar, *Adv. Mater*, 2015, **27**, 7261–7284.
16. T. B. Nezakati, G. Cousins and A. M. Seifalian, *Arch. Toxicol.*, 2014, **88**, 1987–2012.
17. F. M. Tonelli, V. A. Goulart, K. N. Gomes, M. S. Ladeira, A. K. Santos, E. Lorencon, L. O. Ladeira and R. R. Resende, *Nanomedicine*, 2015, **10**, 2423–2450.
18. Z. Singh, *Nanotechnol. Sci. Appl.*, 2016, **9**, 15–28.
19. V. C. Sanchez, A. Jachak, R. H. Hurt and A. B. Kane, *Chem. Res. Toxicol.*, 2012, **25**, 15–34.
20. W. C. Su, B. K. Ku, P. Kulkarni and Y. S. Cheng, *J. Occup. Environ. Hyg.*, 2016, **13**, 48–59.
21. L. Ou, B. Song, H. Liang, J. Liu, X. Feng, B. Deng, T. Sun and L. Shao, *Part. Fibre Toxicol.*, 2016, **13**, 57.
22. A. C. Jachak, M. Creighton, Y. Qiu, A. B. Kane and R. H. Hurt, *MRS Bull.*, 2012, **37**, 1307–1313.
23. G. Oberdorster, E. Oberdorster and J. Oberdorster. *Environ. Health Perspect*, 2005, **113**, 823–839.

24. N. R. Yacobi, F. Fazllolahi, Y. H. Kim, A. Sipos, Z. Borok, K. J. Kim and E. D. Crandall, *Air Qual. Atmos. Health,* 2011, **4**, 65–78.
25. E. Goldstein, *Rev. Infect. Dis.*, 1983, **5**, 1078–1092.
26. Y. Zhang, Y. H. Bai, J. B. Jia, N. N. Gao, Y. Li, R. N. Zhang, G. Jiang and B. Yan, *Chem. Soc. Rev.*, 2014, **43**, 3762–3809.
27. X. Zhang, J. Yin, C. Peng, W. Hu, Z. Zhu and W. Li, *Carbon*, 2011, **49**, 986–95.
28. B. Li, J. Yang, Q. Huang, Y. Zhang, C. Peng, Y. Zhang, Y. He, J. Shi, W. Li, J. Hu and C. Fan, *NPG Asia Mater.*, 2013, **5**, e44.
29. S. Zhang, K. Yang, L. Feng and Z. Liu, *Carbon*, 2011, **49**, 4040–4049.
30. D. Jasim, H. Boutin, M. Fairclough, C. Menard-Moyon, C. Prenant, A. Bianco and K. Kostarelos, *Applied Materials Today,* 2016, **4**, 24–30.
31. P. Borm, J. D. Robbins, S. Haubold, T. Kuhlbusch, H. Fissan, K. Donaldson, R. Schins, V. Stone, W. Kreyling, J. Lademann, J. Krutmann, D. Warheit and E. Oberdorster, *Part. Fibre Toxicol.*, 2006, **3**, 11.
32. M. C. Mendonca, E. S. Soares, M. B. de Jesus, H. J. Ceragioli, M. S. Ferreira, R. R. Catharino and M. A. da Cruz-Hofling, *J. Nanobiotechnol.*, 2015, **13**, 78.
33. C. Buzea, II. Pacheco and K. Robbie, *Biointerphases*, 2007, **2**, MR17–71.
34. M. Pelin, L. Fusco, V. Leon, C. Martin, A. Criado, S. Sosa, E. Vazquez, A. Tubaro and M. Prato, *Sci. Rep.*, 2017, **7**, 40572.
35. A. Sasidharan, P. Chandran, D. Menon, S. Raman, S. Nair and M. Koyakutty, *Nanoscale*, 2011, **3**, 3657–3669.
36. J. Liu and R. H. Hurt, *Environ. Sci. Technol.*, 2010, **44**, 2169–2175.
37. Z. Wang, A. von dem Bussche, P. K. Kabadi, A. B. Kane and R. H. Hurt, *ACS Nano*, 2013, **7**, 8715–8727.
38. J. Liu, Z. Wang, F. D. Liu, A. B. Kane and R. H. Hurt, *ACS Nano*, 2012, **6**, 9887–9899.
39. H. S. Su, Y. T. Hsu, Y. H. Chang, M. H. Chiu, C. L. Hsu, W. T. Hsu, W. H. Chang, J. H. He and L. J. Li, *Small*, 2014, **10**, 2589–2594.
40. T. Kim, K. Lee, M. S. Gong and S. W. Joo, *Langmuir*, 2005, **21**, 9524–9528.
41. I. Lynch and K. A. Dawson, *Nano Today*, 2008, **3**, 40–47.
42. Y. Yong, L. Zhou, Z. Gu, L. Yan, G. Tian, X. Zheng, X. Liu, X. Zhang, J. Shi, W. Cong, W. Yin and Y. Zhao, *Nanoscale,* 2014, **6**, 10394–10403.
43. G. Wang, Y. Ling, H. Wang, X. Yang, C. Wang, J. Z. Zhang and Y. Li, *Energy Environ. Sci.*, 2012, **5**, 6180–6187.
44. V. A. Petrochenkov, I. G. Gorichev, V. V. Batrakov, A. D. Izotov and A. M. Kutepov, *Theor. Found. Chem. Eng.*, 2004, **38**, 386–393.
45. J. W. Boclair and P. S. Braterman, *Chem. Mater.*, 1999, **11**, 298–302.
46. W. Stumm and R. Wollast, *Rev. Geophys.*, 1990, **28**, 53–69.
47. J. S. O. Evans and D. O'Hare, *Adv. Mater.*, 1994, **6**, 646–648.
48. R. Brec, pp. 93–124. *In*: M. Dresselhaus (ed.). Intercalation in Layered Materials, *NATO Advanced Study Institute, Series B, Plenum, New York,* 1986, vol. 148.
49. D. Ruiz-Leo n, V. Manr quez, J. Kasaneva and R. E. Avila, *Mater. Res. Bull.*, 2002, **37**, 981–989.
50. N. V. Venkataraman, L. Mohanambe and S. Vasudevan, *J. Mater. Chem.*, 2003, **13**, 170–171.
51. J. Das, B. S. Patra, N. Baliarsingh and K. M. Parida, *Appl. Clay Sci.*, 2006, **32**, 252–260.
52. S. Mandal and S. Mayadevi, *Appl. Clay Sci.*, 2008, **40**, 54–62.
53. A. Legrouri, M. Lakraimi, A. Barroug, A. De Roy and J. P. Besse, *Water Res.,* 2005, **39**, 3441–3448.
54. G. Abate and J. C. Masini, *Colloids Surf. A*, 2005, **262**, 33–39.
55. A. A. El-Bayaa, N. A. Badawy and E. A. AlKhalik, *J. Hazard. Mater.,* 2009, **170**, 1204–1209.
56. M. Malandrino, O. Abollino, A. Giacomino, M. Aceto and E. Mentasti, *J. Colloid Interface Sci.*, 2006, **299**, 537–546.
57. D. Yang, Z. Zheng, H. Liu, H. Zhu, X. Ke, Y. Xu, D. Wu and Y. Sun, *J. Phys. Chem. C*, 2008, **112**, 16275–16280.
58. S. Sarina, A. Bo, D. Liu, H. Liu, D. Yang, C. Zhou, N. Maes, S. Komarneni and H. Zhu, *Chem. Mater.,* 2014, **26**, 4788–4795.
59. S. Ross and A. Sussman, *J. Phys. Chem.*, 1955, **59**, 889–892.
60. R. H. Z. Wang, presented in part at the SETAC North America 36th Annual Meeting, Salt Lake City, Utah, November 1–5, 2015.

61. E. Parzinger, B. Miller, B. Blaschke, J. A. Garrido, J. W. Ager, A. Holleitner and U. Wurstbauer, *ACS Nano*, 2015, **9**, 11302–11309.
62. H. Tributsch and J. C. Bennett, *J. Chem. Technol. Biotechnol.*, 1981, **31**, 627–635.
63. L. Bryner and R. Anderson, *Ind. Eng. Chem.*, 1957, **49**, 1721–1724.
64. K. Rosecker, *FEMS Microbiol. Rev.*, 1997, **20**, 591–604.
65. B. A. Goodman and M. V. Cheshire, *Nature*, 1982, **299**, 618–620.
66. W. G. Sunda, S. A. Huntsman and G. R. Harvey, *Nature*, 1983, **301**, 234–236.
67. M. Yamamoto, T. L. Einstein, M. S. Fuhrer and W. G. Cullen, *J. Phys. Chem. C.*, 2013, **117**, 25643–25649.
68. S. S. Chou, M. De, J. Kim, S. Byun, C. Dykstra, J. Yu, J. Huang and V. P. Dravid, *J. Am. Chem. Soc.*, 2013, **135**, 4584–4587.
69. T. Liu, C. Wang, X. Gu, H. Gong, L. Cheng, X. Shi, L. Feng, B. Sun and Z. Liu, *Adv. Mater.*, 2014, **26**, 3433–3440.
70. D. Voiry, A. Goswami, R. Kappera, C. e Silva, D. Kaplan, T. Fujita, M. Chen, T. Asefa and M. Chhowalla, *Nat. Chem.*, 2015, **7**, 45–49.
71. R. Kurapati, L. Muzi, A. P. R. de Garibay, J. Russier, D. Voiry, I. A. Vacchi, M. Chhowalla and A. Bianco, *Adv. Funct. Mater.*, 2017, **27**, 1605176.
72. Z. Liu, Y. Gong, W. Zhou, L. Ma, J. Yu, J. C. Idrobo, J. Jung, A. H. MacDonald, R. Vajtai, J. Lou and P. M. Ajayan, *Nat. Commun.*, 2013, **4**, 2541.
73. R. Kurapati, C. Backes, C. Menard-Moyon, J. N. Coleman and A. Bianco, *Angew. Chem.*, 2016, **55**, 5506–5511.
74. M. A. Creighton, J. R. Rangel-Mendez, J. Huang, A. B. Kane and R. H. Hurt, *Small*, 2013, **9**, 1921–1927.
75. S. Y. Choi, M. Mamak, E. Cordola and U. Stadler, *J. Mater. Chem.*, 2011, **21**, 5142–5147.
76. C. Y. Zhi, Y. Bando, T. Terao, C. C. Tang, H. Kuwahara and D. Golberg, *Chem. Asian J.*, 2009, **4**, 1536–1540.
77. T. Sainsbury, A. Satti, P. May, Z. Wang, I. McGovern, Y. K. Gun'ko and J. Coleman, *J. Am. Chem. Soc.*, 2012, **134**, 18758–18771.
78. Y. Lin, T. V. Williams, W. Cao, H. E. Elsayed-Ali and J. W. Connell, *J. Phys. Chem. C.*, 2010, **114**, 17434–17439.
79. S. Presolski and M. Pumera, *Mater. Today*, 2016, **19**, 140–145.
80. B. Zhang, P. Wei, Z. Zhou and T. Wei, *Adv. Drug Delivery Rev.*, 2016, **105**, 145–162.
81. N. Chatterjee, H. J. Eom and J. Choi, *Biomaterials*, 2014, **35**, 1109–1127.
82. Y. Li, H. Yuan, A. von dem Bussche, M. Creighton, R. H. Hurt, A. B. Kane and H. Gao, *Proc. Natl. Acad. Sci. U. S. A.*, 2013, **110**, 12295–12300.
83. Y. Tu, M. Lv, P. Xiu, T. Huynh, M. Zhang, M. Castelli, Z. Liu, Q. Huang, C. Fan, H. Fang and R. Zhou, *Nat. Nanotechnol.*, 2013, **8**, 594–601.
84. O. Akhavan and E. Ghaderi, *ACS Nano*, 2010, **4**, 5731–5736.
85. K.-H. Liao, Y.-S. Lin, C. W. Macosko and C. L. Haynes, *ACS Appl. Mater. Interfaces*, 2011, **3**, 2607–2615.
86. A. Sasidharan, L. S. Panchakarla, P. Chandran, D. Menon, S. Nair, C. N. R. Rao and M. Koyakutty, *Nanoscale*, 2011, **3**, 2461–2464.
87. Y. Zhang, S. F. Ali, E. Dervishi, Y. Xu, Z. Li, D. Casciano and A. S. Biris, *ACS Nano*, 2010, **4**, 3181–3186.
88. R. Zhou and H. Gao, *Wiley Interdiscip. Rev.: Nanomed. Nanobiotechnol.*, 2014, **6**, 452–474.
89. Y. Li, Y. Liu, Y. Fu, T. Wei, L. Le Guyader, G. Gao, R.-S. Liu, Y.-Z. Chang and C. Chen, *Biomaterials*, 2012, **33**, 402–411.
90. K. Yang, Y. Li, X. Tan, R. Peng and Z. Liu, *Small*, 2013 **99-10**, 1492–1503.
91. H. Yue, W. Wei, Z. Yue, B. Wang, N. Luo, Y. Gao, D. Ma, G. Ma and Z. Su, *Biomaterials*, 2012, **33**, 4013–4021.
92. Y. Chang, S. T. Yang, J. H. Liu, E. Dong, Y. Wang, A. Cao, Y. Liu and H. Wang, *Toxicol. Lett.*, 2011, **200**, 201–210.
93. B. Luan, T. Huynh, L. Zhao and R. Zhou, *ACS Nano*, 2015, **91**, 663–669.
94. O. Akhavan, E. Ghaderi and A. Akhavan, *Biomaterials*, 2012, **33**, 8017–8025.
95. J. Yuan, H. Gao, J. Sui, H. Duan, W. N. Chen and C. B. Ching, *Toxicol. Sci.*, 2012, **126**, 149–161.
96. S. Liu, L. Wei, L. Hao, N. Fang, M. W. Chang, R. Xu, Y. Yang and Y. Chen, *ACS Nano*, 2009, **22**, 3891–902.

97. Q. Mu, D. L. Broughton and B. Yan, *Nano Lett.*, 2009, **9**, 4370–4375.
98. Z. Wang, W. Zhu, Y. Qiu, X. Yi, A. von dem Bussche, A. Kane, H. Gao, K. Koski and R. Hurt, *Chem. Soc. Rev.*, 2016, **45**, 1750–1780.
99. W. Zhu, A. von dem Bussche, X. Yi, Y. Qiu, Z. Wang, P. Weston, R. H. Hurt, A. B. Kane and H. Gao, *PNAS*, 2016, **113**, 12374–12379.
100. J. Wang, Y. Wei, X. Shi and H. Gao, *RSC Adv.*, 2013, **3**, 15776–15782.
101. A. V. Titov, P. Kr´al and R. Pearson, *ACS Nano*, 2010, **4**, 229–234.
102. C. M. Girish, A. Sasidharan, G. S. Gowd, S. Nair and M. Koyakutty, *Adv. Healthc. Mater.*, 2013, **2**, 1489–1500.
103. M. Geiser, *J. Aerosol. Med. Pulm. Drug Deliv.*, 2010, **23**, 207–217.
104. S. De Koker, B. G. De Geest, C. Cuvelier, L. Ferdinande, W. Deckers, W. E. Hennink, S. C. DeSmedt and N. Mertens, *Adv. Funct. Mater.*, 2007, **17**, 3754–3763.
105. M. D. Witmer-Pack, M. T. Crowley, K. Inaba and R. M. Steinman, *J. Cell Sci.*, 1993, **105**, 965–973.
106. B. Godin, C. Chiappini, S. Srinivasan, J. F. Alexander, K. Yokoi, M. Ferrari, P. Decuzzi and X. Liu, *Adv. Funct. Mater*, 2012, **22**, 4225–4235.
107. L. Kuo, J. Sun, Y. Zhai and Z. He, *Asian J. Pharm. Sci.*, 2013, **8**, 1–10.
108. G. Sahay, D. Y. Alakhova and A. V. Kabanov, *J. Controlled Relea*se, 2010, **145**, 182–195.
109. J. Linares, M. C. Matesanz, M. Vila, M. J. Feito, G. Goncalves, M. Vallet-Regi, P. A. Marques and M. T. Portoles, *ACS Appl. Mater. Interfaces*, 2014, **6**, 13697–13706.
110. H. Kettiger, A. Schipanski, P. Wick and J. Huwyler, *Int. J. Nanomed.*, 2013, **8**, 3255–3269.
111. Q. Mu, G. Su, L. Li, B. O. Gilbertson, L. H. Yu, Q. Zhang, Y.-A. Sun and B. Yan, *ACS Appl. Mater. Interfaces*, 2012, **4**, 2259–2266.
112. Z. Wang, A. von dem Bussche, Y. Qiu, T. M. Valentin, K. Gion, A. B. Kane and R. H. Hurt, *Environ. Sci. Technol.*, 2016, **50**, 7208–7217.
113. R. S. Flannagan, V. Jaumouille and S. Grinstein, *Annu. Rev. Pathol.*, 2012, **7**, 61–98.
114. G. P. Kotchey, B. L. Allen, H. Vedala, N. Yanamala, A. A. Kapralov, Y. Y. Tyurina, J. Klein-Seetharaman, V. E. Kagan and A. Star, *ACS Nano*, 2011, **5**, 2098–2108.
115. V. E. Kagan, N. V. Konduru, W. Feng, B. L. Allen, J. Conroy, Y. Volkov, I. I. Vlasova, N. A. Belikova, N. Yanamala, A. Kapralov, Y. Y. Tyurina, J. Shi, E. R. Kisin, A. R. Murray, J. Franks, D. Stolz, P. Gou, J. Klein-Seetharaman, B. Fadeel, A. Star and A. A. Shvedova, *Nat. Nanotechnol.*, 2010, **5**, 354–359.
116. R. Kurapati, J. Russier, M. A. Squillaci, E. Treossi, C. Menard-Moyon, A. E. Del Rio-Castillo, E. Vazquez, P. Samori, V. Palermo and A. Bianco, *Small*, 2015, **11**, 3985–3994.
117. D. M. Brown, I. A. Kinloch, U. Bangert, A. H. Windle, D. M. Walter, G. S. Walker, C. A. Scotchford, K. Donaldson and V. Stone, *Carbon*, 2007, **45**, 1743–1756.
118. H. Ali-Boucetta, K. T. Al-Jamal and K. Kostarelos, *Methods Mol. Biol.*, 2011, **726**, 299–312.
119. G. Duan, S. G. Kang, X. Tian, J. A. Garate, L. Zhao, C. Ge and R. Zhou, *Nanoscale*, 2011, **7**, 15214–15224.
120. X. Tan, L. Feng, J. Zhang, K. Yang, S. Zhang, Z. Liu and R. Peng, *ACS Appl. Mater Interfaces*, 2013, **5**, 1370–1377.
121. C. McCallion, J. Burthem, K. Rees-Unwin, A. Golovanov and A. Pluen, *Eur. J. Pharm. Biopharm.*, 2016, **104**, 235–250.
122. B. Li, X. Y. Zhang, J. Z. Yang, Y. J. Zhang, W. X. Li, C. H. Fan and Q. Huang, *Int. J. Nanomed.*, 2014, **9**, 4697–4707.
123. S. Tenzer, D. Docter, J. Kuharev, A. Musyanovych, V. Fetz, R. Hecht, F. Schenk, D. Fischer, K. Kioptsi, C. Reinhardt, K. Landfester, H. Schild, M. Maskos, S. K. Knauer and R. H. Stauber, *Nat. Nanotechnol.*, 2013, **8**, 772–781.
124. M. Xu, J. Zhu, F. Wang, Y. Xiong, Y. Wu, Q. Wang, J. Weng, Z. Zhang, W. Chen and S. Liu, *ACS Nano*, 2016, **10**, 3267–3281.
125. S. Liu, T. H. Zeng, M. Hofmann, E. Burcombe, J. Wei, R. Jiang, J. Kong and Y. Chen, *ACS Nano*, 2011, **5**, 6971–6980.
126. S. K. Manna, S. Sarkar, J. Barr, K. Wise, E. V. Barrera, O. Jejelowo, A. C. Rice-Ficht and G. T. Ramesh, *Nano Lett.*, 2005, **5**, 1676–1684.
127. X. Liu, S. Sen, J. Liu, I. Kulaots, D. Geohegan, A. Kane, A. A. Puretzky, C. M. Rouleau, K. L. More, G. T. Palmore and R. H. Hurt, *Small*, **7**, 2775–2785.
128. M. I. E. Mejias-Carpio, C. M. Santos, X. Wei and D. F. Rodrigues, *Nanoscale*, 2012, **4**, 4746–4756.
129. J. Chen, H. Peng, X. Wang, F. Shao, Z. Yuan and H. Han, *Nanoscale*, 2014, **6**, 1879–1889.

130. L. Hui, J. G. Piao, J. Auletta, K. Hu, Y. Zhu, T. Meyer, H. Liu and L. Yang, *ACS Appl. Mater Interfaces*, 2014, **6**, 13183–13190.
131. F. Perreault, A. F. de Faria, S. Nejati and M. Elimelech, *ACS Nano*, 2015, **9**, 7226–7236.
132. M. Sametband, I. Kalt, A. Gedanken and R. Sarid, *ACS Appl. Mater. Interfaces*, 2014, **6**, 1228–1235.
133. S. Ye, K. Shao, Z. Li, N. Guo, Y. Zuo, Q. Li, Z. Lu, L. Chen, Q. He and H. Han, *ACS Appl. Mater Interfaces*, 2015, **7**, 21571–21579.
134. M. Sawangphruk, P. Srimuk, P. Chiochan, T. Sangsri and P. Siwayaprahm, *Carbon*, 2012, **50**, 5156–5161.
135. X. Wang, X. Liu, J. Chen, H. Han and Z. Yuan, *Carbon*, 2014, **68**, 798–806.
136. R. Kempaiah, S. Salgado, W. L. Chung and V. Maheshwari, *Chem. Commun.*, 2011, **47**, 11480–11482.
137. B. Wang, P. Liu, W. Jiang, H. Pan, X. Xu and R. Tang, *Angew. Chem., Int. Ed.*, 2008, **47**, 3560–3564.
138. S. H. Yang, T. Lee, E. Seo, E. H. Ko, I. S. Choi and B.-S. Kim, *Macromol. Biosci.*, 2012, **12**, 61–66.
139. M. H. Wahid, E. Eroglu, X. Chen, S. M. Smith and C. L. Raston, *RSC Adv.*, 2013, **3**, 8180–8183.
140. P. F. M. Nogueira, D. Nakabayashi and V. Zucolotto, *Aquat. Toxicol.*, 2015, **166**, 29–35.
141. X. Hu, K. Lu, L. Mu, J. Kang and Q. Zhou, *Carbon*, 2014, **80**, 665–676.
142. C. Pretti, M. Oliva, R. Di Pietro, G. Monni, G. Cevasco, F. Chiellini, C. Pomelli and C. Chiappe, *Ecotoxicol. Environ. Saf.*, 2014, **101**, 138–145.
143. Z. Chen, Q. Wang, M. Asmani, Y. Li, C. Liu, C. Li, M. J. Lippmann, Y. Wu and R. Zhao, *Sci. Rep.*, 2016, **6**, 31304.
144. A. Kermanizadeh, M. Lohr, M. Roursgaard, S. Messner, P. Gunness, J. M. Kelm, P. Moller, V. Stone and S. Loft, *Part. Fibre Toxicol.*, 2014, **11**, 56.
145. Y. Luo, C. Wang, M. Hossain, Y. Qiao, L. Ma, J. An and M. Su, *Anal. Chem.*, 2012, **84**, 6731–6738.
146. E. R. Shamir and A. J. Ewald, *Nat. Rev. Mol. Cell Biol.*, 2014, **15**, 647–664.
147. R. Edmondson, J. J. Broglie, A. F. Adcock and L. Yang, *Assay Drug Dev. Technol.*, 2014, **12**, 207–218.
148. F. Sambale, A. Lavrentieva, F. Stahl, C. Blume, M. Stiesch, C. Kasper, D. Bahnemann and T. Scheper, *J. Biotechnol.* 2015, **205**, 120–129.
149. C. A. McConkey, E. Delorme-Axford, C. A. Nickerson, K. S. Kim, Y. Sadovsky, J. P. Boyle and C. B. Coyne, *Sci. Adv.*, 2016, **2**, e1501462.
150. P. Gunness, D. Mueller, V. Shevchenko, E. Heinzle, M. Ingelman-Sundberg and F. Noor, *Toxicol. Sci.*, 2013, **133**, 67–78.
151. S. Messner, I. Agarkova, W. Moritz and J. M. Kelm, *Arch. Toxicol.*, 2013, **87**, 209–213.
152. S. P. Mukherjee, N. Lozano, M. Kucki, A. E. Del Rio-Castillo, L. Newman, E. Vazquez, K. Kostarelos, P. Wick and B. Fadeel, *PLoS One*, 2016, **11**, e0166816.
153. Z. M. Markovic, L. M. Harhaji-Trajkovic, B. M. Todorovic-Markovic, D. P. Kepic, K. M. Arsikin, S. P. Jovanovic, A. C. Pantovic, M. D. Dramicanin and V. S. Trajkovic, *Biomaterials*, 2011, **32**, 1121–1129.
154. B. Tian, C. Wang, S. Zhang, L. Feng and Z. Liu, *ACS Nano*, 2011, **5**, 7000–7009.
155. D. Sebastian, 2012, Material safety data sheets-Graphenea, Online, https://www.Graphenea.com/pages/msds.
156. A. B. Seabra, A. J. Paula, R. de Lima, O. L. Alves and N. Duran, *Chem. Res. Toxicol.*, 2014, **27**, 159–168.
157. S. Park, J. An, I. Jung, R. D. Piner, S. J. An, X. Li, A. Velamakanni and R. S. Ruoff, *Nano Lett.*, 2009, **9**, 1593–1597.
158. X. Guo and N. Mei, *J. Food Drug Anal.*, 2014, **22**, 105–115.
159. M. Orecchioni, C. Menard-Moyon, L. G. Delogu and A. Bianco, *Adv. Drug Delivery Rev.*, 2016, **105Pt B**, 163–175.
160. A. Sasidharan, L. S. Panchakarla, A. R. Sadanandan, A. Ashokan, P. Chandran, C. M. Girish, D. Menon, S. V. Nair, C. N. Rao and M. Koyakutty, *Small*, 2012, **8**, 1251–1263.
161. H. Zhou, K. Zhao, W. Li, N. Yang, Y. Liu, C. Chen and T. Wei, *Biomaterials*, 2012, **33**, 6933–6942.
162. G. Y. Chen, H. J. Yang, C. H. Lu, Y. C. Chao, S. M. Hwang, C. L. Chen, K. W. Lo, L. Y. Sung, W. Y. Luo, H. Y. Tuan and Y. C. Hu, *Biomaterials*, 2012, **33**, 6559–6569.
163. S. Das, S. Singh, V. Singh, D. Joung, J. M. Dowding, D. Reid, D. Reid, J. Anderson, L. Zhai, S. I. Khondaker, W. T. Self and S. Seal, *Part. Part. Syst. Charact.*, 2013, **30**, 148–157.
164. H. Ali-Boucetta, D. Bitounis, R. Raveendran-Nair, A. Servant, J. van den Bossche and K. Kostarelos, *Adv. Healthcare Mater.*, 2013, **2**, 433–441.
165. A. Schinwald, F. A. Murphy, A. Jones, W. MacNee and K. Donaldson, *ACS Nano*, 2012, **6**, 736–746.

166. J. R. Roberts, R. R. Mercer, A. B. Stefaniak, M. S. Seehra, U. K. Geddam, I. S. Chaudhuri, A. Kyrlidis, V. K. Kodali, T. Sager, A. Kenyon, S. A. Bilgesu, T. Eye, J. F. Scabilloni, S. S. Leonard, N. R. Fix, D. Schwegler-Berry, B. Y. Farris, M. G. Wolfarth, D. W. Porter, V. Castranova and A. Erdely, *Part. Fibre Toxicol.,* 2016, **13**, 34.
167. A. Schinwald, F. Murphy, A. Askounis, V. Koutsos, K. Sefiane, K. Donaldson and C. J. Campbell, *Nanotoxicology*, 2014, **8**, 824–832.
168. J. H. Shin, S. G. Han, J. K. Kim, B. W. Kim, J. H. Hwang, J. S. Lee, J. H. Lee, J. E. Baek, T. G. Kim, K. S. Kim, H. S. Lee, N. W. Song, K. Ahn and I. J. Yu, *Nanotoxicology*, 2015, **9**, 1023–1031.
169. M. C. Duch, G. R. Budinger, Y. T. Liang, S. Soberanes, D. Urich, S. E. Chiarella, L. A. Campochiaro, A. Gonzalez, N. S. Chandel, M. C. Hersam and G. M. Mutlu, *Nano Lett.,* 2011, **11**, 5201–5207.
170. X. Wang, M. C. Duch, N. Mansukhani, Z. Ji, Y. P. Liao, M. Wang, H. Zhang, B. Sun, C. H. Chang, R. Li, S. Lin, H. Meng, T. Xia, M. C. Hersam and A. E. Nel, *ACS Nano*, 2015, **9**, 3032–3043.
171. N. M. Latiff, Z. Sofer, A. C. Fisher and M. Pumera, *Chemistry*, 2017, **23**, 684–690.
172. P. Miro, M. Audiffred and T. Heine, *Chem. Soc. Rev.*, 2014, **43**, 6537–6554.
173. R. Mas-Balleste, C. Gomez-Navarro, J. Gomez-Herrero and F. Zamora, *Nanoscale*, 2011, **3**, 20–30.
174. T. Liu, S. Shi, C. Liang, S. Shen, L. Cheng, C. Wang, X. Song, S. Goel, T. E. Barnhart, W. Cal and Z. Liu, *ACS Nano*, 2015, **9**, 950–960.
175. M. Jaishankar, T. Tseten, N. Anbalagan, B. B. Mathew and K. N. Beeregowda, *Interdiscip. Toxicol.*, 2014, **7**, 60–72.
176. P. B. Tchounwou, C. G. Yedjou, A. K. Patlolla and D. J. Sutton, *Exs.*, 2012, **101**, 133–164.
177. W. Z. Teo, E. L. K. Chng, Z. Sofer and M. Pumera, *Chem. – Eur. J.*, 2014, **20**, 9627–9632.
178. F. Hazane-Puch, P. Champelovier, J. Arnaud, C. Trocme, C. Garrel, P. Faure and F. Laporte, *Metallomics*, 2014, **6**, 1683–1692.
179. J. Noth, J. Esselborn, J. Guldenhaupt, A. Brunje, A. Sawyer, U. P. Apfel, K. Gerwert, E. Hofmann, M. Winkler and T. Happe, *Angew. Chem.*, 2016, **55**, 8396–8400.
180. X. Wang, N. D. Mansukhani, L. M. Guiney, Z. Ji, C. H. Chang, M. Wang, Y.-P. Liao, T.-B. Song, B. Sun, R. Li, T. Xia, M. C. Hersam and A. E. Nel, *Small*, 2015, **11**, 5079–5087.
181. K. Krishnamoorthy, M. Veerapandian, K. Yun and S. J. Kim, *Colloids Surf., B*, 2013, **112**, 521–524.
182. T. Anh Tran, K. Krishnamoorthy, Y. W. Song, S. K. Cho and S. J. Kim, *ACS Appl. Mater. Interfaces,* 2014, **6**, 2980–2986.
183. S. M. Hussain, K. L. Hess, J. M. Gearhart, K. T. Geiss and J. J. Schlager, *Toxicol. In Vitro*, 2005, **19**, 975–983.
184. M. Park, N. Lee, S. H. Choi, K. An, S.-H. Yu, J. H. Kim, S.-H. Kwon, D. Kim, H. Kim, S.-I. Baek, T.-Y. Ahn, O. K. Park, J. S. Son, Y.-E. Sung, Y.-W. Kim, Z. Wang, N. Pinna and T. Hyeon, *Chem. Mater.*, 2011, **23**, 3318–3324.
185. W. Utembe, K. Potgieter, A. B. Stefaniak and M. Gulumian, *Part. Fibre Toxicol.*, 2015, **12**, 11.

# CHAPTER 4

# Interactions Between 2D Graphene-Based Materials and the Nervous Tissue

*Mattia Bramini,*[1,2] *Giulio Alberini,*[1,3] *Fabio Benfenati,*[1,2,3] *Luca Maragliano*[1,*] *and Fabrizia Cesca*[1,2,*]

## Introduction

Graphene (G) is a single- or few-layered sheet of $Sp^2$-bonded carbon atoms tightly packed to a two-dimensional (2D) honeycomb lattice.[1] Each carbon atom has three µ-bonds and an out-of-plane π-bond that can bind with neighboring atoms, conferring graphene unique chemical and physical properties.[1] The family members of graphene-related materials (GRMs) include: single- and few-layered graphene (1–10 layers; GR), graphene oxide (single layer, 1:1 C/O ratio; GO), reduced graphene oxide (rGO), graphite nano- and micro-platelets (more than 10 layers, but < 100 nm thickness and average lateral size in the order of the nm and µm, respectively), graphene and graphene oxide quantum dots (GQDs and GOQDs, respectively), and a variety of hybridized graphene nanocomposites.[2–4] The diversity among the GRMs is mainly dependent on the graphene production method; chemical vapor deposition (CVD),[5,6] mechanical cleavage,[7] and electrochemical exfoliation of graphite,[8] can in fact give the material a wide range of specific properties based on the number of layers, lateral dimension, purity and defect density, conductivity, surface chemistry and shape.[4,9–12]

[1] Center for Synaptic Neuroscience & Technology (NSYN@UniGe), Istituto Italiano di Tecnologia, Largo Rosanna Benzi, 10, 16132 – Genova, Italy.
[2] Graphene Labs, Istituto Italiano di Tecnologia, Via Morego 30, 16163 – Genova, Italy.
[3] Department of Experimental Medicine, University of Genova, Viale Benedetto XV 3, 16132, Genova, Italy.
[*] Corresponding authors: luca.maragliano@iit.it; fabrizia.cesca@iit.it

G and GRMs also possess tunable and extreme mechanical strength, exceptionally high electronic and thermal conductivities, flexibility and transparency.[7] Therefore, G has all the characteristics to play a key role in many applications, opening new advantageous opportunities in supercapacitors,[13,14,15] flexible electronics,[16,17] printable inks,[18,19] batteries,[20,21] optical and electrochemical sensors,[22–24] and energy storage.[25–27]

In the last few years, biomedical applications of G have attracted an ever-increasing interest, including the use of G and GRMs for bioelectrodes, bioimaging, drug/gene/peptide delivery, nanopore-based DNA-sequencing devices, stem cell differentiation and tissue engineering (Fig. 1).[28,29] Moreover, GRMs have generated great interests for the design of nanocarriers and nanoimaging tools, tissue scaffolds (both two- and three-dimensional), anti-bacterial coatings and biosensors.[30,31]

In this chapter we discuss G and GRMs biomedical applications to neuroscience, distinguishing between G nanosheets and G substrates, and include a future perspective on the role of computational modeling in predicting graphene-biological system interactions. For each approach, we describe the state of art in terms of bio-applications and outline the biocompatibility issues raised by nano-toxicological studies performed *in vitro* and *in vivo*.

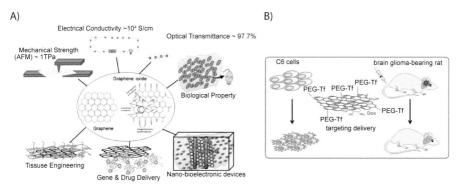

**Fig. 1.** *Biomedical applications of graphene-related materials (GRMs).* (A) Schematic illustrating some properties of G (high mechanical strength and conductivity) and possible biomedical applications, ranging from drug/gene delivery, to biosensing, imaging, phototherapy and tissue engineering. Modified with permission from Goenka et al.[50] (B) Example of G-based drug delivery system to the CNS. Transferrin (Tf) was chosen as the ligand to develop Tf-conjugated PEGylated graphene oxide (GO) nanosheets for loading and glioma targeting delivery of anticancer drug doxorubicin (Dox; Tf-PEG-GO-Dox). Tf-PEG-GO-Dox displayed greater intracellular delivery efficiency and stronger cytotoxicity against C6 glioma cells *in vitro* and *in vivo*. Modified with permission from Liu et al.[51]

## Graphene and the Brain

Neurological diseases, such as neurodegenerative diseases (i.e., Parkinson's disease, Alzheimer) and brain tumors (glioblastoma, neuroblastoma), are characterized by abnormal cell functionality and cell deterioration. Developing new tools that could outperform the current state of the art technologies for imaging, drug delivery, neuronal regeneration and electrical recording and sensing is one of the main challenges of modern medicine and neuroscience.[32] With the development of carbon nanomaterials, and ultimately of graphene, nanotechnology is likely to play a major role and impact on (i) drug, gene and protein delivery, to cross the blood-brain barrier (BBB) and reach brain compromised areas; (ii) neuro-regenerative techniques, to restore cell-cell communication upon damage; and (iii) highly specific and reliable diagnostic tools, for *in vivo* sensing of disease biomarkers.[33,34] Moreover, the combination of different forms and states of the material, diverse chemical functionalization and the possible association with other biomaterials to form G-based composites, may allow to devise an all-in one tool for both diagnosis and therapy, thus effectively building a powerful theranostic device.

To date, the possible bio-applications of graphene that have been proposed and explored at the central nervous system (CNS) level include: (i) cell labeling and real-time live-cell monitoring;[35,36] (ii) delivery to the brain of molecules that are usually rejected by the BBB;[37,38] (iii) G-based scaffolds for cell culture;[39–41] and (iv) cell analysis based on G-electrodes (Fig. 1).[42,43] Additionally, interfacing graphene with neural cells could be extremely advantageous for exploring their electrical behavior or facilitating neuronal regeneration by promoting controlled elongation of neuronal processes.[44–46] All these applications open up new research lines in neuro-therapeutics, including neuro-oncology, neuro-imaging, neuro-regeneration, functional neuro-surgery and peripheral nerve surgery.[34]

One major concern about graphene biomedical applications relies upon its biocompatibility and biosafety. The different biological impact of G nanosheets and GRMs depends on many features such as layer number, lateral dimension, surface chemistry, contact area and material purity,[2,47–49] and has to be clearly assessed when investigating possible toxic effects. In this perspective, a detailed *in vitro/in vivo* toxicity profile is still not available for every G/GRM material; however several groups have contributed to this topic and a general picture is beginning to emerge. Here, we tried to summarize the main findings related to both G- and G-scaffold biocompatibility with nerve tissues.

### *Graphene nanosheet biocompatibility in the central nervous system*

As for many other nanomaterials, the G toxicity profile is highly dependent on the size, surface functionalization, exposure time and aggregation state in biological fluids.[2,52] The capacity of G to adsorb biomolecules onto its surface needs to be taken into consideration. In fact, when using these materials both *in vitro* and *in vivo*, the flake surface gets covered by various biomolecules (proteins, lipids, etc.), drastically changing most of the properties of G towards the biological systems such as hydrophobicity, topography, size and surface charges, influencing the immune reaction and the interaction between the material and the organism.[53,54,55] The fact that G can reach the CNS and get in contact with neuronal cells, poses further concerns about its safety.[56,57] Published data suggest that GO is less toxic than G, rGO and hydrogenated-G. Additional factors shown to affect toxicity include surface functionalization, size (with smaller nanosheets being less toxic than large flakes) and solubility (where highly dispersible G solutions are safer than aggregating ones). G shows very little biological degradation, although carboxylated and oxidated derivatives may form under some conditions.[58,59]

The importance of assessing toxicological studies on nerve tissues comes together with the evidence that G nanosheets, and especially GO flakes, are able to accumulate (even if in small quantities) in the CNS after intravenous (*i.v.*) injection without prior surface functionalization.[56,57] In a recent study, rGO was detected in brain tissues, particularly in the thalamus and hippocampus, after *i.v.* injection that was accompanied by BBB disruption, raising additional concerns on using G nanosheets for biomedical applications.[57] Interestingly, rats treated with rGO flakes were screened for signs of tremor, convulsions, salivation, lacrimation, dyspnea and motor abnormalities and did not show any clinical signs of neurotoxicity. These findings are in contrast with the work carry out by Zhang et al.,[60] whose results displayed a short-term decrease in locomotor activity and neuromuscular coordination in mice orally administered with rGO nanosheets. This discrepancy underlines that the route of administration is another key parameter in determining G biocompatibility. Thus the portal of entry of G into the organism, together with its dose, size, functionalization and aggregation, will determine the final biological effects.

The interactions of GRMs with neurons and astrocytes, the two main cell populations of the CNS, have been investigated at the cellular level. No changes in neuronal and glial cell viability were detected upon G exposure, both *in vivo* and *in vitro*,[57,61,62] However, primary neuronal cultures exposed to GO nanosheets displayed clear alterations in a number of physiological pathways, such as calcium and lipid homeostasis, synaptic connectivity and plasticity (Fig. 2C).[61,62] These effects were observed upon chronic G exposure, stressing the need of urgent and further biocompatibility assessment of the material with nerve tissues in long-term studies, hopefully linking *in vivo* effects to *in vitro* cellular and molecular interactions.

Once internalized in cells, G flakes were seen to preferentially accumulate in lysosomes (Fig. 2B), as well as to damage mitochondria, endoplasmic reticulum and, in some cases, nuclei.[33] Another study also suggested that irregular protrusions and sharp edges can damage the plasma membrane, thus letting G nanosheets enter the cell by piercing the phospholipid-bilayer (Fig. 2A).[63] These features raise additional safety concerns, as free GRMs in the cytoplasm may lead to disruption of the cytoskeleton, impaired cell motility and blockade of the cell-cycle, in a carbon nanotube-like cytotoxic fashion.

A first evidence of G-induced CNS toxicity came from a recent *in vivo* study.[64] To recreate a situation of G environmental pollution, researchers dispersed GO in water in the presence of zebrafish larvae. In this scenario, GO was found to reach the CNS and, most importantly, to induce Parkinson's disease-like symptoms such as disturbance of locomotor activity, dopaminergic neurons loss and formation of Lewy bodies. These effects could be due to mitochondrial damage and apoptosis induction through the caspase 8 pathway, in the presence of a more general metabolic disturbance.

In summary, while toxicity assays showed a good biocompatibility of graphene scaffolds with neuronal cells (see next section), graphene nanosheets may have some toxic effects compared to other nanomaterials, leaving open the debate whether use them or not as drug-delivery system.[62] It is important to notice that the variable and sometimes contrasting results about graphene biocompatibility available in the literature most likely reflect the high heterogeneity of materials present on the market and the large variety of synthesis methods. Thus, depending on the graphite source

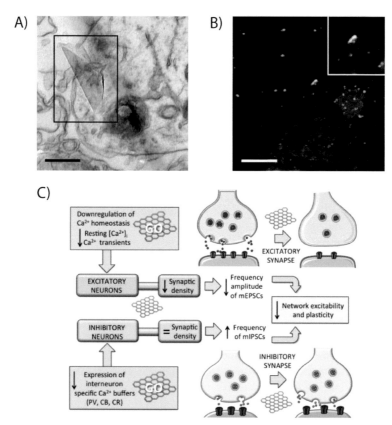

**Fig. 2.** *Graphene nanosheet biocompatibility with neuronal cells.* (A) Example of GO nanosheet piercing the cell membrane of a neuron and entering the cell through lipid-bilayer disruption (scale bar 200 nm). (B) Neurons exposed to GO nanosheets were fixed and stained with antibodies anti-EEA1 (green, for early endosomes) and anti-LAMP1 (red, for lysosomes). Cell nuclei are in blue (DAPI) and graphene in white (by reflection light acquisition). Neurons were imaged by fluorescence confocal microscopy and GO accumulation in lysosomes was noticed over time (scale bar 20 μm). (C) Schematic representation of the mechanisms of the effects of chronic exposure of GO on the activity of neuronal networks. Modified with permission from Bramini et al.,[62] copyright (2016) American Chemical Society.

(starting material), the synthesis method used, the chemical (if needed) adopted and the final product dispersion form (solution or powder), graphene can present different sizes, thickness, chemical surface and aggregation state, which all affect to various extent its interaction with biological systems. Thus, it will be necessary to conduct a full and detailed study of the biocompatibility and biodistribution of each type of material in use.

## Graphene functionalization for biomolecule delivery to the central nervous system

The term 'G nanosheets' includes GRMs with a lateral dimension in the order of nanometers.[2] Moreover, to be considered graphene (and not graphite), the sheets have to be composed of no more than 10 layers. Finally, depending on the surface chemistry composition, G nanosheets can be classified as pristine graphene (GR), graphene oxide (GO) and reduced graphene oxide (rGO).[2,32,65]

The large surface area available and the possibility of conjugating different biomolecules onto its surface, make G a suitable material for holding and carrying small-molecules, drugs, genes, antibodies, proteins, DNA and RNA sequences (including siRNA and miRNA).[66] In addition, it is also possible to modify its chemical structure by functional groups such as amino, carboxyl, hydroxyl, alkyl halogen, or azide groups.[33] All these functionalizations have a dual goal: (i) loading high quantity of biomolecules and specifically deliver them to target cells; (ii) allowing a more homogenous dispersion of the material, since pure G is highly hydrophobic and tends to aggregate in aqueous solution, including biological fluids containing salts and proteins.[33,34] As already mentioned, the high surface area, the π–π stacking, and the electrostatic and/or hydrophobic interactions of G can be exploited to achieve high drug loading of otherwise poorly soluble drugs, without compromising and affecting their efficacy. Additionally, functionalized G nanosheets could be applied in systemic, targeted, and local delivery systems.[67–69] Because of its unique fluorescent, photoacustic and magnetic resonance profiles, several studies have also explored the possibility of incorporating G-based nanoparticles to enhance the *in vivo* visualization of brain tumors and improve tumor targeting of molecular anticancer strategies.[69,70] As a matter of fact, G surface is endowed with bi-functional units, capable of loading drugs (or other biomolecules) and simultaneously assessing drug delivery through target protein recognition.[71,72] *In vivo* studies revealed that GO, more than GR, has good potential for neuroscience applications. In fact, systemically administered radiolabeled GO ($^{188}$Re-GO) could reach the brain parenchyma, although in a small amount (0.04%).[73]

### Blood-brain barrier crossing

As mentioned above, the main limitation for G application in neuroscience is its very low accumulation in the brain parenchyma upon intravenous injection. Among the various approaches to cross the BBB, ultrasounds were applied to mice to physically open the tights junctions of the BBB endothelium layer and permit the drug delivery system (GO nanosheets grafted with Gd-DTPA and poly(amidoamine) dendrimer, loaded with EPI and *Let-7*, a tumor suppressor miRNA) to reach the brain upon tail vein injection. The main advantage of this approach is the reversibility of the BBB opening. Interestingly, G could at the same time also make possible high contrast MRI analysis and a quantification of the distribution of the delivery system inside the brain tissue.[74] These results are promising, however in-depth pharmacokinetics and toxicological studies are needed, especially in the long-term, keeping in mind that, with respect to what has been obtained so far, this technique achieves a much higher G accumulation in the CNS.

Alternatively, the G surface can be functionalized with specific biomolecules that enable the material to cross the BBB.[33,50,75] Researchers created G-PEG nanoparticles conjugated to transferrin, known to cross the BBB through transcytosis (Fig. 1B). After loading them with doxorobucin (DOX) they demonstrated that, after intravenous injection, the particles migrated from the bloodstream to glioma cells that had previously been implanted in the rat brain.[51] The results showed higher nanoparticles concentration in the CNS compared to other organs, and a higher efficiency in tumor regression compared to the control of DOX alone. Following the same approach with similar promising results, Yang et al.[76] functionalized PEG-GO nanosheets with the Tat protein of the Human Immunodeficiency Virus (HIV), which allowed the drug-loaded PEG-GO system to cross the BBB by transcytosis, while leaving the barrier endothelium fully preserved.[76]

Another promising strategy to challenge the BBB filter is the coating with surfactants.[77,78] Kanakia et al.[79] improved GO delivery to the CNS by functionalizing the nanosheets with dextran; the material was found to cross the BBB and reach the brain without exerting toxic effects. Surprisingly, the GO concentration in the CNS increased with time, being almost absent in other organs. Thus, the study suggests a slow accumulation of G in the CNS and long-term persistency of the material, that is encouraging from the point of view of the drug delivery system, but also raises safety concern on long-term toxicity of G nanosheets.[32]

### *Drug and gene delivery to the central nervous system*

The key goal of any drug delivery system is to create a smart tool that recognizes specific targets and releases the drug in a controlled way.[75] Because of its features, G could play a major role in the development of novel drug delivery systems. For example, water insoluble hydrophobic bioactive reagents can attach to the surface of G nanosheets through hydrophobic, Van der Waals or $\pi$–$\pi$ interactions. Moreover, G can be further modified to be soluble in the aqueous environment, either by grafting water-soluble molecules onto it, or by increasing its oxidation state (thus obtaining GO).[80]

Researchers have already successfully linked some drugs to G, confirming its potential as drug delivery system. For example, GO nanosheets functionalized with branched, biocompatible PEG, become more hydrophilic under physiological conditions.[81] Liu and colleagues[81] showed that GO-PEG flakes can be decorated with the water insoluble aromatic molecule 7-ethyl-10-hydroxy-camptothecin (SN38), via non-covalent van der Waals interactions. Similarly, other drugs, including different camptothecin analogues,[81] Iressa (gefitinib),[81] and DOX,[82] were successfully attached onto the GO-PEG complex by simple non-covalent binding. Moreover, PEG functionalization renders of G nanosheets less cytotoxic for the CNS. rGO-PEG particles were able to cross the endothelial layer of the BBB without disrupting the tight junctions, in both *in vitro* and *in vivo* studies.[56,57]

One of the main applications of G-based drug delivery systems is anticancer therapy, by linking G composites with chemotherapeutics. Given their strong optical absorbance in the near-infrared (NIR) region, G-based hybrid materials are also intensively studied for their promising applications in cancer phototherapy.[83,84,70,85] The rationale beyond this approach is to exploit the heat produced by the G accumulated in tumor regions upon NIR laser stimulation to kill cancer cells. This technique was successfully applied *in vitro* using U251 glioma cells.[86] Such experimental approaches are of special interest, as they might help overcoming the limitations imposed by the BBB.[87] Moreover, as previously described, G can be modified and functionalized for carrying multiple drugs, potentially enhancing their cytotoxicity to the cancer cells.[88] Such innovative strategy is very promising, especially for the treatment of very resistant and aggressive tumors, such as the glioblastoma.

In addition, the intrinsic properties of G in the visible (VIS) and NIR range make it an attractive tool for bio-imaging,[71] both *in vitro* and *in vivo*.[89] For example, aptamer-carboxyfluorescein/GO complexes were employed for intracellular monitoring and *in situ* molecular probing of specific clusters of living cells, such as artificially implanted tumors in mice. Moreover, GO nanosheets were used for photo-acoustic imaging, which relies on the acoustic response on heat expansion following optical energy absorption.[90–92] Specifically for CNS applications, *in vivo* studies showed that intracranial administered PEG-GO and its derivatives can be imaged in the brain by two-photon microscopy.[91] Through this imaging technique, a 3D distribution map can be reconstructed in the brain parenchyma due to the high tissue penetration of the fluorescence signal of PEG-GO composites.

These promising results could lead to the use of G as a diagnostic tool for imaging brain cancerous lesions, especially if the material is engineered with biomolecules that specifically target tumorigenic cells. Furthermore, once the targeting is achieved, G properties can be optimized accordingly to the specific application, i.e., the size and oxidation state might be changed to shift the emission wavelength from VIS to NIR, which has a deeper tissue penetration, thus improving the depth of the diagnostic imaging device. By combining the optical properties of G with other biodegradable and functional materials, it will be possible to create G-based composites and hybrids suitable for several live-imaging applications. So far, most of the tools have been tested *in vitro* on cancer cell lines, and *in vivo* for cancer detection and diagnosis, leaving unaddressed the possibility of using them to explore and image the CNS.[3,71]

Similarly to drug delivery, also genetic engineering can exploit G properties and open new opportunities in biomedicine. The concept in this case would be to deliver nucleic acids (DNA or various types of RNA molecules, including miRNA and shRNA) to specific target cell populations, to restore physiological conditions.[3] For this purpose, carbon allotrope G has been investigated as a vector for gene, gene-drug and protein delivery.[33,69] Being the technique of functionalization the same, both drugs and genes can be delivered simultaneously using G-based hybrid materials.[93] The latter option would exhibit a synergic effect, as it would bring a significant enhancement of drug efficacy as well as transfection efficiency, making G a promising tool for future applications in non-viral based gene therapy. On this line, G-nanosheets were functionalized with the cationic polymer PEI, a non-viral gene vector that forms strong electrostatic interactions with the negatively charged phosphate groups of both RNA and DNA.[68] A step further was taken by Chen et al., that used PEI-functionalized GO for gene delivery yielding a high transfection efficiency in the absence of any cytotoxic effect.[94]

In summary, G-based delivery systems, when conveniently functionalized or associated with complementary technologies, represent promising candidates for both diagnostics (i.e., imaging) and therapeutics (i.e., drug and gene delivery) neuroscience applications. Moreover, in spite of few studies showing toxic effects of exposure the nervous system to bare G and rGO,[57,61,62] to date there is no solid evidence that functionalized-G is harmful to neuronal cells and the BBB. Since G-based technologies for biomedical applications are constantly and rapidly evolving, the near future may see the development of new safe and highly neurocompatible materials.

### *Graphene biosensing in the brain*

Electrical conductivity is one of the most interesting properties of G, which can potentially be exploited to amplify biological signals in innovative diagnostic tools.[71] This new arena is however still under development, and further research is needed, especially for applications in the neuroscience field. The general idea is to generate 2D G-based composite structures with the ability to recognize specific biomacromolecules (antibodies, peptides, drugs, etc.) for biomarker detection. For this purpose, both 2D planar G and G nanosheets are very promising for the development of the diagnostic tools. In several instances, G was introduced in previously developed biosensing tools, to improve their sensitivity and, in some cases, to improve the biosensor's specificity.[95]

One of the first successful attempts to exploit G properties for biosensing consisted in the detection of PSA, a prostate cancer marker. Specific anti-PSA antibodies were conjugated with bovine serum albumin (BSA) and G nanosheets. The resulting composite exploits the specificity of the antibody-antigen binding to recognize the cancer marker (PSA), while taking advantage of G conductivity, which changes depending on the biomolecules attached onto its surface, to detect and quantify the signal.[96] Theoretically, this can be done with many other tumor and non-tumor biomarkers.[71] Indeed, G-based sensors showed excellent performances for detection of various markers *in vitro*, with high accuracy and reproducibility; unfortunately, few studies have been carried out *in vivo* and the systems still have to find an application for CNS-related diseases. In this regard, initial standard testing has demonstrated that new biosensors based on graphene technology can detect with high accuracy important enzymes involved in the synthesis and metabolism of neurotransmitters, such as acetylcholinesterase.[97]

Very recently, graphene has been used to implement both *in vitro* and *in vivo* neuronal biosensing by coupling to electrodes for chemical and electrophysiology recordings.[98,99] These devices aim to a continuous monitoring of the brain microenvironment, especially under pathological conditions. Graphene has been incorporated to create new generation of electrochemical sensors, with robust stability, flexibility, high strength and low electrical resistance, that are able to detect extracellular concentrations of various molecules (such as glucose or dopamine) and simultaneously record electrical activity of the neuronal network.[34,100] The latest development of new molecular and electrochemical tools[101] has special importance and impact for experimental neuroscience research, opening to promising, innovative and effective technologies for fast and accurate neuro-pathologies diagnosis, monitoring and therapy.

# Graphene Substrates as Neuronal Interfaces
## *in vitro* and *in vivo*

## 2D Supports

### *CVD-graphene*

A number of studies characterized the biocompatibility of 2 dimensional G substrates, mostly CVD-G, with stem cells of different origins,[102–105] including neural stem cells (NSCs).[106–111] A general shortfall of these studies is that, because of technical limitations, in many cases the precise number of G layers composing the substrates was not precisely defined, although the number of graphene layers is known to have a great influence on the physico-chemical properties of a graphene-based platform.[112] Notwithstanding these technical considerations, G devices have shown very interesting properties when used as neuronal interfaces, in that the specific physico-chemical features of G substrates can drive stem cell differentiation towards different lineages.[104,113,114] One possible explanation for this observation is that G substrates act as concentration platforms for trophic factors. For example insulin, an inducer of adipogenic differentiation, is denatured on G, but not denatured on GO supports, with different effects on the differentiation of bone marrow-derived mesenchymal stem cells.[103] On the other hand, the aromatic scaffold of G-materials can increase the local concentrations of extracellular matrix (ECM) proteins such as collagen, laminin and fibronectin by non-covalent binding, thus affecting cell adhesion and differentiation.[115] Besides the interaction with bioactive molecules, the supports themselves can be functionalized to improve cell adhesion. On this line, fluorinated graphene was shown to drive stem cell to the neuronal lineage in the absence of chemical inducers.[115] NSCs grown on CVD-G displayed enhanced electrical activity,[107] and moreover, the G film itself could be used as stimulating electrode to enhance differentiation of human neural stem cells into neurons.[106] Modeling studies suggested that the electric field produced by the electronegative cell membrane is higher on G-substrates than on standard devices, thus leading to a more negative resting membrane potential. This was suggested to contribute to the faster differentiation of NSCs grown on G into neurons.[111] The conductive properties of G substrates can be also externally modulated. Indeed, flash photo-stimulation of human NSCs on G/TiO$_2$ heterojunction films[108] or G nanogrids[109] could induce differentiation of the neuronal lineage.

   Besides stem cells, the biocompatibility of G substrates with primary neurons (see discussion below) and with microglial cells[116] has been investigated (Fig. 3). Peptide-free CVD-G layers support the growth of primary retinal and hippocampal neurons, which maintained physiological excitability and network activity.[45,117] Interestingly, the crystalline quality of the graphene substrates directly affects their biocompatibility, with low crystallinity being fully repellent and high crystallinity being highly adhesive for neurons.[118] Primary rodent neurons on PLL-coated CVD-G displayed increased neurite sprouting and branching, spine density, synapse formation and activity.[44,119,120] Interestingly, PDL can be patterned on CVD-G to achieve directional neuronal growth.[118,121]

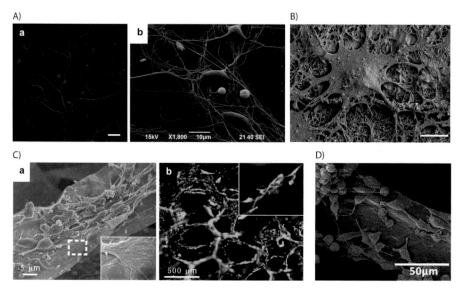

**Fig. 3.** *Neural cells grow on two- and three-dimensional G scaffolds.* (A) Immunofluorescence (a) and SEM (b) images of primary cortical neurons grown on 2D CVD-G supports. Neurons form a healthy network, in strict adhesion with the substrate. In a: β-III tubulin in red, DAPI-stained nuclei in blue. Scale bar, 50 μm (Bramini et al., unpublished). (B) Field emission-SEM of differentiated NSCs cultured on PCL/graphene nanofiber hybrid scaffolds after six days of culture. The differentiated cells (pseudo-colored blue for contrast) on the graphene nanofiber hybrid scaffolds show a clear oligodendrocyte-like differentiated morphology. Scale bar: 10 μm. Modified with permission from Shah et al.[122] (C) (a) SEM images of NSCs cultured on 3D-G foams under proliferation conditions. The insets illustrate the interaction between the cell filopodia and surface. (b) Fluorescence images of NSCs cultured on 3D-G foams for 5 days. Nestin (green) is a marker for neural stem cells, and DAPI (blue) identifies nuclei. Modified with permission from Li et al.[39] (D) Representative SEM image of BV2 microglial cells on 3D-graphene substrates. Modified with permission from Song et al.[116]

## *Graphene oxide*

Stem cells growth has been examined also on GO/rGO substrates.[103,105,123–126] Microcontact printing of GO grid patterns on a variety of substrates could drive the efficient conversion of mesodermal stem cells to ectodermal neuronal cells, due to the ability of the grid patterns to mimic interconnected/elongated neuronal networks.[127] Similarly, laminin-coated arrays of hybrid silica/GO nanoparticles could drive axonal alignment and enhanced neuronal differentiation of NSCs.[128] The surface of GO/PEDOT devices was functionalized by cross-linking of interferon-γ (IFNγ) and platelet-derived growth factor (PDGF), which selectively drove NSCs toward either neuronal or oligodendrocyte lineage.[129] Similar to the above-described approach for G substrates, the conductive properties of GO nanomesh platforms were challenged by near infrared laser stimulation, and exploited to drive human NSCs toward neuronal differentiation.[130]

Neuronal growth is supported also by PDL-coated rGO films, even if the functional interface between these substrates and primary neurons has not been thoroughly investigated yet.[131] With the aim of improving the adhesion of primary neuronal cultures, GO surfaces were functionalized by covalently bonding acetylcholine-like (dimethylaminoethyl methacrylate, DMAEMA) or phosphorylcholine-like (2-methacryloyloxyethyl phosphorylcholine, MPC) units[132] or amino-($-NH_2$), poly-*m*-aminobenzene sulfonic acid-($-NH_2/-SO_3H$), or methoxyl-($-OCH_3$) functional groups. In the latter case, positively charged GO was found to be more beneficial for neurite outgrowth and branching.[46]

## *Fibers*

In the attempt to build 3D supports more closely mimicking the extracellular environment, PCL nanofiber meshes were coated with GO at various concentrations, and were shown to support the growth of both primary cortical neurons[133] and stem cells (Fig. 3).[134] In the latter case, higher GO concentrations promoted the differentiation of stem cells into oligodendrocytes, increasing the expression of oligodendrocyte-specific integrin-related molecules and signaling.[122] The conductive properties of GO could also be exploited in the nanomesh structure. Indeed, pulsed electrical stimulation was applied to poly(3,4-ethylenedioxythiophene) (PEDOT)-rGO hybrid microfibers to promote neuronal differentiation of mesenchymal stem cells,[134] and to GO-coated electrospun poly(vinyl chloride) nanofibers to drive growth and development of primary motor neurons.[135]

## *3D foams*

*In vivo* neuroscience applications of G-based materials will only be possible upon development of three-dimensional scaffolds able to support nerve regeneration across the injured/lesioned site (Fig. 4). Various three-dimensional G and GO foams were shown to be compatible substrates for stem cells.[39,136–139] Interestingly, the features of the G scaffolds (i.e., stiff *vs.* soft) differentially affected cell adhesion and proliferation and could drive neural stem cell differentiation toward the astrocyte and neuronal lineages, respectively.[140] Microglial cells were also grown on 3D G foams (Fig. 3). In this case, the 3D structure of the scaffolds affected the neuroinflammatory response of the cultured cells, probably because of spatial constraints due to the 3D topographic features.[116] Similar to what described for 2D materials, also 3D G/GO scaffolds were used as cell stimulating electrodes, to drive neuronal growth and differentiation of NSCs.[39,141]

Several 3D scaffolds have been generated and tested *in vitro*, however so far only a very limited number of them have also been implanted *in vivo*. Some examples include G-coated electrospun PCL microfiber scaffolds, which were implanted in the striatum or the subventricular zone of adult rats (Fig. 4). G-coated implants were associated with a lower microglia/macrophage infiltration when compared to bare scaffolds, while supporting astrocytes and neuroblast migration from the SVZ.[142] Free-standing 3D GO porous scaffolds were implanted in the injured rat spinal cord, showing no local or systemic toxicity and a good biocompatibility also in the case of chronic implantation. Of note, long-term (30 days) implants were able to promote angiogenesis and partial axonal regeneration.[143]

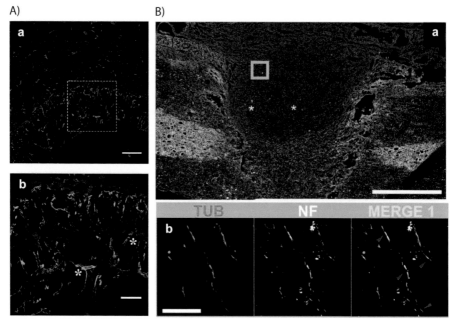

**Fig. 4.** *3D G scaffolds implanted in vivo.* (A) (a,b) Brain astrocyte/G-scaffolds interaction and astrocyte process infiltration 3 weeks after scaffold implantation. Green: GFAP-positive astrocytes, blue: DAPI-stained nucleus, red: surface-functionalized scaffolds. (b) Detailed astrocyte morphology of the dash-box indicated area in (a). * indicate astrocytes that bridge a gap between two scaffold layers. Scale bar, 50 (a) and 20 (b) μm. Modified with permission from Zhou et al.[142] (B) Representative immunofluorescence images for axon growth detection inside rGO scaffolds implanted in the injured spinal cord and analysed at 30 days post-injury. (a) Panoramic view, axons inside the scaffold structure marked with white asterisks. (b) Higher magnification images corresponding to the area indicated by the gray square in the panoramic view. Axons were specifically detected by expression of b-III tubulin (TUB, green) and neurofilaments (NF, white). Red arrowheads indicate areas with coexistence of TUB and NF. Scale bars: 1 μm (top) and 100 μm (bottom). Modified with permission from Lopez-Dolado et al.[144]

## Modeling and Classical Simulations of Graphene Biosystems

The comprehension of the subtle details underlying interactions between biomolecules and inorganic surfaces is pivotal to the rational design of new classes of tools for a wide range of applications in nanomedicine. Although relevant experimental results about the dynamics of these interactions have been recently reported, many topological details of the mechanisms involved are still elusive. To overcome this gap between the structural properties and the associated functions, molecular modeling and classical molecular simulations can give an irreplaceable contribution, obtaining details not accessible by experimental techniques and improving the knowledge of the possible biological safety and applications of these materials.[145]

With its many properties, graphene has shown great potential in various applications, and its interest in computational simulations grows incredibly.[146] Here, we introduce a brief description of the applications of classical simulations of biomolecules-graphene complexes, broadly studied at a multiscale level. The most important problem referring to classical simulations of these systems, currently not fully resolved, is the formulation of an appropriate model, in terms of force fields (FFs) parameters. Indeed, although the computational approaches employed to study biomolecules and graphene are quite similar, the models and FFs used for the description of the physical properties can differ. Today, different compromises in the allocation of the FFs parameters of graphene exist. Specifically, in the context of classical all atom molecular dynamics (MD) simulations, the choices used by the Patra group[147,148] or similar variants, and the treatment of carbon atoms in graphene as uncharged Lennard-Jones spheres[149] are commonly accepted. A description of the choices of the parameters of graphene in difference FFs has been recently reported.[150]

Several computational results about graphene-biomolecules systems have been recently published, focusing on:

1. the adsorption of proteins/peptides on graphene substrates, to elucidate the structural stability of these systems and the evolution of the secondary and tertiary structure of the proteins, in the context of substrate self-organization of proteins into functional architectures for biomedical applications.
2. the interactions of graphene with biomembranes, to assess biological safety/ toxicity of graphene as well as its promising function as a vector of new classes of antibiotics or pharmaceutical products able to disrupt pathogenic cells. In this respect, the migration of graphene nanosheets inside the cellular environment is of particular interest.
3. the computational investigation of DNA detection by graphene nanopores, a promising class of nanosensors that are less sensitive than biological pores to various factors such as the temperature and pH.

All these processes span a wide range of time- and length-scales, rendering the use of multi-scale approaches unavoidable, by alternating the use of classical all atom MD simulations with Coarse Grained MD (CGMD) simulations. CGMD methods are based, with respect to all atom MD, on the reduction of the number of degrees of freedom, the use of shorter-range potential functions and of a larger time-step. This strategy makes possible the production of simulations with longer time scales and/ or larger size of the system, supporting the fascinating challenge of a combination of different methods, in which CGMD simulations with medium resolution are combined with the topological details allowed by all atom MD simulations.

We will focus separately on the first two macro-areas, by discussing the details of the system studied, the models involved and the main results obtained. The reader is directed to Schulten works (Sathe et al.[151,152] and Qiu et al.[153]) for all the details of the last topic.

## Adsorption of Biomolecules onto Graphene

### Pristine graphene substrates

Classical all atom MD simulations have been widely used to investigate the adsorption of different biomolecules such as proteins or peptides onto graphene. One of the first efforts is described in Zuo et al.,[154] where the authors studied the adsorption of the headpiece (HP35) of villin, a large protein broadly used as a model for protein folding, onto a graphene layer. The simulations showed unequivocally a rapid adsorption of HP35 by the substrate with relevant conformational changes in both the secondary and tertiary structures (Fig. 5). The $\pi/\pi$ stacking interactions between aromatic residues and graphene dominate the protein-graphene interaction differently from other HP35-curved carbon nanostructures such as single-wall carbon nanotube (SWCNT) and fullerene (C60) where the interactions are mostly controlled by the aliphatic side chains, with a lower attitude to forming flat $\pi/\pi$ stackings with aromatic residues and with a resulting lower binding affinity. Given this background, it is clear that the dispersion interaction between aromatic residues and the nanomaterials depends on the surface

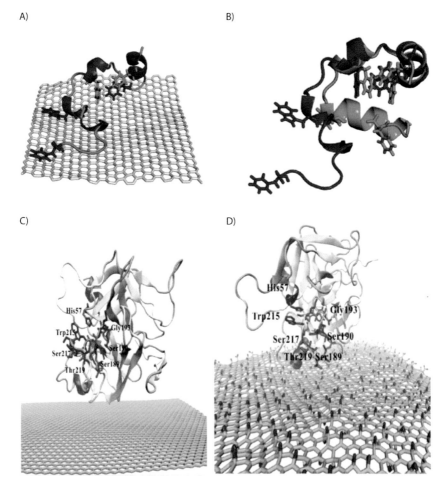

**Fig. 5.** (A) Representative structure of HP35 adsorbed onto a graphene layer. HP35 is shown with red helix and green loop cartoons, graphene as cyan lines. (B) Superposition of the adsorbed HP35 structure on graphene (red) with its native structure (green) to highlight the structural differences. Modified with permission from Zuo et al.[154] (C) The S1 active site of ChT (highlighted in red) is far away from the surface of pristine graphene. (D) The S1 active site is adsorbed onto GO. Modified with permission form Sun et al., copyright (2014) American Chemical Society.[158]

curvature, since the number of carbon atoms of the nanomaterial in contact with a residue decreases from graphene to the other two systems. These results suggest that the surface curvature and hardness affect the interaction with proteins.

At a later time, Zhou and collaborators (in Gu et al.)[155] performed MD simulations able to show how blood proteins such as bovine fibrinogen (BFG) can rapidly adsorb onto the graphene surface to form a biomolecular corona complex. Markedly, in addition to the aforementioned effect of strong $\pi/\pi$ stacking interactions, basic residues, like arginine, play a relevant role during the process because of the strong dispersion interactions between the side-chains of these residues and the substrate.

### *Graphene oxides substrates*

Because of its water solubility and functional groups, GO can serve as an ideal substrate for adsorbing biomolecules without any surface modification. The adhesion of biomolecules to GO and reduced graphene oxide (rGO) layers has been further investigated in Chong et al.,[156] to test the different advantages of GO due to the presence of the oxygen atoms. The interactions of serum proteins with GO nanosheets were investigated with a large set of experimental techniques (fluorescence, circular dichroism, atomic force microscopy, surface plasmon resonance) and MD simulations showing that the adsorption capacities of GO and rGO are higher than those of SWCNTs with a reduction of cytotoxicity. However, it is important point out that, while GO and rGO were used in the experiments, pristine graphene was chosen to simulate the relevant non-oxidized regions (up to $\sim 60\%$ for GO and $\sim 97\%$ for rGO) of the surface present on GO nanosheets (the so-called "$sp^2$-domain").

The action of GO was investigated more explicitly, representing the substrate with the Lerf-Klinowski model[157] based on the molecular formula $C_{10}O_1(OH)_1(COOH)_{0.5}$ to mimic the behavior of a standard oxidation process. Using this approach, two paradigmatic papers[158,159] demonstrated that GO displays an enhanced adsorption of the attached protein.

Firstly, in the work of Sun et al.,[158] an atomistic description of the inhibitory action of GO on the activity of $\alpha$-chymotrypsin (ChT),[160] has been provided based on the deformation of its binding sites and supporting the hypothesis that GO can be considered as an ideal artificial receptor for enzyme inhibition. The distinct binding modalities of ChT to GR and GO substrates is shown in Fig. 5. Secondly, Zeng et al.[159] show the details of the binding of GO to Vpr13-33, a fragment of the viral protein R (Vpr), that plays a key role in regulating nuclear import of HIV. In this process, the Vpr13-33 fragment can modify its conformation after the adsorbing process. The binding energy of the protein domain on GO was obtained using the potential of mean force (PMF) calculations with umbrella sampling calculations.[161] The self aggregation of multiple copies of the Vpr13-33 peptide was strongly affected by the adsorption on GO substrate, with a weaker interaction between two of these peptides, while it is shown that a stable dimer can be obtained in water or on GR substrate.

## Interactions of Graphene with Biomembranes

The interaction of graphene with biomolecular complexes is crucial to understand its biological safety and potential toxicity. The computational results displayed in this section show that graphene can strongly interact with, and induce relevant structural modifications in, different biological targets. Thus, we have to consider both the potential toxicity of graphene and its promising function of vector able to disrupt pathogenic cells.

In 2013, Tu et al.[162] showed that GR and GO nanosheets (described according to the Lerf-Klinowski model) induce the degradation of the inner and outer cell membranes of *Escherichia coli*. Specifically, MD simulations showed that graphene is able to extract actively phospholipid molecules from a lipid bilayer and fixing them on its surface, taking advantage of the dispersion interactions. These results introduce graphene as a convenient tool able to kill bacteria, and therefore as a suitable substitute of antibiotics.

On the other hand in 2014, Luan et al.[163] demonstrated that the hydrophobic protein-protein interactions, essential to diverse biological functions, can be disrupted by a graphene nanosheet, inducing the dissociation of the complex with possible irreversible effects on the cell's activities.

The comprehension of the mechanisms of the internalization of graphene layers inside the cellular environment, in particular through the plasma membrane, is crucial. To this aim, the use of classical all atom-MD still remain limited by its algorithmic efficiency and the available computing power. Therefore, a controlled reduction of the complexity of the system is required. Coarse Grained MD (CGMD) simulations[164] can be appropriate for this task. Comprehensive studies investigating the above-mentioned concepts have been carried out by various groups.

Titov et al.[165] used CGMD with the Martini force field[166] to study the interaction of small graphene and few-layer graphene (FLG) nanosheets with phospholipid bilayers formed by 1-palmitoyl-2-oleoyl-sn-glycero-3-phosphocholine (POPC membrane). They demonstrated that graphene sheets could be hosted in the hydrophobic interior of the membrane, forming stable graphene-lipid structures. In this work, the structure of pristine graphene is reduced to a triangular lattice of CG particles, where every three carbons are modeled as a single particle.

Guo and collaborators,[167] by using the CGMD approach of Dissipative Particle Dynamics (DPD),[168–170] explored the translocation of small graphene nanosheets across a membrane bilayer and the influence of graphene size and edge on this process. Using the same approach, they also investigated the membrane interactions of graphene nanosheets in various oxidation states[171] showing that graphene with a higher oxidation degree can produce a stronger perturbation of the membrane, and that this effect increases with the edge length of nanosheet.

Additionally, Li et al.[63] have explored the cell membrane interaction with graphene microsheets, using both all atom steered MD (SMD)[172] and CGMD-DPD simulations in combination with analytical modeling, confocal fluorescence imaging, and electron microscopy imaging. The results displayed that the entry of graphene microsheets is initiated at corners or at asperities along the edges. CGMD simulations illustrated the process of graphene-bilayer interaction, whereas the SMD simulations allowed to determine the energy barriers associated with the initial graphene penetration and the influence of the edge shape on the lipid penetration energy barrier.

In all these studies, however, lipid extraction or membrane damage is not observed, in contrast to the results of Tu et al. on the bacterial wall.[162] More recently, computational simulations were used to elucidate whether graphene causes cell membrane damage.[173] All atom MD simulations investigating of the behavior of both G and GO with respect to a dipalmitoylphosphatidylcholine (DPPC) bilayer revealed that G quickly entered into the membrane by assuming a position parallel to the lipid tails, while GO could not enter the membrane in the simulation timescale and localized at the solvent-membrane interface due to the hydrophilicity of the oxygen-containing groups. However, considering that GO, when placed onto the membrane, can enter the cell through endocytosis, stripping of membrane phospholipids to the GO surface and rupture of the bilayer can occur, resulting in the formation of pores in the plasma membrane.

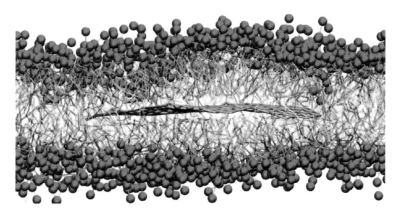

**Fig. 6.** Equilibrated superstructure of a graphene sheet inside the phospholipid bilayer formed by POPC lipids. Polar heads of the POPC lipids are depicted as green beads, hydrophobic hydrocarbon chains as thick blue lines; the graphene sheet is shown with brown lines (water molecules not shown; modified with permission from Titov et al., copyright (2010) American Chemical Society).[165]

# Conclusions

In the past few years, GRMs have been widely studied and used in a wide range of technological fields, including biomedical applications. More recently, graphene has also been adopted in CNS diagnostic and therapeutic tools development, due to its electrical properties and the promising potential of crossing the BBB.

In this chapter we reviewed the major areas of investigation for graphene and GRMs utilization in the CNS, from diagnostic to therapeutic usage. Neuro-oncology may profit form the development of graphene nanosheets and graphene nanoparticles for tumor-targeted imaging, photothermal therapy, anticancer drug delivery and gene therapy. New electrical, chemical and optical sensors may have great impact for neuro-intensive care and neuro-monitoring. Finally, tissue engineering research is expected to develop novel brain-implant interfaces based on graphene, to exploit the material electrical conductivity and enhance cell-cell communication and repair. The advancement in developing new tools for CNS applications is also taking advantages from molecular dynamics studies, which are providing extremely precise indications and predictions on G/cell and G/protein interactions, guiding the researcher to design more powerful G-based devices.

Nevertheless, despite the fact that initial studies have already demonstrated the graphene biocompatibility (especially when conjugated with other materials, forming 2D and 3D scaffolds), only few systems were demonstrated to be successful in *in vivo* neuro-applications. Future studies, in particular long-term biological effects, are still required before the neuroscience community can properly take advantage of the technological applications that such a new and promising material could offer to CNS biomedical applications.

# References

1.  A. K. Geim, *Science*, 2009, **324**, 1530–1534.
2.  P. Wick, A. E. Louw-Gaume, M. Kucki, H. F. Krug, K. Kostarelos, B. Fadeel, K. A. Dawson, A. Salvati, E. Vazquez, L. Ballerini, M. Tretiach, F. Benfenati, E. Flahaut, L. Gauthier, M. Prato and A. Bianco, *Angewandte Chemie*, 2014, **53**, 7714–7718.
3.  C. Cheng, S. Li, A. Thomas, N.A. Kotov and Rainer Haag, *Chem. Rev.*, 2017, **117**, 1826–1914.
4.  A. Bianco, *Angewandte Chemie*, 2013, **52**, 4986–4997.
5.  L. Gomez De Arco, Y. Zhang, C. W. Schlenker, K. Ryu, M. E. Thompson and C. Zhou, *ACS Nano*, 2010, **4**, 2865–2873.
6.  R. Munoz, C. Munuera, J. I. Martinez, J. Azpeitia, C. Gomez-Aleixandre and M. Garcia-Hernandez, *2D Materials*, 2017, **4**.
7.  K. S. Novoselov, A. K. Geim, S. V. Morozov, D. Jiang, Y. Zhang, S. V. Dubonos, I. V. Grigorieva and A. A. Firsov, *Science*, 2004, **306**, 666–669.
8.  B. H. Wee, T. F. Wu and J. D. Hong, *ACS Applied Materials & Interfaces*, 2017, **9**, 4548–4557.
9.  J. Lu, J. X. Yang, J. Wang, A. Lim, S. Wang and K. P. Loh, *ACS Nano*, 2009, **3**, 2367–2375.
10. K. Chen, D. Xue and S. Komarneni, *Journal of Colloid and Interface Science*, 2017, **487**, 156–161.
11. A. K. Geim and K. S. Novoselov, *Nature Materials*, 2007, **6**, 183–191.
12. R. Kurapati, K. Kostarelos, M. Prato and A. Bianco, *Advanced Materials*, 2017, **29**.
13. S. Sahoo, P. Bhattacharya, S. Dhibar, G. Hatui, T. Das and C. K. Das, *Journal of Nanoscience and Nanotechnology*, 2015, **15**, 6931–6941.
14. S. Casaluci, M. Gemmi, V. Pellegrini, A. Di Carlo and F. Bonaccorso, *Nanoscale*, 2016, **8**, 5368–5378.
15. L. H. Hess, M. Jansen, V. Maybeck, M. V. Hauf, M. Seifert, M. Stutzmann, I. D. Sharp, A. Offenhausser and J. A. Garrido, *Advanced Materials*, 2011, **23**, 5045–5049, 4968.
16. G. Eda, G. Fanchini and M. Chhowalla, *Nature Nanotechnology*, 2008, **3**, 270–274.
17. I. Meric, M. Y. Han, A. F. Young, B. Ozyilmaz, P. Kim and K. L. Shepard, *Nature Nanotechnology*, 2008, **3**, 654–659.
18. C. Zhu, T. Y. Han, E. B. Duoss, A. M. Golobic, J. D. Kuntz, C. M. Spadaccini and M. A. Worsley, *Nature Communications*, 2015, **6**, 6962.
19. F. Bonaccorso, A. Bartolotta, J. N. Coleman and C. Backes, *Advanced Materials*, 2016, **28**, 6136–6166.
20. M. K. Dufficy, M. T. Geiger, C. A. Bonino and S. A. Khan, *Langmuir: The ACS Journal of Surfaces and Colloids*, 2015, **31**, 12455–12463.
21. J. Hassoun, F. Bonaccorso, M. Agostini, M. Angelucci, M. G. Betti, R. Cingolani, M. Gemmi, C. Mariani, S. Panero, V. Pellegrini and B. Scrosati, *Nano Letters*, 2014, **14**, 4901–4906.
22. D. Du, Z. Zou, Y. Shin, J. Wang, H. Wu, M. H. Engelhard, J. Liu, I. A. Aksay and Y. Lin, *Analytical Chemistry*, 2010, **82**, 2989–2995.
23. X. Kang, J. Wang, H. Wu, J. Liu, I. A. Aksay and Y. Lin, *Talanta*, 2010, **81**, 754–759.
24. M. Pumera, *Chemical Record*, 2009, **9**, 211–223.

25. M. F. El-Kady and R. B. Kaner, *Nature Communications*, 2013, **4**, 1475.
26. A. Ambrosi and M. Pumera, *Chemistry*, 2016, **22**, 153–159.
27. F. Bonaccorso, L. Colombo, G. Yu, M. Stoller, V. Tozzini, A. C. Ferrari, R. S. Ruoff and V. Pellegrini, *Science*, 2015, **347**, 1246501.
28. K. Yang, L. Feng, X. Shi and Z. Liu, *Chem. Soc. Rev.*, 2013, **42**, 530–547.
29. L. Feng, L. Wu and X. Qu, *Advanced Materials*, 2013, **25**, 168–186.
30. D. Bitounis, H. Ali-Boucetta, B. H. Hong, D. H. Min and K. Kostarelos, *Advanced Materials*, 2013, **25**, 2258–2268.
31. X. Ding, H. Liu and Y. Fan, *Advanced Healthcare Materials*, 2015, **4**, 1451–1468.
32. M. Baldrighi, M. Trusel, R. Tonini and S. Giordani, *Frontiers in Neuroscience*, 2016, **10**, 250.
33. A. A. John, A. P. Subramanian, M. V. Vellayappan, A. Balaji, H. Mohandas and S. K. Jaganathan, *International Journal of Nanomedicine*, 2015, **10**, 4267–4277.
34. T. A. Mattei and A. A. Rehman, *Neurosurgery*, 2014, **74**, 499–516; discussion 516.
35. L. Zuccaro, C. Tesauro, T. Kurkina, P. Fiorani, H. K. Yu, B. R. Knudsen, K. Kern, A. Desideri and K. Balasubramanian, *ACS Nano*, 2015, **9**, 11166–11176.
36. X. Wang, X. Sun, J. Lao, H. He, T. Cheng, M. Wang, S. Wang and F. Huang, *Colloids and Surfaces. B, Biointerfaces*, 2014, **122**, 638–644.
37. F. M. Tonelli, V. A. Goulart, K. N. Gomes, M. S. Ladeira, A. K. Santos, E. Lorencon, L. O. Ladeira and R. R. Resende, *Nanomedicine*, 2015, **10**, 2423–2450.
38. H. Dong, M. Jin, Z. Liu, H. Xiong, X. Qiu, W. Zhang and Z. Guo, *Lasers in Medical Science*, 2016, **31**, 1123–1131.
39. N. Li, Q. Zhang, S. Gao, Q. Song, R. Huang, L. Wang, L. Liu, J. Dai, M. Tang and G. Cheng, *Scientific Reports*, 2013, **3**, 1604.
40. F. Menaa, A. Abdelghani and B. Menaa, *Journal of Tissue Engineering and Regenerative Medicine*, 2015, **9**, 1321–1338.
41. C. Defterali, R. Verdejo, L. Peponi, E. D. Martin, R. Martinez-Murillo, M. A. Lopez-Manchado and C. Vicario-Abejon, *Biomaterials*, 2016, **82**, 84–93.
42. M. Medina-Sanchez, S. Miserere and A. Merkoci, *Lab on a Chip*, 2012, **12**, 1932–1943.
43. N. Li, T. Xiao, Z. Zhang, R. He, D. Wen, Y. Cao, W. Zhang and Y. Chen, *Nanoscale*, 2015, **7**, 16354–16360.
44. N. Li, X. Zhang, Q. Song, R. Su, Q. Zhang, T. Kong, L. Liu, G. Jin, M. Tang and G. Cheng, *Biomaterials*, 2011, **32**, 9374–9382.
45. A. Fabbro, D. Scaini, V. Leon, E. Vazquez, G. Cellot, G. Privitera, L. Lombardi, F. Torrisi, F. Tomarchio, F. Bonaccorso, S. Bosi, A. C. Ferrari, L. Ballerini and M. Prato, *ACS Nano*, 2016, **10**, 615–623.
46. Q. Tu, L. Pang, Y. Chen, Y. Zhang, R. Zhang, B. Lu and J. Wang, *The Analyst*, 2014, **139**, 105–115.
47. V. C. Sanchez, A. Jachak, R. H. Hurt and A. B. Kane, *Chemical Research in Toxicology*, 2012, **25**, 15–34.
48. J. Russier, E. Treossi, A. Scarsi, F. Perrozzi, H. Dumortier, L. Ottaviano, M. Meneghetti, V. Palermo and A. Bianco, *Nanoscale*, 2013, **5**, 11234–11247.
49. H. Yue, W. Wei, Z. Yue, B. Wang, N. Luo, Y. Gao, D. Ma, G. Ma and Z. Su, *Biomaterials*, 2012, **33**, 4013–4021.
50. S. Goenka, V. Sant and S. Sant, *Journal of Controlled Release: Official Journal of the Controlled Release Society*, 2014, **173**, 75–88.
51. G. Liu, H. Shen, J. Mao, L. Zhang, Z. Jiang, T. Sun, Q. Lan and Z. Zhang, *ACS Applied Materials & Interfaces*, 2013, **5**, 6909–6914.
52. S. Gurunathan and J. H. Kim, *International Journal of Nanomedicine*, 2016, **11**, 1927–1945.
53. M. P. Monopoli, C. Aberg, A. Salvati and K. A. Dawson, *Nature Nanotechnology*, 2012, **7**, 779–786.
54. M. P. Monopoli, F. B. Bombelli and K. A. Dawson, *Nature Nanotechnology*, 2011, **6**, 11–12.
55. A. Sasidharan, L. S. Panchakarla, P. Chandran, D. Menon, S. Nair, C. N. Rao and M. Koyakutty, *Nanoscale*, 2011, **3**, 2461–2464.
56. M. C. Mendonca, E. S. Soares, M. B. de Jesus, H. J. Ceragioli, A. G. Batista, A. Nyul-Toth, J. Molnar, I. Wilhelm, M. R. Marostica, Jr., I. Krizbai and M. A. da Cruz-Hofling, *Molecular Pharmaceutics*, 2016, **13**, 3913–3924.
57. M. C. Mendonca, E. S. Soares, M. B. de Jesus, H. J. Ceragioli, S. P. Irazusta, A. G. Batista, M. A. Vinolo, M. R. Marostica Junior and M. A. da Cruz-Hofling, *Journal of Nanobiotechnology*, 2016, **14**, 53.

58. K. Donaldson, R. Aitken, L. Tran, V. Stone, R. Duffin, G. Forrest and A. Alexander, *Toxicological Sciences: An Official Journal of the Society of Toxicology*, 2006, **92**, 5–22.

59. R. Kurapati, C. Backes, C. Menard-Moyon, J. N. Coleman and A. Bianco, *Angewandte Chemie*, 2016, **55**, 5506–5511.

60. D. Zhang, Z. Zhang, Y. Liu, M. Chu, C. Yang, W. Li, Y. Shao, Y. Yue and R. Xu, *Biomaterials*, 2015, **68**, 100–113.

61. R. Rauti, N. Lozano, V. Leon, D. Scaini, M. Musto, I. Rago, F. P. Ulloa Severino, A. Fabbro, L. Casalis, E. Vazquez, K. Kostarelos, M. Prato and L. Ballerini, *ACS Nano*, 2016, **10**, 4459–4471.

62. M. Bramini, S. Sacchetti, A. Armirotti, A. Rocchi, E. Vazquez, V. Leon Castellanos, T. Bandiera, F. Cesca and F. Benfenati, *ACS Nano*, 2016, **10**, 7154–7171.

63. Y. Li, H. Yuan, A. von dem Bussche, M. Creighton, R. H. Hurt, A. B. Kane and H. Gao, *Proceedings of the National Academy of Sciences of the United States of America*, 2013, **110**, 12295–12300.

64. C. Ren, X. Hu, X. Li and Q. Zhou, *Biomaterials*, 2016, **93**, 83–94.

65. S. R. Shin, Y. C. Li, H. L. Jang, P. Khoshakhlagh, M. Akbari, A. Nasajpour, Y. S. Zhang, A. Tamayol and A. Khademhosseini, *Advanced Drug Delivery Reviews*, 2016, **105**, 255–274.

66. Y. Chen, A. Star and S. Vidal, *Chemical Society Reviews*, 2013, **42**, 4532–4542.

67. J. Liu, L. Cui and D. Losic, *Acta Biomaterialia*, 2013, **9**, 9243–9257.

68. L. Feng, S. Zhang and Z. Liu, *Nanoscale*, 2011, **3**, 1252–1257.

69. H. Kim, R. Namgung, K. Singha, I. K. Oh and W. J. Kim, *Bioconjugate Chemistry*, 2011, **22**, 2558–2567.

70. K. Yang, L. Hu, X. Ma, S. Ye, L. Cheng, X. Shi, C. Li, Y. Li and Z. Liu, *Advanced Materials*, 2012, **24**, 1868–1872.

71. H. Zhang, G. Grüner and Y. Zhao, *Journal of Materials Chemistry B*, 2013, **1**, 2542–2567.

72. T. Y. Hsieh, W. C. Huang, Y. D. Kang, C. Y. Chu, W. L. Liao, Y. Y. Chen and S. Y. Chen, *Advanced Healthcare Materials*, 2016, **5**, 3016–3026.

73. J. Y. X. Zhang, C. Peng, W. Hu, Z. Zhu, W. Li et al., *Carbon*, 2011, **49**, 986–995.

74. H. W. Yang, C. Y. Huang, C. W. Lin, H. L. Liu, C. W. Huang, S. S. Liao, P. Y. Chen, Y. J. Lu, K. C. Wei and C. C. Ma, *Biomaterials*, 2014, **35**, 6534–6542.

75. T. M. Allen and P. R. Cullis, *Science*, 2004, **303**, 1818–1822.

76. L. Yang, F. Wang, H. Han, L. Yang, G. Zhang and Z. Fan, *Colloids and Surfaces. B, Biointerfaces*, 2015, **129**, 21–29.

77. S. Gelperina, O. Maksimenko, A. Khalansky, L. Vanchugova, E. Shipulo, K. Abbasova, R. Berdiev, S. Wohlfart, N. Chepurnova and J. Kreuter, *European Journal of Pharmaceutics and Biopharmaceutics: Official Journal of Arbeitsgemeinschaft fur Pharmazeutische Verfahrenstechnik e.V*, 2010, **74**, 157–163.

78. J. Kreuter, P. Ramge, V. Petrov, S. Hamm, S. E. Gelperina, B. Engelhardt, R. Alyautdin, H. von Briesen and D. J. Begley, *Pharmaceutical Research*, 2003, **20**, 409–416.

79. S. Kanakia, J. D. Toussaint, S. Mullick Chowdhury, T. Tembulkar, S. Lee, Y. P. Jiang, R. Z. Lin, K. R. Shroyer, W. Moore and B. Sitharaman, *Biomaterials*, 2014, **35**, 7022–7031.

80. H. K. Sang Kyu Lee and Bong Sup Shim, *Carbon Letters*, 2013, **14**, 63–75.

81. Z. Liu, J. T. Robinson, X. Sun and H. Dai, *Journal of the American Chemical Society*, 2008, **130**, 10876–10877.

82. X. Sun, Z. Liu, K. Welsher, J. T. Robinson, A. Goodwin, S. Zaric and H. Dai, *Nano Research*, 2008, **1**, 203–212.

83. Z. Liu, J. T. Robinson, S. M. Tabakman, K. Yang and H. Dai, *Mater. Today*, 2011, **14**, 316–323.

84. H. Honigsmann, *Photochemical & Photobiological Sciences: Official Journal of the European Photochemistry Association and the European Society for Photobiology*, 2013, **12**, 16–21.

85. J. T. Robinson, S. M. Tabakman, Y. Liang, H. Wang, H. S. Casalongue, D. Vinh and H. Dai, *Journal of the American Chemical Society*, 2011, **133**, 6825–6831.

86. Z. M. Markovic, L. M. Harhaji-Trajkovic, B. M. Todorovic-Markovic, D. P. Kepic, K. M. Arsikin, S. P. Jovanovic, A. C. Pantovic, M. D. Dramicanin and V. S. Trajkovic, *Biomaterials*, 2011, **32**, 1121–1129.

87. N. J. Abbott, *Journal of Inherited Metabolic Disease*, 2013, **36**, 437–449.

88. L. Zhang, J. Xia, Q. Zhao, L. Liu and Z. Zhang, *Small*, 2010, **6**, 537–544.

89. G. Gollavelli and Y. C. Ling, *Biomaterials*, 2012, **33**, 2532–2545.

90. Y. Wang, Z. Li, D. Hu, C. T. Lin, J. Li and Y. Lin, *Journal of the American Chemical Society*, 2010, **132**, 9274–9276.

91. J. Qian, D. Wang, F. H. Cai, W. Xi, L. Peng, Z. F. Zhu, H. He, M. L. Hu and S. He, *Angewandte Chemie*, 2012, **51**, 10570–10575.
92. K. Yang, S. Zhang, G. Zhang, X. Sun, S. T. Lee and Z. Liu, *Nano Letters*, 2010, **10**, 3318–3323.
93. L. Zhang, Z. Lu, Q. Zhao, J. Huang, H. Shen and Z. Zhang, *Small*, 2011, **7**, 460–464.
94. B. Chen, M. Liu, L. Zhang, J. Huang, J. Yao and Z. Zhang, *J. Mater. Chem.*, 2011, **11**, 7736–7741.
95. D. Du, L. Wang, Y. Shao, J. Wang, M. H. Engelhard and Y. Lin, *Analytical Chemistry*, 2011, **83**, 746–752.
96. M. Yang and S. Gong, *Chemical Communications*, 2010, **46**, 5796–5798.
97. C. Zhai, Y. Guo, X. Sun, Y. Zheng and X. Wang, *Enzyme and Microbial Technology*, 2014, **58-59**, 8–13.
98. B. M. Blaschke, M. Lottner, S. Drieschner, A. Bonaccini Calia, K. Stoiber, L. Rousseau, G. Lissourges and J. A. Garrido, *2D Materials*, 2016, **3**, 025007.
99. T. C. Liu, M. C. Chuang, C. Y. Chu, W. C. Huang, H. Y. Lai, C. T. Wang, W. L. Chu, S. Y. Chen and Y. Y. Chen, *ACS Applied Materials & Interfaces*, 2016, **8**, 187–196.
100. H. Gu, Y. Yu, X. Liu, B. Ni, T. Zhou and G. Shi, *Biosensors & Bioelectronics*, 2012, **32**, 118–126.
101. A. Bonanni and M. Pumera, *ACS Nano*, 2011, **5**, 2356–2361.
102. M. Kalbacova, A. Broz, J. Kong and M. Kalbac, *Carbon*, 2010, **48**, 4323–4329.
103. W. C. Lee, C. H. Lim, H. Shi, L. A. Tang, Y. Wang, C. T. Lim and K. P. Loh, *ACS Nano*, 2011, **5**, 7334–7341.
104. T. R. Nayak, H. Andersen, V. S. Makam, C. Khaw, S. Bae, X. Xu, P. L. Ee, J. H. Ahn, B. H. Hong, G. Pastorin and B. Ozyilmaz, *ACS Nano*, 2011, **5**, 4670–4678.
105. G. Y. Chen, D. W. Pang, S. M. Hwang, H. Y. Tuan and Y. C. Hu, *Biomaterials*, 2012, **33**, 418–427.
106. S. Y. Park, J. Park, S. H. Sim, M. G. Sung, K. S. Kim, B. H. Hong and S. Hong, *Advanced Materials*, 2011, **23**, H263–267.
107. M. Tang, Q. Song, N. Li, Z. Jiang, R. Huang and G. Cheng, *Biomaterials*, 2013, **34**, 6402–6411.
108. O. Akhavan and E. Ghaderi, *Nanoscale*, 2013, **5**, 10316–10326.
109. O. Akhavan and E. Ghaderi, *Journal of Materials Chemistry B*, 2013, **1**, 6291–6301.
110. C. Defterali, R. Verdejo, S. Majeed, A. Boschetti-de-Fierro, H. R. Mendez-Gomez, E. Diaz-Guerra, D. Fierro, K. Buhr, C. Abetz, R. Martinez-Murillo, D. Vuluga, M. Alexandre, J. M. Thomassin, C. Detrembleur, C. Jerome, V. Abetz, M. A. Lopez-Manchado and C. Vicario-Abejon, *Front. Bioeng. Biotechnol.*, 2016, **4**, 94.
111. R. Guo, S. Zhang, M. Xiao, F. Qian, Z. He, D. Li, X. Zhang, H. Li, X. Yang, M. Wang, R. Chai and M. Tang, *Biomaterials*, 2016, **106**, 193–204.
112. S. Bae, H. Kim, Y. Lee, X. Xu, J. S. Park, Y. Zheng, J. Balakrishnan, T. Lei, H. R. Kim, Y. I. Song, Y. J. Kim, K. S. Kim, B. Ozyilmaz, J. H. Ahn, B. H. Hong and S. Iijima, *Nature Nanotechnology*, 2010, **5**, 574–578.
113. J. Park, S. Park, S. Ryu, S. H. Bhang, J. Kim, J. K. Yoon, Y. H. Park, S. P. Cho, S. Lee, B. H. Hong and B. S. Kim, *Advanced Healthcare Materials*, 2014, **3**, 176–181.
114. J. Kim, S. Park, Y. J. Kim, C. S. Jeon, K. T. Lim, H. Seonwoo, S. P. Cho, T. D. Chung, P. H. Choung, Y. H. Choung, B. H. Hong and J. H. Chung, *Journal of Biomedical Nanotechnology*, 2015, **11**, 2024–2033.
115. Y. Wang, W. C. Lee, K. K. Manga, P. K. Ang, J. Lu, Y. P. Liu, C. T. Lim and K. P. Loh, *Advanced Materials*, 2012, **24**, 4285–4290.
116. Q. Song, Z. Jiang, N. Li, P. Liu, L. Liu, M. Tang and G. Cheng, *Biomaterials*, 2014, **35**, 6930–6940.
117. A. Bendali, L. H. Hess, M. Seifert, V. Forster, A. F. Stephan, J. A. Garrido and S. Picaud, *Advanced Healthcare Materials*, 2013, **2**, 929–933.
118. F. Veliev, A. Briancon-Marjollet, V. Bouchiat and C. Delacour, *Biomaterials*, 2016, **86**, 33–41.
119. D. Sahni, A. Jea, J. A. Mata, D. C. Marcano, A. Sivaganesan, J. M. Berlin, C. E. Tatsui, Z. Sun, T. G. Luerssen, S. Meng, T. A. Kent and J. M. Tour, *J. Neurosurg. Pediatr.*, 2013, **11**, 575–583.
120. Z. He, S. Zhang, Q. Song, W. Li, D. Liu, H. Li, M. Tang and R. Chai, *Colloids and Surfaces. B, Biointerfaces*, 2016, **146**, 442–451.
121. M. Lorenzoni, F. Brandi, S. Dante, A. Giugni and B. Torre, *Scientific Reports*, 2013, **3**, 1954.
122. S. Shah, P. T. Yin, T. M. Uehara, S. T. Chueng, L. Yang and K. B. Lee, *Advanced Materials*, 2014, **26**, 3673–3680.
123. J. Kim, K. S. Choi, Y. Kim, K. T. Lim, H. Seonwoo, Y. Park, D. H. Kim, P. H. Choung, C. S. Cho, S. Y. Kim, Y. H. Choung and J. H. Chung, *J. Biomed. Mater. Res. A.*, 2013, **101**, 3520–3530.

124. J. Kim, Y. R. Kim, Y. Kim, K. T. Lim, H. Seonwoo, S. Park, S. P. Cho, B. H. Hong, P. H. Choung, T. D. Chung, Y. H. Choung and J. H. Chung, *Journal of Materials Chemistry B.*, 2013, **1**, 933–938.
125. O. Akhavan, E. Ghaderi, E. Abouei, S. Hatamy and E. Ghasemi, *Carbon*, 2013, **66**, 395–406.
126. O. Akhavan and E. Ghaderi, *J. Mater. Chem. B.*, 2014, **2**, 5602–5611.
127. T. H. Kim, S. Shah, L. Yang, P. T. Yin, M. K. Hossain, B. Conley, J. W. Choi and K. B. Lee, *ACS Nano*, 2015, **9**, 3780–3790.
128. A. Solanki, S. T. Chueng, P. T. Yin, R. Kappera, M. Chhowalla and K. B. Lee, *Advanced Materials*, 2013, **25**, 5477–5482.
129. C. L. Weaver and X. T. Cui, *Advanced Healthcare Materials*, 2015, **4**, 1408–1416.
130. O. Akhavan, E. Ghaderi and S. A. Shirazian, *Colloids and Surfaces. B, Biointerfaces*, 2015, **126**, 313–321.
131. S. M. Kim, P. Joo, G. Ahn, I. H. Cho, D. H. Kim, W. K. Song, B. S. Kim and M. H. Yoon, *Journal of Biomedical Nanotechnology*, 2013, **9**, 403–408.
132. Q. Tu, L. Pang, L. Wang, Y. Zhang, R. Zhang and J. Wang, *ACS Applied Materials & Interfaces*, 2013, **5**, 13188–13197.
133. K. Zhou, G. A. Thouas, C. C. Bernard, D. R. Nisbet, D. I. Finkelstein, D. Li and J. S. Forsythe, *ACS Applied Materials & Interfaces*, 2012, **4**, 4524–4531.
134. W. Guo, X. Zhang, X. Yu, S. Wang, J. Qiu, W. Tang, L. Li, H. Liu and Z. L. Wang, *ACS Nano*, 2016, **10**, 5086–5095.
135. Z. Q. Feng, T. Wang, B. Zhao, J. Li and L. Jin, *Advanced Materials*, 2015, **27**, 6462–6468.
136. S. W. Crowder, D. Prasai, R. Rath, D. A. Balikov, H. Bae, K. I. Bolotin and H. J. Sung, *Nanoscale*, 2013, **5**, 4171–4176.
137. S. Sayyar, M. Bjorninen, S. Haimi, S. Miettinen, K. Gilmore, D. Grijpma and G. Wallace, *ACS Applied Materials & Interfaces*, 2016, **8**, 31916–31925.
138. W. Guo, S. Wang, X. Yu, J. Qiu, J. Li, W. Tang, Z. Li, X. Mou, H. Liu and Z. Wang, *Nanoscale*, 2016, **8**, 1897–1904.
139. M. C. Serrano, J. Patiño, C. García-Rama, M. L. Ferrer, J. L. G. Fierro, A. Tamayo, J. E. Collazos-Castro, F. del Monte and M. C. Gutiérrez, *J. Mater. Chem. B.*, 2014, **2**, 5698–5706.
140. Q. Ma, L. Yang, Z. Jiang, Q. Song, M. Xiao, D. Zhang, X. Ma, T. Wen and G. Cheng, *ACS Applied Materials & Interfaces*, 2016, **8**, 34227–34233.
141. O. Akhavan, E. Ghaderi, S. A. Shirazian and R. Rahighi, *Carbon*, 2016, **97**, 71–77.
142. K. Zhou, S. Motamed, G. A. Thouas, C. C. Bernard, D. Li, H. C. Parkington, H. A. Coleman, D. I. Finkelstein and J. S. Forsythe, *PloS One*, 2016, **11**, e0151589.
143. E. Lopez-Dolado, A. Gonzalez-Mayorga, M. T. Portoles, M. J. Feito, M. L. Ferrer, F. Del Monte, M. C. Gutierrez and M. C. Serrano, *Advanced Healthcare Materials*, 2015, **4**, 1861–1868.
144. E. Lopez-Dolado, A. Gonzalez-Mayorga, M. C. Gutierrez and M. C. Serrano, *Biomaterials*, 2016, **99**, 72–81.
145. M. Ozboyaci, D. B. Kokh, S. Corni and R. C. Wade, *Quarterly Reviews of Biophysics*, 2016, **49**, e4.
146. T. Cavallucci, K. Kakhiani, R. Farchioni and V. Tozzini, *arXiv:1609.07871*, 2016.
147. N. Patra, Y. Song and P. Kral, *ACS Nano*, 2011, **5**, 1798–1804.
148. N. Patra, B. Wang and P. Kral, *Nano Letters*, 2009, **9**, 3766–3771.
149. G. Hummer, J. C. Rasaiah and J. P. Noworyta, *Nature*, 2001, **414**, 188–190.
150. M. Pykal, P. Jurecka, F. Karlicky and M. Otyepka, *Physical Chemistry Chemical Physics: PCCP*, 2016, **18**, 6351–6372.
151. C. Sathe, A. Girdhar, J. P. Leburton and K. Schulten, *Nanotechnology*, 2014, **25**, 445105.
152. C. Sathe, X. Zou, J. P. Leburton and K. Schulten, *ACS Nano*, 2011, **5**, 8842–8851.
153. H. Qiu, A. Sarathy, J. P. Leburton and K. Schulten, *Nano Letters*, 2015, **15**, 8322–8330.
154. G. Zuo, X. Zhou, Q. Huang, H. Fang and R. Zhou, *J. Phys. Chem. C.*, 2011, **115**, 23323–23328.
155. Z. Gu, Z. Yang, L. Wang, H. Zhou, C. A. Jimenez-Cruz and R. Zhou, *Scientific Reports*, 2015, **5**, 10873.
156. Y. Chong, C. Ge, Z. Yang, J. A. Garate, Z. Gu, J. K. Weber, J. Liu and R. Zhou, *ACS Nano*, 2015, **9**, 5713–5724.
157. A. Lerf, H. He, M. Forster and J. Klinowsky, *J. Phys. Chem. B.*, 1998, **102**, 4477–4482.
158. X. Sun, Z. Feng, T. Hou and Y. Li, *ACS Applied Materials & Interfaces*, 2014, **6**, 7153–7163.
159. S. Zeng, G. Zhou, J. Guo, F. Zhou and J. Chen, *Scientific Reports*, 2016, **6**, 24906.

160. M. De, S. S. Chou and V. P. Dravid, *Journal of the American Chemical Society*, 2011, **133**, 17524–17527.
161. J. Kästner, *Adv. Rev.*, 2011, **1**.
162. Y. Tu, M. Lv, P. Xiu, T. Huynh, M. Zhang, M. Castelli, Z. Liu, Q. Huang, C. Fan, H. Fang and R. Zhou, *Nature Nanotechnology*, 2013, **8**, 594–601.
163. B. Luan, T. Huynh, L. Zhao and R. Zhou, *ACS Nano*, 2015, **9**, 663–669.
164. S. Kmiecik, D. Gront, M. Kolinski, L. Wieteska, A. E. Dawid and A. Kolinski, *Chemical Reviews*, 2016, **116**, 7898–7936.
165. A. V. Titov, P. Kral and R. Pearson, *ACS Nano*, 2010, **4**, 229–234.
166. S. J. Marrink, H. J. Risselada, S. Yefimov, D. P. Tieleman and A. H. de Vries, *The Journal of Physical Chemistry. B*, 2007, **111**, 7812–7824.
167. R. Guo, J. Mao and L. T. Yan, *Biomaterials*, 2013, **34**, 4296–4301.
168. P. J. Hoogerbrugge and J. M. V. A. Koelman, *Europhys. Lett.*, 1992, **19**, 155.
169. J. M. V. A. Koelman and P. J. Hoogerbrugge, *Europhys. Lett.*, 1993, **21**, 363.
170. P. W. Español, P., *Europhys. Lett.*, 1995, **30**, 191.
171. J. Mao, R. Guo and L. T. Yan, *Biomaterials*, 2014, **35**, 6069–6077.
172. B. G. Isralewitz, M.; Schulten, K., *Curr. Opin. Struct. Biol.*, 2001, **11**, 224–230.
173. J. Chen, G. Zhou, L. Chen, Y. Wang, X. Wang and S. Zeng, *J. Phys. Chem. C.*, 2016, **120**, 6225–6231.

CHAPTER 5

# Antimicrobial Properties of Modified Graphene and other Advanced 2D Material Coated Surfaces

*Anthony J. Slate,*[1,2] *Nathalie Karaky*[1] and *Kathryn A. Whitehead*[1,*]

## Introduction

Over recent decades, antimicrobial resistance (AMR) has emerged as an enormous global issue. The rise in observed AMR bacteria can be attributed to the continuous overuse of antimicrobial therapies, not limited to healthcare but within a wide range of areas including in the veterinary services and agriculture.[1] The problems with the overuse of antibiotics has been coupled with the lack of new antibiotic therapies. A number of pharmaceutical companies have abandoned research in the antibiotic field which has led to a lack of compounds able to effectively treat AMR bacteria.[2] The continual over reliance placed upon antibiotics, along with a better infrastructure in society has led to geographical barriers no longer being an issue. This has resulted in the unrestricted movement of people, products and their microbial counterparts. Coinciding with this increase in movement has been the evolution of bacteria, which has allowed genes coding for antimicrobial resistance to be developed, shared and expressed by a number of different bacterial species. Once a resistance gene is produced, it is able to be transferred to other bacteria (both of the same and different genus) by horizontal

[1] Microbiology at Interfaces Group, Manchester Metropolitan University, Manchester M1 5GD, UK.
[2] Division of Chemistry and Environmental Science, Manchester Metropolitan University, Manchester M1 5GD, UK.
* Corresponding author: K.A.Whitehead@mmu.ac.uk

gene transfer, therefore the potential to spread through the bacterial population is high.[3] The result has been specific antimicrobial agents becoming ineffective and therefore impractical against certain bacterial strains.[4] In light of this, emphasis must be placed on the development of alternative antimicrobial agents in order to reduce the transmission of microbial agents and the burden placed upon conventional (antibiotic) therapies.

Antimicrobial resistance is an issue within a wide range of infectious organisms. The "ESKAPE" pathogens include *Enterococcus* spp., *Staphylococcus aureus*, *Klebsiella pneumoniae*, *Acinetobacter baumannii*, *Pseudomonas aeruginosa* and *Enterobacter* spp. These pathogens demonstrate high levels of multidrug resistance, and are responsible for a substantial percentage of nosocomial infections.[5] The magnitude of this worldwide problem and the impact of AMR on human health, results in increased costs for the health-care sector. Further, the wider societal impact of such infections are of key concerns for governments. A report by the World Health Organisation (WHO) reported than an average of 8.7% of hospital patients had nosocomial infections. The costs associated with this include an increased length of stay for infected patients, the increased use of drugs, additional laboratory costs, the possibility of the need for isolation of the patient and increased morbidity and mortality rates. Since the issues associated with AMR include a number of areas outside the health sector, WHO estimated over 10 years ago that this phenomenon would result in a fall in real gross domestic product (GDP) of 0.4% to 1.6% (which is equivalent to several billions of today's dollars).[6] Being a complex global public health challenge, no single or simple strategy will suffice to fully contain the emergence and spread of nosocomial and community acquired infectious organisms that become resistant to available antimicrobial drugs.

Within nature, cells living freely in bulk solution usually become attached to a surface, and if retained, can then form a biofilm. A biofilm is a matrix-enclosed bacterial populations that are attached to a surface or an interface. Biofilms demonstrate a wide array of resistance mechanisms, including persistent dormant cells, hyper-mutability, quorum sensing and efflux pumps making them extremely tolerant/resistant to antibiotics and antimicrobials and thus greater antimicrobial resistance has been demonstrated in biofilms when compared to planktonic cells.[7,8] Thus, there is a need for advanced antimicrobial surfaces to be developed. A range of 2D-nanomaterials that may be exploited is antimicrobial surface coatings, which could be used in areas with a population that is of increased risk of potential microbial/bacterial transfer, e.g., nosocomial settings. These include the carbon based materials such as graphite, graphite oxide, reduced graphite oxide, graphene, graphene oxide and reduced graphene oxide and non-carbon based materials for example boron nitrite, tungsten diselenide and molybdenum disulphide.

## Carbon-Based Nanomaterials

The graphite and graphene derivatives have specific definitions (Table 1). There are a number of graphite and graphene derivatives (Table 2) that have been used in nanoparticulate and 2D form to determine their antimicrobial activity.

**Table 1.** Definitions of the graphene/graphite derivatives.

| Graphite/graphene type | Definition |
|---|---|
| Graphite | In graphite, adjacent graphene layers overlap due to $p_z$ orbitals, producing bulk graphite.[9] |
| Graphite oxide | Graphite oxide refers to graphite with functional groups containing oxygen attached. It is prepared by treating graphite with strong aqueous oxidizing agents.[10] |
| Reduced graphite oxide | Reduced graphite oxide is produced by reducing graphite oxide, thereby removing attached oxygenated groups, allowing the honeycomb lattice to be achieved—which restores electrochemical properties.[11] |
| Graphene | Graphene sheets comprise of a 2D layer of $sp^2$-hybridized carbon atoms, arranged in a hexagonal (honeycomb) lattice.[9] |
| Graphene oxide | Graphene oxide refers to single-atom layers of carbon (graphene) with functional groups containing oxygen attached. The oxygen groups allow the molecular to become polar and therefore soluble.[12] |
| Reduced graphene oxide | In order to create reduced graphene oxide, graphene oxide can be reduced, often by thermal mechanisms allowing for the removal of attached oxygenated groups.[12] |

## Graphite

Graphite is one of the oldest and most widely used of the carbon-based materials.[30] Graphite is routinely used as a starting material in the production of a variety of carbon-based nanomaterials including fullerenes, nanodiamonds, single and multi-walled nanotubes and the synthesis of graphene.[31] Graphite has been used in a variety of biomedical applications[32,33] including drug delivery,[34,35,36] photothermal anticancer activity,[37,38,39] biosensors, biofunctionalisation,[34] disease diagnostics[40,41,42] and antimicrobial therapies.[43]

The antimicrobial activity of graphite has been demonstrated, whereby the interaction between graphite nanoplatelets and *Pseudomonas aeruginosa* (which can cause chronic infections in the lungs of patients with cystic fibrosis) was investigated. The results of a bacterial cell viability assay, performed after 5 h of incubation with graphite nanoplatelets, demonstrated a viability loss of up to 69.5%, as opposed to the control with no graphite nanoplatelets.[44] Work in our laboratories has demonstrated that following minimal bactericidal concentration assays using particulate compounds, when tested against Gram negative *Escherichia coli*, graphite demonstrated greater antimicrobial efficacy than graphene and was comparable with the antimicrobial efficacy of zinc oxide (Fig. 1). However against Gram positive *S. aureus,* although graphite again demonstrated greater antimicrobial activity than graphene, its antimicrobial efficacy was not greater than that of zinc oxide. This may be explained in part due to the differences in the chemical composition of bacterial cell walls of Gram negative and Gram positive bacteria.

The use of graphite as an antimicrobial surface coating is in its infancy, however, early results using carbon thin films have been promising.[45] In a study whereby carbon thin films (previously known as graphite) were used as a antimicrobial coatings on

**Table 2.** Features and antimicrobial efficiency of different graphene and graphite derivatives.

| Graphene/metal | Bacterial strain | Methods used | Concentrations | Suggested mechanisms | References |
|---|---|---|---|---|---|
| Graphite Oxide (GtO) | E. coli | - Plate assay method<br>- Shake flask test in saline<br>- ZOI | | - Membrane stress or Oxidative stress: production of reactive oxygen species (ROS) | 13,14 |
| GtO–Ag | E. coli | Plate assay method<br>- Shake flask test in saline<br>- ZOI | 1.5 mol/L | Production of ROS (harm DNA/ proteins) | 13,15 |
| GtO–Sand | E. coli | Plate assay method<br>- Shake flask test in saline<br>- ZOI | NA | Membrane stress | 13 |
| Ag nanoparticle | E. coli | Plate assay method<br>- Shake flask test in saline<br>- ZOI | 12 µg/mL | Production of ROS (harm DNA/ proteins) | 13 |
| Graphene Oxide (GO) | E. coli | - Agar diffusion method<br>- Viable count<br>- Scanning electron microscopy | 80 µg/mL | Insertion/cutting of cell membrane and extraction of phospholipids | 16 |
| | P. aeruginosa | | 150 µg/mL | ROS generation | 14 |
| | S. aureus | | ≤ 10 µg/mL | Wrapping of bacterial cells | 14 |
| | S. fecalis | | ≤ 10 µg/mL | Membrane stress | 14 |
| Reduced GO (rGO) | E. coli | - Cytotoxicity test<br>- Cell viability test<br>- Metabolic activity assay | 100 µg/mL | Cell membrane damage due to contact interaction | 14,17 |
| | P. aeruginosa | - Cell viability test | 150 µg/mL | NA | 18 |
| | S. aureus | - MTT assay | | Photothermal ablation | 19 |

*Table 2 contd....*

*...Table 2 contd.*

| Graphene/metal | Bacterial strain | Methods used | Concentrations | Suggested mechanisms | References |
|---|---|---|---|---|---|
| GO–Ag | E. coli | - ZOI<br>- Colorimetric methods<br>- UV visible spectroscopy | 2,000 mg/L | Cell wall breakdown and cytoplasm release | 20 |
| | P. aeruginosa | | NA | Cell wall breakdown and cytoplasm release | 20 |
| | B. subtilis | - ZOI<br>- MIC<br>- MBC<br>- Colorimetric method for kinetics | NA | Cell wall breakdown and cytoplasm release | 15,21 |
| rGO–Au | S. aureus | - MIC<br>- MBC<br>- ZOI | NA | Oxidative stress | 22 |
| | E. coli | | 2,000 mg/L | Oxidative stress | 22 |
| | B. subtilis | | NA | Oxidative stress | 22 |
| Graphene oxide-poly-N-vinylcarbazole (PVK-GO) coating | E. coli | - Growth Curves<br>- Epifluorescence microscopy | 3% GO on surface | Oxidative stress | 23 |
| Graphene and Poly-N-vinylcarbazole (PVK) | E. coli<br>B. subtilis | - AFM<br>- Metabolic assay<br>- Live dead assay | 97.3 w/w% | NA | 24 |
| Silver nanoparticles in combination with graphene oxide sheets | E. coli | - Agar well diffusion assay<br>- ZOI | 60 µL<br><br>26 mm | ROS generation and inhibition of respiratory enzymes | 20 |
| | P. aeruginosa | | 18 mm | | |

| Material | Organism | Method | Value | Mechanism | Ref |
|---|---|---|---|---|---|
| Graphene nanosheets | E. coli | - MIC | 1 µg/mL | Production of ROS and lipid peroxidation | 25 |
| | Salmonella typhimurium | | 1 µg/mL | | |
| | Enterococcus faecalis | | 8 µg/mL | | |
| | B. subtilis | | 4 µg/mL | | |
| GO–Nitinol surface coating | E. coli | - Colony forming unit (CFU) counts<br>- Live/dead fluorescent staining<br>- SEM | 0.2 mg/ml<br>NA<br>NA | ROS and sharp edge plane penetration of cell membrane | 26 |
| GO nanosheets | E. coli | - Viability assays<br><br>- Suspension assays | $0.01\ \mu m^2$–$0.65\ \mu m^2$ | ROS production and higher defect density of smaller GO particles<br>Cell entrapment | 27 |
| Graphene film | E. coli<br>S. aureus | - CFU<br>- Live/dead fluorescent staining | 60 µl/mL | Loss of cell membrane integrity and cell membrane leakage | 28 |
| Graphene oxide nanowalls on stainless steel | E. coli<br>S. aureus | - Cytotoxicity test<br>- RNA measurement | NA<br>30 ng/ML<br>38 ng/mL | Loss of cell membrane integrity due to sharp edge penetration | 29 |
| rGO–stainless steel | E. coli<br>S. aureus | - Cytotoxicity test<br>- RNA measurement | NA<br>43 ng/mL<br>56 ng/mL | Loss of cell membrane integrity due to sharp edge penetration | 29 |
| GO nanosheets–paper surface coatings | E. coli | - Luciferase-based ATP assay kit<br>- TEM | 85 µg/mL<br>NA | Loss of cell membrane integrity | 17 |

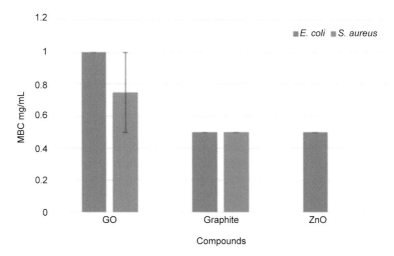

**Fig. 1.** Minimal bactericidal concentrations (MBC) of graphite, graphene oxide and zinc oxide demonstrating that the graphite demonstrates antimicrobial efficacy against *E. coli* and *S. aureus* ($n = 3$). Note, where error bars are not present there was no variation in standard error. (Work courtesy of Daniel Brown, MMU, UK).

a polyethylene terephthalate (PET) surface, the results demonstrated that the carbon thin films reduced bacterial adherence capabilities by 65% and 86% for both *S. aureus* and *Staphylococcus epidermidis,* respectively.[45]

Shteynle (2012), compared the absorbency and fluid retention of two variants of nanostructured graphite wound dressings and a variety of "gold standard" wound dressings. The nanostructured graphite wound dressings demonstrated the greatest adsorption potential, but no antimicrobial testing was carried out. The nanostructured graphite wound dressings were able to adsorb the wound exudate due to a variety of mechanisms including covalent bonding, due to a system of numerous interweaving fibres that produced a high surface area,[46] and a matrix consisting of strong fluoride oxidisers, resulting in large quantities of free radicals. Furthermore, these oxidisers added oxygenated species to the surface of the graphite, increasing its hydrophobicity.[47] It has been reported that the absence of excessive wound exudate (i.e., by adsorption), can reduce the risk of nosocomial/opportunistic infection by making the conditions for the commensal microflora less than favourable.[48] Graphite is proposed to be able to act as an antimicrobial in a variety of ways, including physiochemical responses leading to membrane stress, sharp nanosheets which result in disruption of the cell membrane and via the production of reactive oxygen species, due to the oxidation of glutathione which acts as a redox mediator in bacterial cells.[49]

## Graphite Oxide

Graphite oxide is produced via the oxidation of natural graphite. Characterisation studies identify graphite oxide as a lamellar solid containing phenolic, carboxyl and epoxide groups. This makes graphite oxide hydrophilic, with the production of monolayer colloidal dispersions upon interaction with water.[50] Delamination of graphite oxide solids often forms the basis for the synthesis of graphite oxide nanomaterials. Graphite oxide has unique properties, namely enhanced electrochemical activities in the form of capacitor materials due to the presence of hydroxyl, epoxy and carbonyl groups.[51,52,53] Graphite oxide is often only ever intended to be used as a by-product in the synthesis of graphene-based nanosheets produced by the chemical reduction of exfoliated graphite oxide and the antimicrobial efficacy of graphite oxide has yet to be fully documented. However, in one study,[54] the antimicrobial efficacy of graphite oxide was compared to that of other carbon-based materials (graphite, graphene oxide and reduced graphene oxide) against *E. coli*. The results showed that graphene oxide had the greatest antimicrobial activity (inactivating 69% of bacterial cells after 2 hours of incubation) followed by reduced graphene oxide, graphite followed by graphite oxide which inactivated 15% of bacterial cells at the same dispersion concentration.[54] Das et al. (2011) synthesised a silver nanoparticle and graphite oxide nanosheet composite. Using X-ray diffraction and transmission electron microscopy, the results indicated that the silver nanoparticles decorated the graphite oxide sheets and following antimicrobial testing of the nanosheets, antibacterial activity was indicated against *E. coli* and *P. aeruginosa* using zone of inhibition assys.[20] The underlying antimicrobial mechanism of graphite oxide remains unclear, however its large surface area could allow for cell wrapping. Cell wrapping can be attributed to the hydrophobicity of graphene materials, since the graphene sheets are hydrophobic. This can lead to the phospholipid bilayers of microorganisms (in direct contact with the graphene) to become inversed, therefore leading to the loss of membrane integrity and cell lysis.[55] Other antimicrobial mechanisms for graphite oxide include its oxygen containing surface functionalities such as carboxylic, carbonyl, hydroxyl and epoxide groups as well as its increased water solubility which provides a basis for ion or nanoparticle intercalation, which can in turn have a detrimental effect on bacterial cells.[56]

## Reduced Graphite Oxide

Reduced graphite oxide is often used as a starting product in the production of graphene. Graphite oxide is reduced by a two-step system, with the first step being the removal of oxygen groups via the use of sodium tetrahydridoborate followed by the second step which uses concentrated sulphuric acid to dehydrate and therefore restore the chemical structure.[57] Due to its reported unique electrochemical properties (as demonstrated by all graphene-based materials), an increasing amount of research has been recently targeted towards understanding this material, especially with the application of reduced graphite oxide for use in supercapacitors and batteries.[58,59] Research into reduced graphite oxide as a standalone antimicrobial material is relatively novel, and as of yet has not yet been fully elucidated.

In 2011, Dai et al., produced a novel reduced graphite oxide-silver nanocomposite which showed a synergistic antimicrobial effect. The reduced graphite alone displayed no apparent antimicrobial effect however, when used in combination with silver ions an improved synergistic antimicrobial effect was seen.[60] The authors hypothesized that the reduced graphite oxide acted as a supporting structure, whilst the silver ions possessed the overall antimicrobial ability.[60] This was emphasised by Gerasymchuk et al. (2016) who carried out antimicrobial testing on a reduced graphite oxide, silver nanoparticle and bis(lysinato)zirconium(IV) phthalocyanine complexes.[61] The complex was tested for its antimicrobial efficacy against *S. aureus*, *E. coli* and *P. aeruginosa*. The results showed a prolonged synergistic antimicrobial effect after near-infrared irradiation, with a four-fold decrease observed in the Gram-negative strains (*P. aeruginosa* and *E. coli*) viable cells.[61] The authors suggested that the complex should be tested as an antimicrobial surface coating for use in the field of dentistry or as a wound dressing due to its long lasting properties.[61] Therefore, the use of reduced graphite oxide in combination with another antimicrobial agent could potentially demonstrate a synergistically antimicrobial effect.

## Graphene

Since the discovery and isolation of graphene by Geim and Novoselov in 2004, research into this material and field has increased exponentially.[62] Graphene is a one-layer atom thick sheet of hexagonally arranged carbon. In recent years, graphene has attracted a lot of interest in a wide variety of industries, due to its unique properties such as excellent thermal conductivity (up to 5,000 W m$^{-1}$ K$^{-1}$), electrical conductivity, high electron mobility of up to 200,000 cm$^2$ V$^{-1}$ s$^{-1}$, permeability to gases, excellent tensile strength (42 N m$^{-1}$), and its high surface area (2630 m$^2$ g$^{-1}$).[63] Graphene has a variety of proposed antimicrobial mechanisms including damage due to physiochemical interactions (hydrophobicity), or due to physical interactions (size/sharp edges) (Fig. 2). The lateral size of graphene-materials is essential when determining the antimicrobial

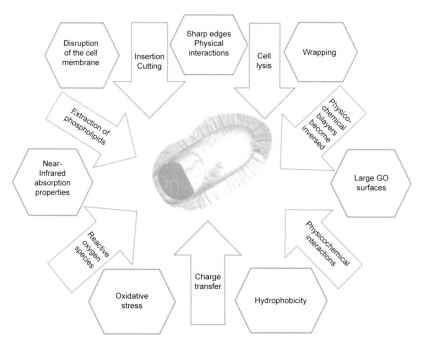

**Fig. 2.** Potential mechanisms of antibacterial activity of graphite, graphene and other carbon materials (Microbial figure courtesy of Dr. Lucia Cabellero, MMU, UK).

activity; this is because lateral size can influence the adsorption, dispersibility and the structure, including sharp edges—which are crucial for antimicrobial physiochemical interactions between the graphene-materials and microorganisms.

Another proposed antimicrobial action suggested to be demonstrated by larger molecules of graphene materials is cell wrapping.[64] Dallavalle et al. (2015) investigated the relationship between graphene material lateral size and antimicrobial mechanisms.[65] It was demonstrated that graphene material, less than 5.2 nm in size, had a predominant physical effect against microorganisms leading to cell lysis by protrusion of cell membranes, which is in turn attributed to the sharp edges of graphene materials.[65] On the other hand, larger graphene molecules (more than 5.2 nm) acted by reducing the microbial activity/cell viability of bacterial microorganisms through "wrapping" process, which did not affect the cell's integrity/shape. This was demonstrated by measuring the amount of nicotinamide adenine dinucleotide hydrogen and/or nicotinamide adenine dinucleotide phosphate, in a cellular metabolic assay.[30]

Large area monolayer graphene foams, have been used as a surface coating of copper conductors, and demonstrated promising properties against both *Staphylococcus aureus* and *Escherichia coli*. However, when the same bacteria were tested on a graphene-coated silicon dioxide insulator, although antimicrobial activity was observed, the morphology of the bacteria remained the same, therefore, charge transfer may be another mechanism of antimicrobial activity.[66] This charge transfer mechanism was described in further detail by Bykkam et al. (2015) which showed graphene foams coated with zinc oxide nanoparticles irradiated high photon energy levels, leading to the transfer of electrons from the valence band to the conduction band of the product material. This was shown to produce holes in the valence band, which were able to react with hydroxyl groups and absorbed water to create hydroxyl radicals (–OH). Electrons trapped in the conduction band by the presence of oxygen were shown to produce superoxide radical ions ($\bullet O^{-2}$), whilst hydrogen peroxide was generated by the combination of the electron pair hole, the production of reactive oxygen species (ROS) then exhibited a detrimental effect on the bacterial cells.[67]

Akhavan and Ghaderi (2010), investigated the antimicrobial activity of graphene layers against both Gram positive (*S. aureus*) and Gram negative bacteria (*E. coli*), however the results demonstrated a greater antimicrobial effect against *S. aureus*, possibly due to lack of an outer membrane.[68] However, antimicrobial activity was also demonstrated when zinc nanoparticles were imbedded into graphene sheets producing zones of inhibition against Gram negative *Salmonella typhi* and Gram negative *E. coli*.[69] In one study, a novel graphene based silver/hydroxyapatite/graphene (Ag/HAP/Gr) composite surface coating was produced by electrophoretic deposition (EPD), and the antimicrobial efficacy was tested against both *S. aureus* and *E. coli*. The results showed that after one hour bacterial growth had been inhibited by 72.9% and 68.4%, respectively, and after 24 hours the samples did not contain any viable cells, thus suggesting that the antimicrobial activity was effective against both planktonic and biofilm-forming strains of bacteria.[70] Due to graphene's low cytotoxic activity against human cells and excellent bacterial toxicity, it is an ideal candidate for application with biomaterials. When graphene was used as a surface coating of poly(N-vinylcarbazole) (PVK), results showed inhibition of up to 80% of biofilm growth, after 24 hour of incubation with *E. coli* and *Bacillus subtilis*, compared against a NIH 3T3 (mouse) cell line where over 80% of the cells were viable after 24 hours.[71] In our laboratories, when 3D graphene foams have been combined with metal ions, using zone of inhibition assays, antimicrobial activity was demonstrated to be increased against Gram positive bacteria (Fig. 3). Antibacterial activity was also demonstrated against the more recalcitrant Gram negative *Klebsiella pneumoniae* and *Acinetobacter baumannii*. Further, our work demonstrated that on a 3D graphene foam substrate that has been soaked in a gallium compound, the physiological structure of the bacterial cells becomes altered due to cellular damaged (Fig. 4). Therefore, with further research, graphene could have the potential to be used as a surface coating for both equipment (i.e., nosocomial setting) and/or in wound dressings, due to its variety of antimicrobial mechanisms, including physiochemical interactions such as cell wall penetration, and cell wrapping depending on the lateral size of the graphene particle.

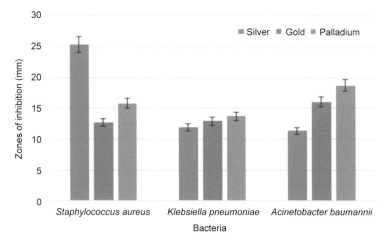

**Fig. 3.** Antimicrobial activity of few layer 3D freestanding graphene foams doped with metals ($n = 3$). Antibacterial activity was also demonstrated against the more recalcitrant Gram negative *Klebsiella pneumoniae* and *Acinetobacter baumannii*.

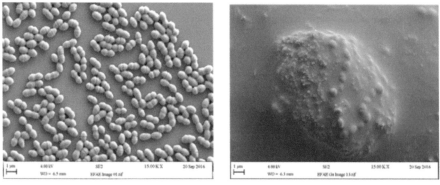

**Fig. 4.** (a) *Enterococcus faecium* cell structure (b) *Enterococcus faecium* following inoculation onto a 3D graphene/gallium structure (Images courtesy of Anthony Slate and Dr Grace Crowther, MMU, UK).

# Graphene Oxide

Much alike graphene, graphene oxide is a 2D-nanomaterial with promising applications in a variety of fields including polymer composites, electrochemical appliances (i.e., electrodes), sensors and biomedical applications[9] due to its excellent electrical, thermal and mechanical properties[12]. Unlike graphene, graphene oxide is hydrophilic due to the oxygen containing groups, allowing it to solubilise in water.[10] Graphene oxide is a promising material for the development of antimicrobial surface coating, due to its reported excellent contact-based antimicrobial activity; however the exact mechanisms are yet to be fully elucidated.[11] Graphene oxide is reported to have demonstrated broad-spectrum antimicrobial activity against both bacteria and fungi, including resistant strains.[13] Graphene oxide has also demonstrated broad spectrum antiviral activity; it was shown to significantly reduce both pseudorabies virus (PRV) and porcine epidemic diarrhoea virus (PEDV), leading to a two log reduction of viral titres.[72] This antimicrobial activity has been attributed to the introduction of oxygen-containing groups present in graphene oxide. It has been shown that altering the surface properties of graphene oxide, such as the edge planes, can dramatically improve antimicrobial activity, leading to a marked improvement in both amphipathy (important for the wrapping mechanism) and physiochemical effects.[68]

Since graphene oxide is water-soluble it has properties that allow for high drug loading and miscibility with polymers, effectively allowing it to be moulded into any desired shape.[14] Bitounis et al. (2013) reported that in a study investigating the antimicrobial effect of graphene based materials, that graphene oxide demonstrated the greatest antimicrobial efficacy followed by reduced graphene oxide, and this was suggested to be due to the production of ROS which led to oxidative stress, leading to more severe membrane damage and therefore loss of cell membrane integrity.[73]

A decrease in graphene oxide sheet size ($0.65\ \mu m^2$ to $0.01\ \mu m^2$), when tested against *E. coli*, also demonstrated a direct correlation of increased antimicrobial activity, due to physiochemical interactions.[11] In contrast, when using a suspension assay, the larger the graphene oxide sheets, the more effective became the antimicrobial activity. This was suggested to be due to cell entrapment, i.e., wrapping, with complete inactivation observed with the $0.65\ \mu m^2$ graphene oxide sheets after 3 hours of exposure. However cell inactivation due to the cell entrapment mechanism could be reversed, when separating the graphene oxide sheets by sonication.[11] Graphene oxide nanosheets have demonstrated strong antimicrobial properties when used in conjunction with thin-film composites as a surface coating. Graphene oxide nanosheets were irreversible attached via amide coupling of carboxyl groups present on both graphene oxide and a polyamide active layer. Microbiological testing of this surface coating against *E. coli* showed 65% bacterial inactivation after 1 hour of incubation.[74] In another study, a graphene oxide-iron oxide nanoparticle-silver nanoparticle (GO-IONP-Ag) complex was evaluated as a surface coating. The results showed that the composite was more effective than silver nanoparticles alone, showing a marked increase in antimicrobial activity against both Gram-positive (*S. aureus*) and Gram-negative (*E. coli*) bacterial strains. This synergistic ability was thought to be due to light being absorbed by the

GO demonstrated a synergistic response, resulting in the photo-thermal killing of the bacteria.[75] Faria et al. (2014) also used graphene oxide in combination with silver nanoparticles as a surface coating, which resulted in 100% growth inhibition of *Pseudomonas aeruginosa*, with a minimum inhibitory concentration (MIC) ranging between 2.5–5.0 mg/L.[76] Work by Whitehead et al. (2017) has demonstrated that when silver-graphene oxide (AgGO) or zinc oxide-graphene oxide (ZnOGO) complexes were tested for their antimicrobial activity against four prominent bacteria which have all demonstrated increased multidrug resistance (*E. coli*, Methicillin Resistant *S. aureus* (MRSA)*, Enterococcus faecium* and *Klebsiella pneumoniae*) that AgGO was the most effective antimicrobial. The addition of Ag enhanced the activity of GO against the bacteria tested, including the generally recalcitrant *K. pneumoniae* and *Enterococcus faecium*. Zhao et al. (2016) developed a gelatin-functionised graphene oxide coating which was impregnated onto a nitinol substrate. The results showed both strong antimicrobial activity against *E. coli* and excellent biocompatibility.[26] Therefore it can be evidenced that alterations to the graphene oxides surface properties as well as its characteristic functional groups (leading to the production of ROS) allows graphene oxide to interact with microorganisms in a dual approach, therefore suggesting an explanation towards the prolific antimicrobial activity demonstrated.

## Reduced Graphene Oxide

In order to obtain reduced graphene oxide (also known as functionalised graphene or reduced graphene), removal of oxygen-containing groups is vital, with the end goal of the reduction protocol being to produce graphene-like materials with similar properties to graphene.[77] Reduction of graphene oxide can be undertaken by a variety of mechanisms including, mechanical/thermal exfoliation, epitaxial growth and chemical vapour deposition.[78,79]

Reduced graphene oxide has demonstrated a vast array of potential applications including electrochemical materials,[15] photocatalysis,[80] industrial lubricants and corrosion protection.[81] The antimicrobial activity of graphene oxide has yet to be fully elucidated, however, studies so far have demonstrated positive results with the potential application of reduced graphene oxide to be used as a viable antimicrobial agent.[79] In one study, a biocompatible reduced graphene oxide-silver nano-hybrid was prepared and tested as an antimicrobial topical agent, i.e., as a surface coating for dressings and bandages. The results found that excellent cytocompatibility was observed towards peripheral blood mononuclear cells and mammalian red blood cells whilst antimicrobial assays against *S. aureus, E. coli* and *Candida albicans* showed synergistic antimicrobial activity compared to reduced graphene oxide and silver nanoparticles as individual antimicrobial therapies.[82] This biocompatibility of reduced graphene oxide has also been described when used in combination with calcium silicate which plays an important role in bone tissue engineering. The use of a calcium silicate and reduced graphene oxide composite increased the fracture toughness by 123% without negatively effecting the attachment of human osteoblast cells (hFOB).[83]

Synergistic antibacterial activity has been demonstrated between reduced graphene oxide and boron-doped diamond anodes, with the results showing a reduction of 0.7 log of *E. coli* after 20 min of incubation with graphene oxide. However, 100%

inactivation of *E. coli* growth was demonstrated when reduced graphene oxide was tested in combination with a three dimensional electrochemical system.[79] Previous studies proved that hydroxyl radicals played a pivotal role in boron doped diamond electrochemical disinfection, and it was hypothesised that the reduced graphene oxide led to an expansion in the electrode area and therefore generated more hydroxyl radicals.[84] Reduced graphene oxide nanosheets doped onto titanium dioxide thin films showed improved antimicrobial activity against *E. coli* under solar light radiation (by 60%) after 0.5 hours. The thin films were demonstrated to be chemically stable and it was suggested that they could be used as an antimicrobial surface coating for hospital equipment, thus potentially reducing the transmission of AMR and other microorganisms.[85] Evidence suggests that the main mechanism of antimicrobial activity produced by reduced graphene oxide is due to the production of reactive oxygen species. This was shown to be the case in one study, where the exposure of graphene oxide and reduced graphene oxide to *P. aeruginosa* induced significant amounts of superoxide radical anions, leading to the loss of cell viability.[16]

## Non-carbon Based 2D Nanomaterials

Other non-carbon based, 2D-nanomaterials with potential applications in antimicrobial surface coatings include boron nitrite, tungsten dioxide and molybdenum disulphide. However, these surfaces have not been investigated in great depth for their antimicrobial activity.

## Boron Nitride

Boron nitride commonly exists in a layered structure of hexagonal honeycombs comprised of equal amounts of boron and nitrogen atoms. These layers are held together by van der Waals forces, with the nitrogen atoms directly above the boron atoms.[17] Due to its similarity to graphene, both in terms of structure and properties it is often referred to as "white graphene".[86] This similarity has been observed by a number of studies, with one example being the use of boron nitride as a electrochemical dopamine biosensor.[87] In recent years, research with boron nitride has shown antimicrobial activity related to the 2D-nanomaterial. It was demonstrated that aqueous dispersions of boron nitride nanotubes exhibited little antimicrobial activity against *E. coli* and *S. aureus*. However, when coated in combination with polyethyleneimine, the boron nitride nanotubes exhibited a significant synergistic antimicrobial activity against *E. coli* and *S. aureus,* with the authors suggesting the potential application of boron nitride nanotubes as surface coatings for use in water purification and food packaging systems.[88]

## Tungsten Diselenide

Another 2D-nanomaterial with great potential is tungsten diselenide ($Se_2W$). When in crystalline monolayers it is found to act as a promising emitter of light at around 750 nm.[89] Tungsten diselenide is a two dimensional metal dichalcogenide, with a semiconducting nature.[90] These properties allow $WSe_2$ to act as a photo-catalyst, leading to the production of hydrogen, which result in the generation of ROS which have the ability to cause damage to a wide array of microorganisms by a variety of mechanisms.[91,92] The research into the antimicrobial effect of tungsten diselenide is in its infancy, however recent studies have demonstrated impressive antimicrobial efficacy.[90,93] One study demonstrated the antimicrobial activity of selenium nanoparticles when used as a surface coasting in conjunction with polymeric medical devices.[94] In another study, tungsten diselenide in combination with single-stranded DNA (ssDNA) and the antimicrobial efficacy of the $WSe_2$-ssDNA nanosheets was evaluated against *E. coli*. The results demonstrated a greater antimicrobial effect compared to the use of graphene oxide.[93] Research into $Se_2W$ may show potential for this to be developed into an alternative antimicrobial surface coating.[95]

## Molybdenum Disulfide

Molybdenum disulphide ($MoS_2$) which is also a two dimensional metal dichalcogenide, is the principle natural source of molybdenum, and is mainly obtained as a secondary product from the mining of copper. Anisotropic properties of $MoS_2$ which arise due to its laminar nature (similar to graphite), allow $MoS_2$ to be used in a variety of applications including industrial lubricants catalysts and electrical energy storage products.[18]

The antimicrobial efficacy of 2D chemically exfoliated $MoS_2$ sheets was evaluated by Yang et al. (2014) against *E. coli*. The results demonstrated that an MIC of 2.5 µg/mL was inhibitory to the bacteria and this antimicrobial activity was attributed to the production of ROS, particularly due to glutathione oxidation which showed a time and concentration dependent trend. It was speculated that the accumulation of ROS led to both membrane and oxidative stress, and eventually loss of cell membrane integrity and death.[96] It has been further suggested that the $MoS_2$ showed a greater amount of antimicrobial activity towards Gram-positive bacteria.[97,98] In agreement with this, work in our laboratories has demonstrated that molybdenum disulphide surfaces do have some antimicrobial activity against Gram positive (*S. aureus*) and Gram negative (*E. coli*) biofilms (Table 3).

Biofilm inhibition has also been demonstrated by molybdenum disulphide by others. In a study, *P. aeruginosa* biofilms were grown with molybdenum disulphide at a concentration of 150 µg/mL and the results showed a decrease in the biofilm growth

**Table 3.** Antimicrobial effect of $MoS_2$ surfaces evaluated against biofilms at $OD_{540}$ ($n = 3$) (Work courtesy of Mohsin Amin, MMU, UK).

| Concentration $MoS_2$ | *Staphylococcus aureus* | *Pseudomonas aeruginosa* |
|:---:|:---:|:---:|
| 0% | 1.30 ± 0.12 | 1.58 ± 0.22 |
| 5% | 1.27 ± 0.23 | 1.53 ± 0.20 |
| 10% | 1.16 ± 0.23 | 1.53 ± 0.16 |
| 15% | 1.15 ± 0.15 | 1.34 ± 0.17 |
| 20% | 0.93 ± 0.12 | 1.03 ± 0.30 |

of up to 40%, in addition to shrinkage in biofilm depth after treatment.[89] These results may prove to be extremely important given the recalcitrant nature of biofilms.

# Conclusion

Throughout this chapter, the antimicrobial potential and applications for the use of 2D-nanomaterials, especially in the case of potential antimicrobial surface coatings has been discussed. Although research into 2D-nanomaterials as surface coatings is in its infancy, research in this area is rapidly progressing. However, more research is needed in order to fully elucidate and characterise all the mechanisms of antimicrobial activity of such materials. This would allow researchers to be able to modify the nanomaterials structures/properties in order to make them more potent and effective as antimicrobial agents. In addition to using these 2D nanomaterials as stand-alone antimicrobial interventions, there is significant interest in utilising these 2D nanomaterials in combinations with other molecules, functional groups and metals. Combinations of such chemical moieties may result in surfaces that demonstrate synergistic antimicrobial effects, that may have the potential to reduce the transmission of nosocomial infections. This in turn may decrease the current burden which is currently placed upon conventional bacterial treatments such as antibiotics. The current literature highlights the huge potential these nanomaterials could have in order to reduce the transmission of antimicrobial resistant bacteria.

# Acknowledgements

The authors would like to thank Dr Lucia Cabellero, Dr Grace Crowther, Mohsin Amin and Daniel Brown for their contributions of the Figures towards this work.

# References

1.  C. L. Ventola, *Pharmacy and Therapeutics*, 2015, **40**, 277–283.
2.  J. G. Bartlett, D. N. Gilbert and B. Spellberg, *Clinical and Infectious Diseases*, 2013, **56**, 1445–1450.
3.  H. Ochman, J. G. Lawrence and E. A. Groisman, *Nature*, 2000, **405**, 299–304.
4.  J. Davies and D. Davies, *Microbiology and Molecular Biology Reviews: MMBR*, 2010, **74**, 417–433.
5.  J. N. Pendleton, S.P. Gorman and B.F. Gilmore, *Expert Review of Anti Infective Therapy*, 2013, **11**, 297–308.
6.  World Health Organisation (2014). http://apps.who.int/iris/bitstream/10665/112642/1/978241564748_eng.pdf
7.  N. Wu, L. He, P. Cui, W. Wang, Y. Yuan, S. Liu, T. Xu, S. Zhang, J. Wu, W. Zhang and Y. Zhang, *Frontiers Microbiology*, 2015, **6**, 1003.
8.  P. Chen, A. K. Seth, J. J. Abercrombie, T. A. Mustoe and K. P. Leung, *Antimicrobial Agents and Chemotherapy*, 2014, **58**, 1208–1213.
9.  O. C. Compton and S. T. Nguyen, *Small*, 2010, **6**, 711–723.
10. H. He, J. Klinowski, M. Forster and A. Lerf, *Chemical Physics Letters*, 1998, **287**, 53–56.
11. M. Eluyemi, M. Eleruja, A. Adedeji, B. Olofinjana, O. Fasakin, O. Akinwunmi, O. Ilori, A. Famojuro, S. Ayinde and E. Ajayi, *Graphene*, 2016, **5**, 143.
12. C. H. An Wong, *Nanoscale*, 2012, **4**, 4972.
13. I. Sheets, H. Holail, Z. Olama, A. Kabbani and M. Hines, *International Journal of Current Microbiology and Applied Sciences*, 2013, **2**, 1–11.

14. S. Liu, T. H. Zeng, M. Hofmann, E. Burcombe, J. Wei, R. Jiang, J. Kong and Y. Chen, *ACS Nano*, 2011, **5**, 6971–6980.
15. Y. Zhou, Y. Kong , S. Kundu, J. D. Cirillo and H. Liang, *Journal of Nanobiotechnology*, 2012, **10**, 1–9.
16. R. F. Al-Thani, N. K. Patan and M. A. Al-Maadeed, *Online Journal of Biological Sciences*, 2014, **14**, 230–239.
17. W. Hu, C. Peng, W. Luo, M. Lv, X. Li, D. Li, Q. Huang and C. Fan, *ACS Nano*, 2010, **4**, 4317–4323.
18. S. Gurunathan, J. W. Han, A. A. Dayem, V. Eppakayala and J. H. Kim, *International Journal of Nanomedicine*, 2012, **7**, 5901–5914.
19. C. Wu, A. R. Deokar, J.-H. Liao, P.-Y. Shih and Y.-C. Ling, *ACS Nano*, 2013, **7**, 1281–1290.
20. M. R. Das, K. Rupak, R. Sarma, R. Saikia, V. S. Kale, M. V. Shelke and P. Sengupta, *Colloids and Surfaces B: Biointerfaces*, 2011, **83**, 16–22.
21. M. R. Das, R. K. Sarma, S. C. Borah, R. Kumari, R. Saikia, A. B. Deshmukh et al., *Colloids and Surfaces B: Biointerfaces* 2013, **105**, 128–136.
22. N. Hussain, A. Gogoi, R. K. Sarma, P. Sharma, A. Barras, R. Boukherroub, R. Saikia, P. Sengupta and M. R. Das, *Chempluschem.*, 2014, **79**, 1774–1784.
23. C. M. Santos, M. C. R. Tria, R. A. M. V. Vergara, F. Ahmed, R. C. Advincula and D. F. Rodrigues, *Chemical Communications*, 2011, **47**, 8892–8894.
24. C. M. Santos, J. Mangadlao, F. Ahmed, A. Leon, R. C. Advincula and D. F. Rodrigues, *Nanotechnology*, 2012, **23**, 395101.
25. K. Krishnamoorthy, M. Veerapandian, L.-H. Zhang, K. Yun and S. J. Kim, *Journal of Physical Chemistry - C*, 2012, **116**, 17280–17287.
26. C. Zhao, S. Pandit, Y. Fu, I. Mijakovic, A. Jesorka and J. Liu, *RSC Advances*, 2016, **6**, 38124–38134.
27. F. Perreault, A. F. de Faria, S. Nejati and M. Elimelech, *ACS Nano*, 2015, **9**, 7226–7236.
28. J. Li, G. Wang, H. Zhu, M. Zhang, X. Zheng, Z. Di, X. Liu and X. Wang, *Scientific Reports*, 2014, **4**, 4359.
29. O. Akhavan and E. Ghaderi, *ACS Nano*, 2010, **4**, 5731–5736.
30. C. Cha, S. R. Shin, N. Annabi, M. R. Dokmeci and A. Khademhosseini, *ACS Nano*, 2013, **7**, 2891–2897.
31. A. Krueger, in *Carbon Materials and Nanotechnology*, Wiley-VCH Verlag GmbH & Co. KGaA, 2010, pp. 1–32.
32. Y. Song, W. Wei, X. Qu, *Advanced Materials*, 2011, **23**, 4215–4236.
33. T. Kuila, S. Bose, A. K. Mishra, P. Khanra, N. H. Kim, J. H. Lee, *Progress in Material Science*, 2012, **57**, 1061–1105.
34. H. Bai, C. Li and G. Shi, *Advanced Materials*, 2011, **23**, 1089–1115.
35. L. Feng and Z. Liu, *Nanomedicine*, 2011, **6**, 317–324.
36. Z. Liu, J. T. Robinson, X. Sun and H. Dai, *Journal of American Chemical Society*, 2008, **130**, 10876–10877.
37. K. Yang, S. Zhang, G. Zhang, X. Sun, S. T. Lee and Z. Liu, *Nano Letters*, 2010, **10**, 3318–3323.
38. J. L. Li, H.-C. Bao, X.-L. Hou, L. Sun, X. G. Wang and M. Gu, *Angewandte Chemie International Edition*, 2012, **51**, 1830–1834.
39. M. Li, X. Yang, J. Ren, K. Qu and X. Qu, *Advanced Materials*, 2012, **24**, 1722–1728.
40. Y. Tao, Y. Lin, Z. Huang, J. Ren and X. Qu, *Advanced Materials*, 2013, **25**, 2594–2599.
41. N. Mohanty and V. Berry, *Nano Letters*, 2008, **8**, 4469–4476.
42. L. Feng, L. Wu and X. Qu, *Advanced Materials*, 2013, **25**, 68–186.
43. S. Liu, T. H. Zeng, M. Hofmann, E. Burcombe, J. Wei, R. Jiang, J. Kong and Y. Chen, *ACS Nano*, 2011, **5**, 6971–6980.
44. E. Zanni, G. De Bellis, M. P. Bracciale, A. Broggi, M. L. Santarelli, M. S. Sarto, C. Palleschi and D. Uccelletti, *Nano Letters*, 2012, **12**, 2740–2744.
45. J. Wang, N. Huang, P. Yang, Y. X. Leng, H. Sun, Z. Y. Liu and P. K. Chu, *Biomaterials*, 2004, **25**, 3163–3170.
46. A. Shteynle, Proceedings on the 7th International Forum on Strategic Technology, 2012.
47. Y. Liu, Z. Ren, Y. Wei, B. Jiang, S. Feng, L. Zhang, W. Zhang and H. Fu, *Journal of Materials Chemistry*, 2010, **20**, 4802–4808.
48. A. Lerf, H. He, M. Forster and J. Klinowski, *The Journal of Physical Chemistry B*, 1998, **102**, 4477–4482.
49. J. L. Hobman and N. L. Brown, *Metal Ions in Biological Systems*, 1997, **34**, 527–568.

50. C. Nethravathi and M. Rajamathi, *Carbon*, 2008, **46**, 1994–1998.
51. M. Ciszewski and A. Mianowski, *Materials Science-Poland*, 2014, **32**, 307–314.
52. S. Stankovich, D. A. Dikin, R. D. Piner, K. A. Kohlhaas, A. Kleinhammes, Y. Jia, Y. Wu, S. T. Nguyen and R. S. Ruoff, *Carbon*, 2007, **45**, 1558–1565.
53. Y. Zhu, S. Murali, W. Cai, X. Li, J. W. Suk, J. R. Potts and R. S. Ruoff, *Advanced Materials*, 2010, **22**, 3906–3924.
54. S. Liu, T. H. Zeng, M. Hofmann, E. Burcombe, J. Wei, R. Jiang, J. Kong and Y. Chen, *ACS Nano*, 2011, **5**, 6971–6980.
55. J. S. Chen, Z. Wang, X. C. Dong, P. Chen and X. W. Lou *Nanoscale*, 2011, **3**, 2158–2161.
56. W. Gao, L. B. Alemany, L. Ci and P. M. Ajayan, *Nature Chemistry*, 2009, **1**, 403–408.
57. A. L. Casey, D. Adams, T. J. Karpanen, P. A. Lambert, B. D. Cookson, P. Nightingale, L. Miruszenko, R. Shillam, P. Christian and T. S. J. Elliott, *Journal of Hospital Infection*, 2010, **74**, 72–77.
58. S. Silver, T. Phung and G. Silver, *Journal of Industrial Microbiology and Biotechnology*, 2006, **33**, 627–634.
59. Z. Liu, J. T. Robinson, X. Sun and H. Dai, *Journal of the American Chemical Society*, 2008, **130**, 10876–10877.
60. C. Dai, X. Yang and H. Xie, *Materials Research Bulletin*, 2011, **46**, 2004–2008.
61. Y. Gerasymchuk, A. Lukowiak, A. Wedzynska, A. Kedziora, G. Bugla-Ploskonska, D. Piatek, T. Bachanek, V. Chernii, L. Tomachynski and W. Strek, *Journal of Inorganic Biochemistry*, 2016, **159**, 142–148.
62. A. K. Geim and K. S. Novoselov, *Nature Materials*, 2007, **6**, 183–191.
63. D. A. C. Brownson, D. K. Kampouris and C. E. Banks, *Chemical Society Reviews*, 2012, **41**, 6944–6976.
64. F. Taherian, V. Marcon, N. F. van der Vegt and F. D. R. Leroy, *Langmuir*, 2013, **29**, 1457–1465.
65. M. Dallavalle, M. Calvaresi, A. Bottoni, M. Melle-Franco and F. Zerbetto, *ACS Applied Materials & Interfaces*, 2015, **7**, 4406–4414.
66. X. Zou, L. Zhang, Z. Wang and Y. Luo, *Journal of the American Chemical Society*, 2016, **138**, 2064–2077.
67. S. Bykkam, S. Narsingam, M. Ahmadipour, T. Dayakar, K. Venkateswara Rao, C. Shilpa Chakra and S. Kalakotla, *Superlattices and Microstructures*, 2015, **83**, 776–784.
68. O. Akhavan and E. Ghaderi, *ACS Nano*, 2010, **4**, 5731–5736.
69. L. Zhong and K. Yun, *International Journal of Nanomedicine*, 2015, **10**, 79–92.
70. A. Janković, S. Eraković, M. Vukašinović-Sekulić, V. Mišković-Stanković, S. J. Park and K. Y. Rhee, *Progress in Organic Coatings*, 2015, **83**, 1–10.
71. C. M. Santos, J. Mangadlao, F. Ahmed, A. Leon, R. C. Advincula and D. F. Rodrigues, *Nanotechnology*, 2012, **23**, 395101.
72. S. Ye, K. Shao, Z. Li, N. Guo, Y. Zuo, Q. Li, Z. Lu, L. Chen, Q. He and H. Han, *ACS Applied Material Interfaces*, 2015, **7**, 21571–21579.
73. D. Bitounis, H. Ali-Boucetta, B. H. Hong, D. H. Min and K. Kostarelos, *Advanced Materials*, 2013, **25**, 2258–2268.
74. F. Perreault, M. E. Tousley and M. Elimelech, *Environmental Science & Technology Letters*, 2014, **1**, 71–76.
75. T. Tian, X. Shi, L. Cheng, Y. Luo, Z. Dong, H. Gong, L. Xu, Z. Zhong, R. Peng and Z. Liu, *ACS Applied Material Interfaces*, 2014, **6**, 8542–8548.
76. A. F. de Faria, D. S. T. Martinez, S. M. M. Meira, A. C. M. de Moraes, A. Brandelli, A. G. S. Filho and O. L. Alves, *Colloids and Surfaces B: Biointerfaces*, 2014, **113**, 115–124.
77. S. Pei and H.-M. Cheng, *Carbon*, 2012, **50**, 3210–3228.
78. V. Loryuenyong, K. Totepvimarn, P. Eimburanapravat, W. Boonchompoo and A. Buasri, *Advances in Materials Science and Engineering*, 2013, **2013**, 5.
79. S.-J. Chang, M. S. Hyun, S. Myung, M.-A. Kang, J. H. Yoo, K. G. Lee, B. G. Choi, Y. Cho, G. Lee and T. J. Park, *Scientific Reports*, 2016, **6**, 22653.
80. A. Iwase, Y. H. Ng, Y. Ishiguro, A. Kudo and R. Amal, *Journal of the American Chemical Society*, 2011, **133**, 11054–11057.
81. C. Wong, C. Lai, K. Lee and S. Hamid, *Materials*, 2015, **8**, 5363.
82. S. Barua, S. Thakur, L. Aidew, A. K. Buragohain, P. Chattopadhyay and N. Karak, *RSC Advances*, 2014, **4**, 9777–9783.

83. M. Mehrali, E. Moghaddam, S. F. S. Shirazi, S. Baradaran, M. Mehrali, S. T. Latibari, H. S. C. Metselaar, N. A. Kadri, K. Zandi and N. A. A. Osman, *ACS Applied Materials & Interfaces*, 2014, **6**, 3947–3962.
84. J. Jeong, J. Y. Kim and J. Yoon, *Environmental Science & Technology*, 2006, **40**, 6117–6122.
85. O. Akhavan and E. Ghaderi, *The Journal of Physical Chemistry C*, 2009, **113**, 20214–20220.
86. T. H. Yuzuriha and D. W. Hess, *Thin Solid Films*, 1986, **140**, 199–207.
87. A. F. Khan, D. A. C. Brownson, E. P. Randviir, G. C. Smith and C. E. Banks, *Analytical Chemistry*, 2016, **88**, 9729–9737.
88. J. S. Maria Nithya and A. Pandurangan, *RSC Advances*, 2014, **4**, 32031–32046.
89. Z. Wang, Z. Dong, Y. Gu, Y.-H. Chang, L. Zhang, L.-J. Li, W. Zhao, G. Eda, W. Zhang, G. Grinblat, S. A. Maier, J. K. W. Yang, C.-W. Qiu and A. T. S. Wee, *Nature Communications*, 2016, **7**, 11283.
90. G. R. Navale, C. S. Rout, K. N. Gohil, M. S. Dharne, D. J. Late and S. S. Shinde, *RSC Advances*, 2015, **5**, 74726–74733.
91. X. Yu, M. S. Prévot, N. Guijarro and K. Sivula, *Nature Communications*, 2015, **6**, 7596.
92. F. Vatansever, W. C. de Melo, P. Avci, D. Vecchio, M. Sadasivam, A. Gupta, R. Chandran, M. Karimi, N. A. Parizotto, R. Yin, G. P. Tegos and M. R. Hamblin, *FEMS Microbiology Reviews*, 2013, **37**, 955–989.
93. G. S. Bang, S. Cho, N. Son, G. W. Shim, B.-K. Cho and S.-Y. Choi, *ACS Applied Materials & Interfaces*, 2016, **8**, 1943–1950.
94. P. A. Tran and T. J. Webster, *Nanotechnology*, 2013, **24**, 155101.
95. B. Liu, M. Fathi, L. Chen, A. Abbas, Y. Ma and C. Zhou, *ACS Nano*, 2015, **9**, 6119–6127.
96. X. Yang, J. Li, T. Liang, C. Ma, Y. Zhang, H. Chen, N. Hanagata, H. Su and M. Xu, *Nanoscale*, 2014, **6**, 10126–10133.
97. N. Qureshi, R. Patil, M. Shinde, G. Umarji, V. Causin, W. Gade, U. Mulik, A. Bhalerao and D. P. Amalnerkar, *Applied Nanoscience*, 2015, **5**, 331–341.
98. W. Zhang, S. Shi, Y. Wang, S. Yu, W. Zhu, X. Zhang, D. Zhang, B. Yang, X. Wang and J. Wang, *Nanoscale*, 2016, **8**, 11642–11648.

# CHAPTER 6

# Graphene Devices for Biomolecule Detection and Sequencing

*Pauline M. G. van Deursen, Hadi Arjmandi-Tash,
Wangyang Fu* and *Grégory F. Schneider**

## Introduction

### Biomolecular Sequencing

The field of biomolecular sequencing entails finding the sequence of biopolymers that make up and encode life: amino acid sequences in proteins and nucleic acids in DNA. Such analyses form a cornerstone of biological engineering and medical care, and fast and on the spot analysis of biopolymers would greatly benefit many areas of application: epidemics in remote and underdeveloped areas, point-of-care medical science, quick determination of the origin of biological substances in customs and transport, and analysis of genetic modification and post-translocation modification of proteins.

In conventional sequencing techniques, speed and convenience have never been prioritized. Fragment analysis; radioactive or fluorescent labeling; purification and amplification by polymerase chain reaction—all such methods are time consuming and require dedicated laboratory resources. Next generation sequencing, in particular single molecular sequencing, seeks to improve on this aspect, widening the availability of sequencing techniques.

Leiden Institute of Chemistry, Leiden University, Einsteinweg 55, 2333 CC Leiden, Leiden, Netherlands.
* Corresponding author: g.f.schneider@chem.leidenuniv.nl

Single molecule sequencing relies on stringing a biopolymer from head-to-tail through a nanoscale aperture sensing device. Direct read-out of single molecules lifts the requirement of amplification or fragmentation. To make this concept reality, a read-out technique is needed sensitive enough to identify the nanometer-size building blocks in the chain as it traverses the aperture. The viability of this principle has been demonstrated using biological pore-forming protein complexes in lipid bilayers and integrated in nanothin solid state membranes.[1] Biological nanopores, however, have a limited lifetime. Nanopores grafted directly in solid state membranes, notably $SiN_x$, offer an alternative system with superior lifetime to biological pores.[2]

A drawback of these systems is found in the dimensions of biological and $SiN_x$ nanopores. The channel length of biological nanopores is usually larger than the thickness of their embedding membranes, such as lipid bilayers or polymeric matrixes. In solid state nanopores, the channel length equals the thickness of the membrane, which reaches a lower limit around 10 nm in $SiN_x$ membranes. Consequently, multiple building blocks will reside in the channel at any moment during biopolymer passage and the signal that arises from such a system is always an average over multiple polymer building blocks (Fig. 1, left panel).

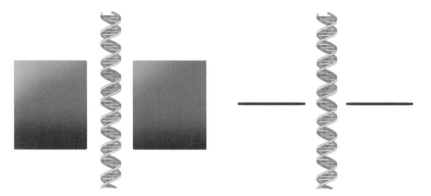

**Fig. 1.** DNA situated in a nanopore of conventional nanothin materials (left) and single layer graphene (right). Graphene, unlike conventional materials, has a thickness of the order of magnitude of single nucleotide building blocks, permitting single nucleotide read-out resolution.

## 2D Materials in Sequencing

The rise of 2D materials, and graphene in particular, opens the road to the true potential of nanopore sequencing. Exploiting the extreme thinness of 2D materials, a spacial resolution can be reached that will reveal single building blocks and enable direct read-out of the sequence, as is displayed in the right panel of Fig. 1.

Turning to the actual read-out mechanism, we can distinguish two primary methods.[3] The first method employs a nanopore suspended in an electrolyte environment. An applied potential over the membrane drives a steady state current through the nanopore. The entrance of a biomolecule into the pore brings about a drop in the current signal. The second technique uses two aligned sheets of graphene separated by a continuous nanoscopic gap. As the two sheets are not in physical contact, a tunneling current from one to the other can be measured. The magnitude of the tunneling current will be sensitive to molecules that temporarily occupy the gap. This chapter will outline the state of art on these techniques, and the developments needed in order to succeed in biopolymer sequencing with 2D materials. In the following, Section 2 will discuss the latest developments in the field of graphene nanopore technology whereas Section 3 will highlight the prospects of nanogap sensing.

Before we continue, we note that the focus of this chapter will be on graphene, only sidestepping to research on other 2D materials such as $MoS_2$. Graphene's electrical conductivity and aptitude for chemical modification pose an invitation to creative solutions to various problems encountered in sensing experiments, including ionic noise, biomolecule dynamics and stability of the nanopore. It is worth noting that many mechanical characteristics of graphene—such as those responsible for noise and spacial resolution—will expectedly apply to other 2D materials in a similar manner, or can be deduced from graphene. For instance, $MoS_2$, being approximately three times thicker than graphene, will rationally give a slight loss of read-out resolution, along with a slight decrease of noise due to mechanical flexibility.

# Nanopore Devices

## Design and Principles

### *Ionic current measurement*

Nanopore devices consist of a flow-cell with two electrolyte compartments connected through a single nanopore. The graphene layer with nanopore is usually supported by a $SiN_x/SiO_2$ support frame, as shown in Fig. 2 (left panel).[4] The nanopore and electrolyte compartments are part of a simple electronic circuit as shown in Fig. 2 (right panel). The read-out mechanism of nanopores relies on the measurement of the ionic current when a potential is applied over the pore.[5] In absence of translocating molecules, the current response to a given potential is determined solely by the dimensions of the pore. When large molecules reside within the pore, they give rise to a current

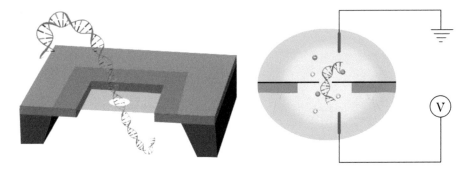

**Fig. 2.** (Left) Schematic of graphene nanopore supported on a frame of thicker material. Incorporated in a flow cell setup, the graphene-support ensemble forms the sensing region for ionic conductivity detection of translocation measurements. (right) The electrical circuit containing the nanopore ensemble during translocation measurement.

blockade proportional to the cross section of the molecule in the pore. Therefore, when charged biomolecules such as DNA or proteins pass the pore—driven by the same potential that drives the background current—a drop in current will occur, signaling a *translocation event* (Fig. 3).

It may be noted at this point that the charge of translocating molecules will make a contribution to the overall current signal as well. However, this contribution will be negligible if the surrounding electrolyte is sufficiently concentrated in salts (typically 1 M KCl is used). Low electrolyte strength measurements, on the other hand, can give information on charge distribution of translocation species—but as such experiments fall beyond the scope of this chapter, we refer the interested reader to the related work.[6]

## *The translocation event*

A typical current record of a translocation measurement in a 22 nm Ø graphene nanopore is presented in Fig. 3.[7] The left image shows the stable current through the pore when +200 mV is applied over the pore in the absence of translocating molecules, i.e., in pure electrolyte. When double stranded DNA is added to the negatively biased compartment, current drops are observed and referred to as 'translocation events'. Information on the length and width of the translocating DNA molecule can be gathered from, respectively, the duration and magnitude of each current blockade. The right-most trace in Fig. 3 zooms in on a single current blockade signal. Through the change in magnitude of the current blockade, the folding state of the translocating DNA strand can be determined (i.e., head-to-tail, doubly folded, triply folded).

Figure 3 (right) shows that measurements with the graphene-nanopore allow to detect the variation in molecular dimensions (i.e., folding of DNA). Furthermore, theoretical works have shown that the size variation of different base pairs in the DNA chain could be discernable using this technique, allowing base-pair sequence read-out with graphene. To move from theory to demonstration of this sequencing technique, two thresholds must be overcome. The first one is straight-forward: the signal-to-noise ratio of graphene nanopores must be increased, as the size variation of different base pairs—and its resulting variation in ionic current blockade—is subtle. Secondly, a biomolecule translocates through graphene nanopores extremely quickly (tens of nanoseconds per nucleotide). Mainly due to their extremely short channel length, graphene nanopores transmit 48 kbp DNA strands within milliseconds.[7] Reduction of

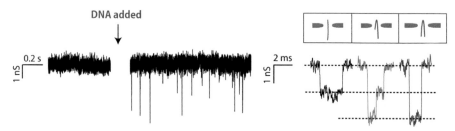

**Fig. 3.** (Left) Ionic current traces of nanopore measurements, showing the signal in absence of DNA and in presence of DNA. Downward current spikes signal the translocation of a DNA strand. (Right) Zooming in on single current spikes reveals information about the structure of the translocating molecule. Adapted with permission from [7].

driving potential may slightly prolong translocation, but also reduces the magnitude of background current therefore also reducing the signal to noise ratio. Viable solutions will consist of enhancing temporal resolution of measurements, or slowing down the passage of the translocating molecule.

## Noise

Although the introduction of 2D materials in nanopore devices raised high hopes for achieving biomolecule sequencing, certain limitations still exist. The signal from a nanopore sculpted in graphene and in $SiN_x$ membranes are compared in Fig. 4. With similar experimental conditions, a graphene nanopore exhibits conductances ~ 14 times higher than $SiN_x$ nanopores which have longer channel lengths. This is mainly a result of the shorter channel length of graphene nanopores. The higher conduction in graphene nanopore is accompanied by a higher current fluctuations, i.e., higher noise. Indeed, similarly, high noise levels have been observed experimentally in nanopores with various membranes of 2D materials and are considered a general drawback of 2D nanopores.

To tell apart various sources contributing to the total noise, we look at the power spectral density (PSD). The PDS reveals the frequency dependence of the current fluctuation. The noise in nanopores exhibits different characteristics at high (> kHz) and low (< kHz) frequency ranges. The origin of the high frequency noise in nanopores is found in the capacitive coupling of the *cis* and *trans* chambers. In this viewpoint, the two electrodes applying the transmembrane potential can be regarded as the parallel plates of a capacitor, separated by the membrane and its nanopore. Basic physics describes the way in which the capacitor conducts current through the membrane (i.e., the noise) with high frequencies. In the context of high frequency noise, the structure of the long channel and 2D nanopores are similar as capacitive coupling is largely governed by the dielectric properties of the $SiN_x$ layer, that is used both as the membrane in long channel nanopores and as the support for 2D materials

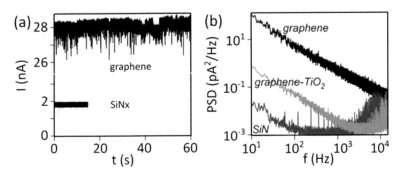

**Fig. 4.** Noise in nanopores in different materials. Ionic conduction through nanopores of similar diameters in graphene (d ≈ 8 nm) and SiNx (d ≈ 6 nm, L = 40 nm) membranes. The experiments were both performed in 1 M KCl, 10 mM Tris (pH 8.1). Numerous translocation events are visible as sharp current dips in both time traces. The figure is adapted from [4]. Power spectral density (PSD) calculated from the current-time traces for nanopores sculpted in bare graphene (d = 8 nm, V = 100 mV), in $TiO_2$-covered graphene (d = 7.5 nm, V = 100 mV) and in SiNx (d = 6 nm, V = 120 mV). The figure is adapted from [4].

in 2D nanopores. Indeed, due to the negligible area of 2D materials compared to the whole surface of the material in contact with the buffer solution, the 2D material does not affect the high frequency noise. Thus, 2D and long channel nanopores exhibit comparable high frequency noise characteristics.

In DNA translocation applications, the signatures of the base pairs lie in the high frequency regime and can be buried within the capacitive noise. Hence, lowering this noise is of great interest. Increasing the thickness of the dielectric layer, e.g., by making $SiN_x$/quartz stacking (instead of using conventional $SiN_x$/Si) can efficiently minimize the high frequency noise.[8]

The characteristics of the experimentally measured low frequency noise in 2D and long channel nanopores are remarkably different.[9–11] Low frequency noise in long channel nanopores follows Hooge's empirical relation[12,13] where the noise amplitude is inversely proportional to the number of ionic charge carriers:[7/]

$$A \propto \frac{1}{N}$$

In 2D nanopores, low frequency noise is typically a few orders of magnitude stronger and does not obey Hooge's relation.[11] While different scenarios have been examined so far, the origin of the high-amplitude low-frequency noise in 2D nanopores is not well understood yet. We shall present here a short review on the state of research in this area.

One possible source of low frequency noise is found in the fluctuations of the charge carriers deposited close to the pore area. In graphene, the random switching between the protonation and deprotonation state of the carboxyl groups at the nanopore rim may cause such fluctuations. If it were in play, the surface charge fluctuations as well as the low frequency noise would be a function of the pH of the environment. However, experimental results ruled out this scenario as no strong dependency of the noise to the pH of the buffer solution was observed.[11]

Another possible cause of low frequency noise is the effect of incomplete wetting of the hydrophobic graphene in buffer solution.[4] Hydrophilicity of the graphene membrane can be improved by rapid oxidation by UV/ozone exposure. The improved wettability of the membrane, however, did not reduce the noise level. In contrast, the evaporation of a thin hydrophilic layer of $TiO_2$ considerably lowered the noise (Fig. 4). Similar results were found in thick membranes of multi-stacked graphene and $Al_2O_3$ layers.[14]

The mechanical instability of the mono-atomically thick membranes was also considered as a potential source of noise, as vibrations of the membrane may induce fluctuations in the ion transport. Long channel nanopores sculpted in thick more than 20 nm) solid state membranes are mechanically more stable than few layer 2D materials, explaining the more prominent low frequency noise in 2D nanopores. Improving the mechanical strength of the free-standing membrane, e.g., by reinforcing the membrane by depositing materials, is a control to assess the contribution of this effect. Indeed, experimental studies have shown trends in lowering the low frequency noise level while increasing the number of layers of graphene and h-BN membranes.[8,11] Linking the low frequency noise to the mechanical instability of the membrane seems a strong hypothesis as it can explain the reduced noise level achieved by covering graphene with $TiO_2$[4] or in stacked systems.[14] Such modifications reinforce the membrane rather than modify its hydrophobicity. Minimizing the area of the free-standing graphene (by using smaller openings in the $SiN_x$ membrane) would be a conclusive control as the surface properties of the membrane remain unaffected.[15] In contrast to multilayer membranes, the spatial resolution of the nanopore—defined by the thickness of the membrane—will be preserved.

## Fabricating Nanopores

### *Suspended graphene*

Device fabrication starts with the preparation of a free-standing single layer of graphene. A limited area of suspended graphene is preferred, to prevent current leakage through inherent defects within the graphene layer, which would obscure the current signal of the nanopore itself. Therefore, graphene is most often suspended over micrometer (or smaller) sized openings in $SiN_x$ film, which is itself a nanothin material with typical thickness of 10 to 200 nm. Figure 2 shows a schematic of such a nanopore system.

The graphene layer is deposited onto the frame by polymer transfer. In this process, a polymer coating is deposited on graphene as a support layer. Then, graphene (either exfoliated or grown by chemical vapor deposition) can be lifted from its native substrate and transferred onto any desired substrate—in this case, the supporting frame. As a last step, the polymer is dissolved and rinsed off. The supporting polymer layer is essential in the fabrication of suspended graphene layers, as suspended graphene is notably more prone to damage than graphene transferred to continuous solid substrates. One drawback of polymer transfer, however, is the presence of polymer traces even after rinsing. Although annealing may help, there will always be traces left, increasing the effective thickness of the graphene layer and therefore decreasing the spacial resolution of the translocation read-out. Cleaner alternatives to polymer transfer are being invented, such as biphasic transfer and lipid-clamping.[16]

Techniques to fabricate free-standing graphene have been established in recent years. The creation of nanopores in suspended graphene is more of a pioneering field. Existing techniques to create nanoscopic openings in conventional nanothin materials, such as electron beam sculpting, are being applied to suspended graphene. The extreme thinness and resulting mutability of graphene challenge us to reach new levels of precision with these techniques.

Furthermore, unconventional ways of nanopore fabrication dedicated to graphene in particular are being developed. The following paragraphs will outline several methods that have been employed to create nanostructures of suspended graphene, nanopores in particular.

### Single pore formation by focused electron beam

The most controlled method to create nanopores in graphene is electron beam induced sculpting. Under the influence of the focused electron beam, carbons atoms are knocked out from the lattice to create apertures. In order to break the graphene lattice, an energy minimum of 150 keV must be used for the collision to overcome the carbon knock-out energy (defined by the summed bonding energy of a single carbon atom to its neighbouring atoms). In a transmission electron microscope (TEM), the electrons from the beam are accelerated readily to energies above this threshold value, allowing direct sculpting. The conductive properties of graphene allow outstanding control over this process, preventing charging effects that often cause loss of accuracy in electron beam applications. Accordingly, the resolution of sculpting is determined completely by the area of the focused electron beam. With a field emission gun—the smallest diameter electron beam source—a sculpting resolution approaching one nanometer can be readily achieved. Using the scanning mode of a TEM, even single carbon atom resolution sculpting has been demonstrated.[17]

Another viable electron beam sculpting technique employs the focused beam of a scanning electron microscope. Here, the electron energy typically lies below the carbon knock-out energy of the graphene lattice. The presence of a oxidative gas in the vacuum chamber assists in breaking the lattice. A sculpting resolution of 7 nm has been reported with this technique.[18]

In addition to the outstanding control over the pore size and shape, electron beam techniques have the advantage of being made primarily for nanoscopic imaging. When spread out, the unfocused beam becomes non-damaging to the graphene lattice, and the sculpted features can be imaged directly after formation.

Conversely, the focused beam may induce unwanted damage, deforming the hexagonal crystallinity of the graphene lattice close to the focusing spot. A solution to this problem is found at elevated temperatures of 500–600°C, where carbon atom mobility enables a self-repair mechanism that regenerates the hexagonal lattice structure up to the edge of created pores.[19] A TEM image of a nanopore sculpted in graphene via this technique is shown in Fig. 5.

**Fig. 5.** TEM image of graphene nanopore drilled under the influence of the TEM focused ion beam at elevated temperature. Reproduced with permission from [17].

## Pore-isolation in porous 2D materials

A different approach is to not create pores in pristine graphene, but isolate pore-defects that are often present in graphene. For instance, graphene grown by chemical vapor deposition (CVD) may show intrinsic nanopores of nanometer and sub-nanometer size. These pores can be exploited by deposition of graphene onto a SiN$_x$ substrate in which an aperture of 10–100 nm randomly isolates a single intrinsic pore-defect in the graphene layer. As the pores are located randomly on the graphene surface, screening can be done on the basis of the ionic current signal. (More details on pore modelling can be found in Paragraph 1.3.) In principle, there is no lower limit to the pore size that can be found with this method. Conductivity measurements have been performed on pores as small as 2 nm found by statistical isolation.[20]

Since no dedicated instrumentation is required, this method is more suitable for upscaling than single pore formation techniques. It offers the possibility to create large sample sets of specific distributions of pore sizes; a distribution that can be controlled by fine-tuning the characteristics of the graphene layer. We will mention a number of methods for acquiring porous 2D membranes and pore-isolation approaches. The first approach is found in the CVD growth of graphene. While this technique, under optimal conditions, can render defect-free monocrystalline graphene, sub-optimal growth parameters result in more defected graphene. Therefore, the growth recipe (partial pressures of precursor gasses; temperature; heating time) of CVD graphene enables the control over the porosity of the resulting graphene layer. Another means of modifying graphene porosity is ultrashort exposure to mild plasma. An important advantage of plasma treatment is that the chemical nature of the pore edge is controlled by the plasma gas type, allowing the exclusive formation of hydrogenated, oxygenated, (etc.) edges.

As an alternative to graphene, other carbon based 2D structures have been produced. For instance, the self-assembly on a gold surface of carbon-rich thiolates forms a versatile precursor for 2D membranes of adjustable thickness and porosity.[21–23] In this process, the self-assembled monolayer is converted to a cross-linked membrane under electron beam irradiation. After subsequent heating, the membrane is oxidized to a pure carbon 2D material. The porosity and thickness of the layer is determined by the precursor thiolates: bulky polycyclic hydrocarbon precursors leading to large pores while biphenyl thiols have been reported to result in a layer of graphene-like quality.[24] Evidently, the versatility of the pore-isolation method comes from the strong influence of the properties of the graphene layer (or any other 2D material). Clever choice of porous 2D film will enable control over the pore length and size, and even the chemical nature of the pore wall.

## Single pore formation by dielectric breakdown

The method of dielectric breakdown is especially convenient as it is performed *in situ* in the flow cell setup used for translocation experiments.[25] In the dielectric breakdown, a voltage in the order of 10 V—much higher than measurement voltages—is applied over the graphene layer suspended in electrolyte. Charge builds up until the graphene barrier breaks down, forming a pore. This process is self-limiting as the presence of a single pore prevents further excessive build-up of charge on the layer. Once formed, a pore can still be enlarged by continued application of a large potential. Indeed, since multiple smaller pores will affect the conductance much more substantially, one can distinguish whether a single pore of the appropriate size is created based on the conductance measurements as a DNA molecule passes.

The dielectric breakdown process gives control over the single pore area, although the actual shape of the pore and the exact mechanisms of pore nucleation and formation remain unknown. Especially given the extremely small amount of material being removed from the graphene membrane, it is unlikely that direct studies of reaction products will be feasible to study the mechanisms of nanopore formation. As a comparison, previously reported controlled dielectric breakdown fabrication on silicon nitride membranes (10–30 nm) applied constant transmembrane voltage (up to 20 V). These applied electric fields (0.4–1 V/nm) are close to the dielectric breakdown strength of low-stress $SiN_x$ films, and are able to induce accumulation of charge traps forming a highly localized conductive path. Physical damage of the conductive path caused by substantial power dissipation and the resulting heating effect are responsible for the nanopore formation through the bulky silicon nitride membranes. The same mechanism cannot describe graphene nanopore nucleation and formation due to the uniqueness of graphene as a two-dimensional membrane ($\sim 0.3$ nm in thickness) compared to the bulky silicon nitride membrane ($\sim 10$–30 nm in thickness). It is speculated that CVD graphene (or even exfoliated graphene) contains defects, including vacancies and grain boundaries. These defective carbon atoms are more inclined to dislocate first in a dielectric breakdown process and may be responsible for nanopore nucleation and growth.

## Modeling Ionic Conductivity

As has become clear from the previous paragraphs, modelling the ionic current of a nanopore is an important means of finding the nanopore size when direct observation thereof is not available. The length of a pore channel is often a known parameter, approximately equal to the thickness of the membrane in which the nanopore is made. The effective diameter is then found from modelling the pore resistance.

The constriction of the nanopore between bulk electrolyte compartments offers resistance to the ionic current when a voltage is applied. For nanometer sized pores this resistance is so large that all other resistance in the circuit can be neglected. In modelling, therefore, it is sufficient to consider the geometry of the constriction met by the electrolyte charge carriers. A simple, yet effective model describes a cylindrical pore in terms of diameter $d$ and channel length $l$.[26] The resistance, normalized for electrolyte conductivity $\sigma_e$, that results from the passage through the channel is given by the inverse of the channel volume:

$$R_{channel} = \frac{1}{\sigma_e} \frac{4l}{\pi d^2}$$

Besides the channel, a significant contribution originates from the space in direct proximity of the pore opening. This resistance can be understood in terms of ions moving through the limited space that gives access to the channel, and is therefore termed 'access resistance'. The access resistance, as the channel resistance, is derived purely from the geometry of the constriction and given here without derivation:

$$R_{access} = \frac{1}{\sigma_e} \frac{1}{2d}$$

As the access resistance is met on either side of the pore, this term appears with a factor two in the formula for the total resistance:

$$R = \frac{1}{\sigma_e}\left(\frac{4l}{\pi d^2} + \frac{1}{d}\right)$$

It is implicitly assumed that the current is carried by a continuum of charge carriers. Therefore, the model breaks down when pore dimensions are scaled down to the dimensions of charge carriers. Sub-continuum ionic conductivity effects have indeed been observed in carbon nanotubes and other sub-nanometer scale systems. In graphene nanopores for translocation purposes, however, the continuum model has shown to be applicable to establish pore size.[7]

## Tunneling Current for Single Molecule Detection

Throughout the last section, we investigated the performance of 2D nanopores as biomolecule sensors. While 2D nanopores raise high hopes for realizing portable and easily accessible biomolecule sequencing devices, the limitations reviewed in the previous section have obstructed their wide-spread application. As a result, alternative schemes for high-fidelity detection and analysis of single molecules are being investigated. One such alternative is found in nanogap devices.[27–31] In a nanogap device, two electrodes are placed facing each other, separated by a nanoscale gap. Due to the nanoscopic dimensions of the gap, an electrical potential over the electrodes will result in a tunneling current through the electrodes and the gap. When molecules are made to translocate through the gap (Fig. 6a), the current signal will be affected. Particularly, the magnitude of the current depends on the local electronic structure and the topology of biomolecules within the gap.[32] Portions of long chain molecules such as polymers, proteins or DNA mark the electronic read-out with specific

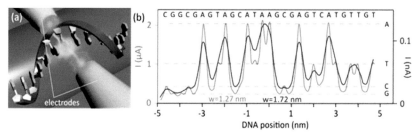

**Fig. 6.** Tunneling current for biomolecule sensing. (a) Schematic representation of a nanogap between two electrodes [27]. (b) Simulation of the tunneling current in a nanogap of two different widths with graphene in-plane electrodes upon translocation of a DNA with the sequence CGGCGA GTAGCATAAGCGAGTCATGTTGT: The tunneling current exhibits unique and experimentally detectible levels corresponding to each nucleotides in the nanogap of 1.27 nm. The current level dramatically drops by increasing the gap up to 1.72 nm [28].

fingerprints of the monomers constituting the macromolecule. This constitutes the read-out principle for nanogap sequencing (Fig. 4b). Compared to ionic conductivity in nanopores, electronic tunneling current measurement offers higher temporal resolution. Moreover, the tunneling current signal directly probes the chemical composition of the nucleotides, removing the influence of capacitive coupling of the *cis* and *trans* cells on the measured signal observed in nanopore setups. It is there prospected that lower levels of high frequency noise will occur.

In practice, various tunneling current setups are being exploited for biomolecular analysis in various measurement setups, including scanning tunneling microscopy (STM) and break junction ensembles. 2D materials offer an enticing alternative. In the following Section, we will first explore the STM and break junction methods, in order to reveal the advantages of the use of 2D materials and graphene in particular.

## Experimental Detection of Single Molecules with Tunneling Current

In STM, a scanning tip is approached to within sub-nanometer distance of a conducting substrate. The approach is controlled by a feedback loop monitoring the onset of tunneling current between tip and substrate. The structure of individual DNA molecules deposited on a conducting substrate has been visualized by mapping the modulations of the tunneling current (Fig. 6c).[33] The resolution attained was high enough to resolve and sequence individual nucleotides within short single strands of DNA.

The mechanical break junction is another technique with confirmed capability of forming nanogap systems. In this method, a mechanical pulling force is applied to a metallic rod, to subtly break it into two pieces. Piezo elements offer the precise control needed to create nanoscale gap as small as ~ 5Å between the two halves.[34] Break junctions has been successfully used to probe different biomolecules. Notably, three nucleotides, namely thymidine monophosphate (TMP), guanosine monophosphate (GMP) and cytidine monophosphate (CMP), revealed unique and distinguishable tunneling current signatures (Fig. 6d).[30] Separately, twelve (out of 20) different amino acids and the post-translational modifications of single peptide molecules have been identified and discriminated using this method.[31]

A drawback of state of the art nanogap methods remains; they rely on the random presence of the molecules in the gap. The lack of a fluidic channel and a driving force to guide the biomolecule hampers the length of the molecules to be investigated. For this reason, the applicability of these methods to full length read-out of DNA molecules is in doubt. Furthermore, these methods perform best under ultrahigh vacuum and thus do not lend themselves well for biological specimens. In fact, in an aqueous solution, parallel conductance channel paths can establish through the buffer, degrading the measurement resolution. In the following Sections we will show that so-called graphene in-plane electrodes can overcome such limitations.

## In-plane graphene nanogap electrodes

The use of 2D materials in the design of nanogap sensors was first suggested by theoretical studies (Fig. 4a).[28,35–37] The implementation of graphene into experimental devices is still a novel field where many practical hurdles have yet to be overcome, but it is prospected, that 2D nanogap devices offer the realization of nanogap sequencing of DNA. As in nanopores, the monoatomic thickness of 2D materials serves as the working electrodes in nanogaps and permits single nucleotide resolution (Fig. 2a). Due to its outstanding electrical properties, graphene is particularly advantageous over other 2D materials. Additionally, the mechanical strength of graphene is enough to realize freestanding membranes[47,38] and to withstand the transmembrane pressure it may experience in applications. The remainder of this section will analyze the effect of different experimental features on the performance of 2D nanogaps, and line out the various experimental approaches that are currently being explored to overcome these difficulties.

## Nanogap width

The current signal is highly sensitive to the width of the  gap, as tunneling current drops exponentially with tunneling path length. Simulations demonstrated that upon the translocation of a short strand composed of a series of nucleotides through a 2D nanogap of 1.27 nm, the tunneling current levels up in certain amplitudes unique for each nucleotide. The tunneling current is in µA range and therefore detectable with commercially available amplifiers. Increasing the width of the nanogap to 1.72 nm, however, lowers the current by few orders of magnitude and smears the unique signatures of the nucleotides, dramatically degrading the sensitivity of the nanogap.

## Conformational variation of DNA in a nanogap

The orientation of the nucleotides and their position with respect to the electrodes are the major sources of uncertainty in translocating DNA through a nanogap. Simulation of the tunneling current[36] showed that even in the presence of such uncertainties, discrimination of the pyrimidine-based (C,T) and purine-based (G,A) nucleotides is feasible by comparing the transmission probability (the probability that an electron passes through the potential barrier), owing to the difference in the size of the nucleotides. Strategies to distinguish between C and T nucleotides and between G and A nucleotides by comparing the position and the width of the transmission peaks have also been documented.[3]

One way of effectively lowering the conformational uncertainties of the gap, is employing chemical functionalization of graphene edges. Particularly hydrogen bonds can establish specific interaction between nucleotides and the graphene electrodes with hydrogenated edges. The main advantages of this approach are the following:[35,39] (a) The stabilization of the nucleotides against thermal fluctuations, lowering measurement noise/error. (b) The slowing down of DNA translocation; prolonged scanning duration improves the fidelity of read-out. (c) The hydrogen-bonded bridges between the electrodes and nucleotides lower the tunneling potential barrier; resulting in up three orders of magnitude improvement of the tunneling current.[35] We note that similar momentary anchoring can be achieved using large bias voltages to provide a transverse electric field exceeding the transmembrane driving potential.[40,41]

## *Sensitivity of the nanogap*

Thermal vibrations of the membrane may decrease the spatial resolution in the devices composed of free-standing graphene.[15,28] Sandwiching graphene electrodes with a self-assembled monolayer or with deposited materials in an elaborate fabrication process can improve the mechanical stability and preserve the resolution. Notably, it was found that the surface coverage improves the sensitivity of the device by blocking any parasitic current that otherwise would establish between the surface of the electrodes in contact with liquid.

## Bilayer Graphene Electrodes

The fabrication of in-plane 2D nanogaps meets important complications, such as the realization of the precise nanogaps of nanometer width and the alignment of the nanogap onto a nanofluidic channel to allow the translocation of biomolecules. The absence of a reproducible fabrication protocol has hampered realization and application of sensors so far. Bilayer graphene with naturally formed sub-nanometer gaps between the layers (Fig. 7) allows to by-pass this problem.[37] Here, the detection signal is established between the edges of layers upon tunneling through the DNA building blocks during their translocation. Simulations demonstrated the viability of this device for DNA sequencing; the electrical coupling of the layers, however, remains a major concern. In the actual device, the conductance via the overlaid surfaces of the two graphene layers can smear the signal established at the edges. Though it is theoretically

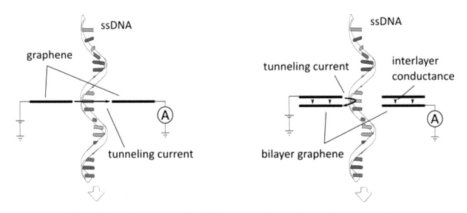

**Fig. 7.** Tunneling current for DNA sequencing with graphene electrodes.

predicted that the orientation angle between the layers can be optimized to minimize the interlayer conductance,[37,42] intercalating a layer of insulating 2D materials, e.g., hexagonal boron nitride, between the graphene layer is a more practical approach.[43]

# References

1. J. J. Kasianowicz, E. Brandin, D. Branton and D. W. Deamer, *Biophysics (Oxf).*, 1996, **93**, 13770–13773.
2. C. Dekker, *Nat. Nanotechnol.*, 2007, **2**, 209–215.
3. H. Arjmandi-Tash, L. A. Belyaeva and G. F. Schneider, *Chem. Soc. Rev.*, 2016, **45**, 476–493.
4. C. A. Merchant, K. Healy, M. Wanunu, V. Ray, N. Peterman, J. Bartel, M. D. Fischbein, K. Venta, Z. Luo, A. T. C. Johnson and M. Drndić, *Nano Lett.*, 2010, **10**, 2915–21.
5. S. van Dorp, U. F. Keyser, N. H. Dekker, C. Dekker and S. G. Lemay, *Nat. Phys.*, 2009, **5**, 347–351.
6. Y. P. Shan, P. B. Tiwari, P. Krishnakumar, I. Vlassiouk, W. Z. Li, X. W. Wang, Y. Darici, S. M. Lindsay, H. D. Wang, S. Smirnov and J. He, *Nanotechnology*, 2013, **24**, 495102.
7. G. F. Schneider, S. W. Kowalczyk, V. E. Calado, G. Pandraud, H. W. Zandbergen, L. M. K. Vandersypen and C. Dekker, *Nano Lett.*, 2010, **10**, 3163–7.
8. A. Kumar, K.-B. Park, H.-M. Kim and K.-B. Kim, *Nanotechnology*, 2013, **24**, 495503.
9. S. Liu, B. Lu, Q. Zhao, J. Li, T. Gao, Y. Chen, Y. Zhang, Z. Liu, Z. Fan, F. Yang, L. You and D. Yu, *Adv. Mater.*, 2013, **25**, 4549–54.
10. K. Liu, J. Feng, A. Kis and A. Radenovic, *ACS Nano*, 2014, **8**, 2504–11.
11. S. J. Heerema, G. F. Schneider, M. Rozemuller, L. Vicarelli and H. W. Zandbergen, *Nanotechnology*, 2014, 1–18.
12. F. N. Hooge, *Phys. Lett. A*, 1969, **29**, 139–140.
13. F. N. Hooge, T. G. M. Kleinpenning and L. K. J. Vandamme, *Reports Prog. Phys.*, 1981, **44**, 479–532.
14. B. M. Venkatesan, D. Estrada, S. Banerjee, X. Jin, V. E. Dorgan, M.-H. Bae, N. R. Aluru, E. Pop and R. Bashir, *ACS Nano*, 2012, **6**, 441–50.
15. S. Garaj, S. Liu, J. A. Golovchenko and D. Branton, *Proceeding Natl. Acad. Sci. PNAS*, 2013, **110**, 12192–12196.
16. L. M. C. Lima, W. Fu, L. Jiang, A. Kros and G. F. Schneider, *Nanoscale*, DOI:10.1039/c6nr05706c.
17. Q. Xu, M. Y. Wu, G. F. Schneider, L. Houben, S. K. Malladi, C. Dekker, E. Yucelen, R. E. Dunin-Borkowski and H. W. Zandbergen, *ACS Nano*, 2013, **7**, 1566–1572.
18. B. Sommer, J. Sonntag, A. Ganczarczyk, D. Braam, G. Prinz, A. Lorke and M. Geller, *Sci. Rep.*, 2015, **5**, 7781.
19. Z. J. Qi, C. Daniels, S. J. Hong, Y. W. Park, V. Meunier, M. Drndic and A. T. C. Johnson, *ACS Nano*, 2015, **9**, 3510–3520.
20. T. Jain, B. C. Rasera, R. J. S. Guerrero, M. S. H. Boutilier, S. C. O'Hern, J.-C. Idrobo and R. Karnik, *Nat. Nanotechnol.*, 2015, **10**, 1–6.
21. J. C. Love, L. A. Estroff, J. K. Kriebel, R. G. Nuzzo and G. M. Whitesides, *Chem. Rev.,* 2005, **105**, 1103–1170.
22. P. Angelova, H. Vieker, N. Weber, D. Matei, O. Reimer, I. Meier, S. Kurasch, J. Biskupek, D. Lorbach, K. Wunderlich, L. Chen, A. Terfort, M. Klapper and K. Mu, *ACS Nano*, 2013, **7**, 6489–6497.
23. A. Turchanin, A. Beyer, C. T. Nottbohm, X. Zhang, R. Stosch, A. Sologubenko, J. Mayer, P. Hinze, T. Weimann and A. Gölzhäuser, *Adv. Mater.*, 2009, **21**, 1233–1237.
24. A. Turchanin, D. Weber, M. Büenfeld, C. Kisielowski, M. V. Fistul, K. B. Efetov, T. Weimann, R. Stosch, J. Mayer and A. Gölzhäuser, *ACS Nano*, 2011, **5**, 3896–3904.
25. H. Kwok, K. Briggs and V. Tabard-Cossa, *PLoS One,* 2014, **9**, e92880.
26. S. W. Kowalczyk, A. Y. Grosberg, Y. Rabin and C. Dekker, *Nanotechnology*, 2011, **22**, 315101.
27. E. Pennisi, *Science*, 2012, **336**, 534–537.
28. H. W. C. Postma, *Nano Lett.*, 2010, **10**, 420–425.
29. T. Ohshiro, K. Matsubara, M. Tsutsui, M. Furuhashi, M. Taniguchi and T. Kawai, *Sci. Rep.*, DOI:10.1038/srep00501.
30. M. Tsutsui, M. Taniguchi, K. Yokota and T. Kawai, *Nat. Nanotechnol.*, 2010, **5**, 286–90.
31. T. Ohshiro, M. Tsutsui, K. Yokota, M. Furuhashi, M. Taniguchi and T. Kawai, *Nat. Nanotechnol.*, 2014, **9**, 835–40.
32. M. Razavy, *Quantum Theory of Tunneling*, 2003.

33. H. Tanaka and T. Kawai, *Surf. Sci.*, 2003, **539**, L531–L536.
34. M. Tsutsui, M. Taniguchi and T. Kawai, *Appl. Phys. Lett.*, 2008, **93**, 163115.
35. Y. He, R. H. Scheicher, A. Grigoriev, R. Ahuja, S. Long, Z. Huo and M. Liu, *Adv. Funct. Mater.*, 2011, **21**, 2674–2679.
36. J. Prasongkit, A. Grigoriev, B. Pathak, R. Ahuja and R. H. Scheicher, *Nano Lett.*, 2011, **11**, 1941–5.
37. Y. He, M. Tsutsui, R. H. Scheicher and M. Taniguchi, *arXiv:1206.4199v1*.
38. S. Garaj, W. Hubbard, A. Reina, J. Kong, D. Branton and J. A. Golovchenko, *Nature*, 2010, **467**, 190–3.
39. H. He, R. H. Scheicher, R. Pandey, A. R. Rocha, S. Sanvito, A. Grigoriev, R. Ahuja and S. P. Karna, *J. Phys. Chem. C.*, 2008, **112**, 3456–3459.
40. J. Lagerqvist, M. Zwolak and M. Di Ventra, *Phys. Rev. E*, 2007, **76**, 013901.
41. J. Lagerqvist, M. Zwolak and M. Di Ventra, *Biophys. J.*, 2007, **93**, 2384–90.
42. R. Bistritzer and A. H. Macdonald, *Phys. Rev. B*, **81**, 245412 .
43. Y. He, M. Tsutsui, S. Ryuzaki, K. Yokota, M. Taniguchi and T. Kawai, *NPG Asia Mater.*, 2014, **6**, e104.

# CHAPTER 7

# New Advances in 2D Electrochemistry—Catalysis and Sensing

*Tharangattu N. Narayanan,\* Ravi K. Biroju[a] and Thazhe Veettil Vineesh[b]*

## Introduction

Atomically thin layered (two-dimensional (2D)) materials can be visualized as extended crystalline planar structures held together by strong in-plane covalent bonds and weak out-of-plane van der Waals interactions.[1] Hence, these individual layers can easily be extracted from their bulk by breaking the van der Waals bonds, with little damage either to the remaining structure or to the extracted layer. This was first exemplified for graphene in 2004 where layers of graphene are peeled away from bulk graphite with an adhesive tape.[2] These extracted layers are found to be stable, transferrable to any substrates with relative ease, and also can be free-floated using suitable substrates. With the discovery that controllable growth of graphene can be achieved on certain metal surfaces by vapor phase deposition from hydrocarbon precursors, scalable graphene-based technologies started looking more feasible.[1] However, defect free graphene crystal, due to its semi-metallic nature, is not an ideal 2D material for many applications, say electronics and electrochemical fields.[3] However, a large number of other layered crystals spanning the entire spectrum of electronic properties from superconducting to insulating are available, and their individual layers

Tata Institute of Fundamental Research, Sy. No. 36/P, Gopanapally Village, Serilingampally Mandal, Hyderabad - 500107, India.
[a] E-mail: birojuravi@gmail.com
[b] E-mail: vineeshchem86@gmail.com
\* Corresponding author: tnn@tifrh.res.in or tn_narayanan@yahoo.com

are found to have exotic physico-chemical properties, enabling them for electronic and electrochemical applications. Presently, this class of layered materials is very rich, and includes metals ($NbSe_2$),[4] semimetals (graphene), semiconductors ($MoS_2$),[5] insulators (hBN),[6] topological insulators ($Bi_2Se_3$),[7] superconductors (e.g., $FeS_2$),[8] and layered heterostructures.[9,10] Further, by exploiting the material properties of the layers in isolation or by mixing and matching them to create new solid structures with atomically thin heterojunctions, researchers are now exploring interesting new physics and interface induced properties.[1] Engineering properties of materials by their sequential stacking or assembly provides another unique opportunity with layered materials to create tailor-made crystals for applications in sensing and catalysis.[11,12]

Since every atom in a mono-layered material is exposed to the environment, 2D materials can be extremely responsive to external stimuli and interactions. Hence, 2D materials/layers can be used as a sensing platform for the detection of biologically important molecules. The vital aim of all sensors is single molecular level sensitivity, but such extreme sensitivity is far from the reach of practical sensors available in the current market.[13] The thermal fluctuations of charges and defects in the sensing platform, leading to increased noise levels, are the root cause for the poor sensitivity of the existing sensors.[14] This can be addressed by the introduction of low electronic noise systems such as graphene, which has the following exceptional features, making the signal to noise ratio high: (i) ideal 2D material where the whole volume is available for interactions with the analyte, (ii) low Johnson–Nyquist noise due to very high electronic conductivity, (iii) low flicker noises, and (iv) possibilities of high precision conductivity/impedance measurements with low contact/charge transfer resistances. Further, it is established that the selectivity in the interaction between 2D materials and analytes can be brought by introducing non-covalent or reversible covalent chemistry between them.[15] Hence, 2D materials can bring selectivity and sensitivity to the sensor, making them ideal sensing platforms.

This chapter discusses the importance of 2D crystals (layers) in electrochemistry (hereafter we have coined it as '2D electrochemistry') and practical guidelines for the successful use of layered materials in catalysis and sensing. Since 'catalysis and sensing' are too big domains to include in a single chapter, emphasis is given to electrocatalysis and electrochemical sensing. In electrocatalysis, four important reactions, namely oxygen evolution reaction (OER), oxygen reduction reaction (ORR), hydrogen evolution reaction (HER), and hydrogen oxidation reaction (HOR) are discussed. The inferences obtaining from the 2D electrochemistry of above mentioned reactions can be extended to other catalytic and sensing reactions.

The initial part of this chapter is dedicated to demonstrate the role of inherent and induced defects in layered crystals on 2D electrochemistry. Later this discussion has been extended to artificial 2D crystals (pseudo 2D layers) called van der Waals solids/ stacks. The chapter concludes with a note on the critical measures to be taken in 2D electrochemistry, and pro and cons of atomic layers in electrochemical applications, such as catalysis and sensing. Most of the discussions on this chapter are centered on the conventional layered crystals such as graphite, hexagonal boron nitride (hBN), transition metal dichalcogenides (TMDCs/TMDs), etc., since these are the well-explored stable materials in catalysis and sensing. However, unconventional layers such as silicene, phosphorene, etc., and anisotropic layered crystals such as $ReS_2$, $NbSe_2$, etc., are not discussed here due to their structural instability (under the exposure of electric field, humidity, air, etc.), substrate specificity (particularly for the growth), issues with large scale productivity, lack of enough fundamental understandings, etc. Also, this discussion does not include some of the crystals like GaSe, $TaS_2$, $Bi_2Se_3$, etc., since these materials are stable in bulk, but once cleaved down to a few layers, start corroding. Instability of graphene (at high temperatures of around 600 K)[12] or TMDs (at elevated temperatures and also in the presence of moist air) are not taken into account since most of the electrochemical reactions discussed here are being held at ambient conditions.

## Synthesis of 2D Materials

Unlike the case of other conventional materials used in electrochemistry, the applicability of a 2D material in electrochemistry highly relies on the method of synthesis adopted, since the method mostly determines the amount and nature of defects in the layer (this includes surface/edge states, vacancy defects, point and line defects, etc.). Numerous physical and chemical methods are reported for the synthesis of layered materials.[16–18] This includes both top-down and bottom-up approaches. Here, some of the important methods are briefed, and later, the layers developed through these methods are discussed for their electrocatalytic applications. A schematic of various methods available for graphene synthesis is shown in Fig. 1,[19] and it is to be noted that many of these techniques are applicable to other conventional layered materials too.

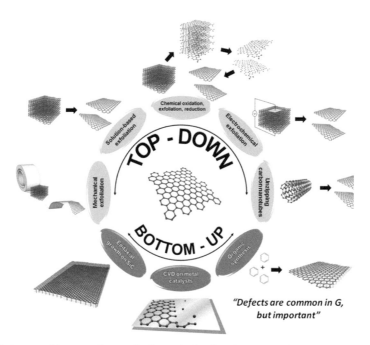

**Fig. 1.** Various possible reported routes for the synthesis of graphene. Reproduced with permission from American Chemical Society.[19]

## Mechanical exfoliation

One of the methods to get high quality (defect-free) atomic layer from their bulk stacked layers is mechanical cleavage and isolation of the layer, as it is first demonstrated for graphene through the scotch tape pealing of highly oriented pyrolytic graphite (HOPG), followed by the washing off of the scotch tape using acetone to isolate the graphene debris.[2] Though this method can result in the production of high quality crystals, the yield of the product (both by weight and area), is very poor and hence it is not a recommendable method for applications such as catalysis, where significant amount of material (either by large area coverage or by weight) is required. Further, since many of the defect-free 2D materials are not electrochemically active, as one can see from the discussions later in this chapter, mechanically exfoliated crystals are not explored much for electrocatalysis or sensing. However, they can still serve as ideal substrates/supports for many other active layers, such as hexagonal boron nitride (h-BN) or transition metal dichalcogenides (TMDCs), which are also important in 2D electrochemistry.

## Chemical/electrochemical exfoliation

Natural graphite has an interplanar cohesive energy of ~ 42.6 meV per atom for an interplanar spacing of 3.35 Å, and hence it needs an exfoliation energy of 42.6 meV to dismantle the stacking.[1] This can be achieved by many means, and a few of the routes follow chemical intercalation and exfoliation, or electrochemical intercalation and electric field assisted exfoliation.[20] Further, other nanostructures such as nanotubes, fullerenes, etc., can also be transformed to layered structures with high aspect ratio (length to breadth)-called nano ribbons (an important candidate in 2D electrochemistry). For example, in the case of carbon nanotubes, longitudinal unzipping via external electric field is found to be resulting to the formation of graphene nano ribbons (GNRs).[21]

Chemically derived graphene or reduced graphene oxide or thermally derived graphene (CDG), where the graphite is oxidative-exfoliated from graphite to form graphene oxide (GO) followed by the reduction of the oxygen functionalities, is one of the extensively researched graphene forms (graphene functional derivative) due to the simplicity of the method and also due to the possibilities of large scale production (both by weight and volume).[18,22] However, one of the best electrical conductivity reported for CDG follows the reduction method mentioned in Fig. 2A. It is apparent from the figure that the conductivity of the final product (CDG/CCG3) is three-orders less than that of the theoretical conductivity of pristine graphene.[23] This indicates that though CDG schematically looks like a perfect crystal, it contains many defects (different forms of defects) resulting in a much lower electrical conductivity. The disordered solid nature of CDG is further verified from the temperature-dependent resistivity and magneto-resistance measurements (Fig. 2B). Two different types of transport mechanisms (as mentioned in Fig. 2B; at low temperatures two-dimensional variable range hopping and at high temperatures Arrhenius-like behavior) are exhibited by most CDGs, indicating the properties of a disordered solid. Further, the negative magneto-resistance observed from these systems (as can

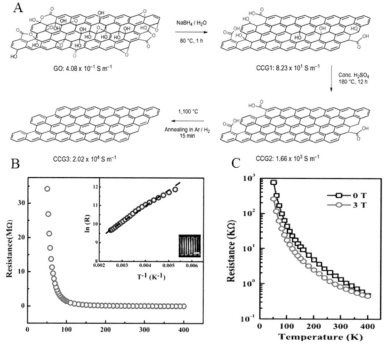

**Fig. 2.** (A) Chemical and thermal reduction of graphene oxide, (B) and (C) electrical resistivity and magneto-resistance measurements on reduced graphene oxide films, respectively.[23,24]

be seen in Fig. 2C) indicates the presence of vacancy induced disorders.[24] However, these defects play a salient role in heterogeneous charge transfer kinetics, and hence in electro-catalysis and sensing. This oxidative exfoliation method can be extended to other layered crystals such as hBN and $MoS_2$ too,[25,26] but they are found to be less effective in comparison to that in graphite.

However, room temperature shear exfoliation of bulk layered crystals to form large scale exfoliated crystals (less defective crystals) by shearing them in low surface tension common solvents such as N-methyl-pyrrolidone (NMP) or dimethyl formamide (DMF) is now becoming a popular protocol for the production of large scale atomic layers.[27] Researchers also speculate this method to be a milestone in 2D research to reach the technology readiness level 9 (TRL9).[28] Plenty of single to a few layered crystals such as graphene, h-BN, $MoS_2$, $WS_2$, $MoSe_2$, $NbSe_2$, $TaSe_2$, $NiTe_2$, $MoTe_2$, etc., are reported by this method from their bulk crystals.[29] A simple modification, by increasing the temperature of shear exfoliation, in the original shear exfoliation method which Coleman et al. (2011) reported, is found to be resulting in edge state manipulation of layered crystals.[30] This oxidative cutting of the sheets also renders smaller sheets or even quantum dots (QDs). This is also an important method for the development of catalytically active layered materials directly from their bulk crystals, since the defective/oxidized edges can act as adsorption centers for many radicals/molecules.

Recently, several other chemical methods are also developed to synthesize large scale 2D layers, such as hydrothermal[31] and solvothermal routes.[32] Another promising method for the synthesis of electrochemically active TMDCs, such as $MoS_2$ layers is the electrochemical exfoliation with the aid of lithium (Li) intercalation.[33] This technique is more popular to achieve '1T' phase (octahedral coordination) 2D TMDCs, which are metallic in nature, and more active in electrochemistry than their stable 2H-phase (trigonal prismatic coordination).[34] It is to be noted that the thermodynamic stable phase of semiconducting TMDCs (STMDs) is '2H', and it is a more favorable phase to grow during chemical vapor deposition (CVD) having controlled number of layers. Though chemical or electrochemical routes can result in a large amount of catalytically active layers, control over defects and disorder is highly difficult in these approaches.

### *Chemical vapor deposition*

CVD is a widely used growth technique to fabricate the thin solid films on the various substrates from the vapor species through chemical reactions.[35] In a conventional CVD reaction, gas precursors are utilized to decompose at relatively higher temperatures and deposited on to transition metal (TM) catalysts.[36] In most of the physical vapor deposition (PVD) processes, the source materials will be solid (powder) and vaporized on to the substrates by transporting through the inert and ambient carrier gases or vacuum.[37] Among all the approaches, CVD growth of graphene on different TM catalysts is most promising for large area crystalline growth with controllable layer thicknesses.[16] However, a clean transfer of graphene onto the alien substrates is the challenge with this growth technique.[16] The residues of transition metals such as polycrystalline nickel (Ni), copper (Cu), etc., used in the CVD based graphene growth often transferred into these alien substrates along with graphene during the transfer process.[38–40] This often complicates the studies using CVD grown graphene samples, particularly when they are used for electrochemical applications—since most of these TMs can participate in the electrochemical reactions of interest. Further, in graphene growth, special attention can be given to control the morphology, domain sizes, shapes, crystal orientations, and lattice defects.[46,41] As one can see later in this chapter, these factors are important in deciding the electrochemical activities of graphene. Further, introducing dopant (such as boron, sulfur, nitrogen, fluorine, phosphorus, etc.) precursors during the CVD growth can be used to modify the graphene honeycomb with these dopants.[42] Multiple dopants can also be incorporated into the graphene lattice by this method. Hence these routes are very important in engineering graphene to obtain functional derivatives of graphene having 'induced defects'.

In case of CVD growth of STMDs, a direct vapor phase reaction of transition metal oxide and sulfur (S)/selenium (Se) has been widely adopted for the growth of layers such as $MoS_2$, $WS_2$, $MoSe_2$, and $WSe_2$. It was found that shape selective—such as triangles, hexagons, truncated triangles, three-point stars, six-point stars, synthesis of TMDs is possible by CVD method with the careful control over the position of the substrate and precursors.[43,44,48] These edge states and dendrimeric surfaces are highly beneficial for many electrochemical reactions where it acts as adsorption centers for radicals/molecules, aiding to an augmented charge transfer mechanism, such as it is recently demonstrated in lithium polysulfide conversion reaction, which is beneficial in Li-Sulfur batteries (Fig. 3).

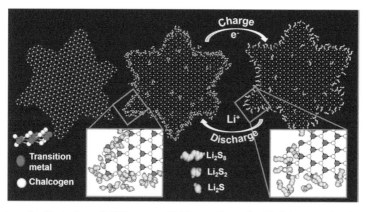

**Fig 3.** Schematic illustration of CVD grown TMD nanosheets for Li–S battery, showing confined deposition of lithium polysulfides at preferential catalytic sites and their conversions during discharge–charge processing in a catholyte solution. Reproduced with permission from American Chemical Society.[45]

It is found that TMDs can be modified with TMs such as Fe, Co, Ni, etc., and also with elements such as Li, Mg, etc.[46] The effect of TMs in TMDs is theoretically also investigated.[47] Further, these modifications of TMDs with external elements can drastically change the structural as well as electrochemical properties, and each element has different effects on TMDs properties.[46]

Carbide-derived carbons include a large group of carbons ranging from extremely disordered to highly ordered structures.[48] Epitaxial growth of graphene/doped graphene from carbides is another promising route to obtain large amount of pristine/doped graphene for applications. It is found that silicon carbide or titanium carbide (amorphous or crystalline) annealing in an inert atmosphere will lead to the sublimation of silicon/titanium atoms, and the remaining carbon atoms will form graphene on the semi-insulating substrate. Further, this method has been extended to boron carbide ($B_4C$) to obtain boron doped graphene.[49] This boron doped graphene can be easily separated from $B_4C$ using a simple water assisted separation method (utilizing the hydrophobicity of graphene flakes), and these doped graphene show bifunctional electrocatalytic activities, which will be discussed in the later section. Recently, doped graphene development was also achieved from bulk carbides, where these single step doped graphene are found to be highly important in various applications.[48] Recent reports show that high quality 2D carbides can be synthesized (such as $\alpha$-$Mo_2C$, WC, and TaC) using CVD method, and crystalline/amorphous carbides can be converted to graphene nanosheets with tunable layers by the thermal-chlorination method, even at relatively low temperatures.[48] This method has now been extended to other metals, such as Cr-, Mo-, Nb-, and V-self-doped graphenes.[48] Both doping and number of layers are important in deciding the electrochemical performance of graphene-like atomic layers, and control over these parameters is highly beneficial in designing materials for 2D electrochemistry.

## Synthesis of 2D Stacks/Hybrids—van der Waals Solids

Recent developments on new stacked layers called van der Waals solids invoke the possibility of new types of quantum heterostructures with atomically sharp interfaces between 2D layers of dissimilar materials.[1] Stacking two or more atomic layers of different compounds in the rich variety of 2D library allows researchers to explore novel and collective quantum phenomena at the interfaces (various types of layers available for the combinatorial stacking is schematically shown in Fig. 4[12]). The basic principle of stacking different layers is simple: for example, take a monolayer, put it on top of another monolayer or few-layered crystal, add another 2D crystal and so on.[12] The resulting stack represents an artificial material assembled in a chosen sequence as Lego blocks defined with one-atomic-plane accuracy. Here, strong covalent bonds provide in-plane stability of 2D crystals, whereas relatively weak, van-der-Waals-like interactions are sufficient to keep the stack together. While making the layer-by-layer transfer, the twisting angle between two 2D materials is an additional degree

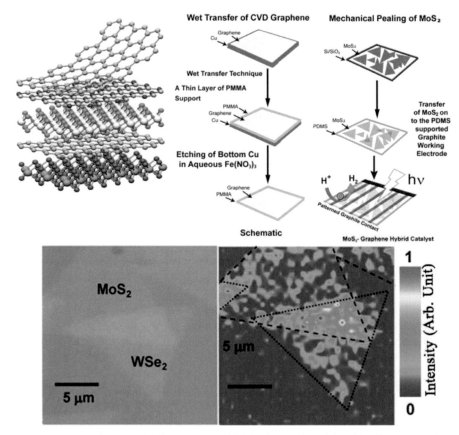

**Fig. 4.** Various atomic layers of different types available for the combinatorial stacking.[12] Schematic of the vertical heterostructure electrode development using $MoS_2$ and graphene,[52] (down) electron microscopic and Raman mapping images of a $MoS_2/WSe_2$ vertical heterostructure.[53]

of freedom.[1] Twisting the layer by layer assembly will result in a new solid. Though two lattices stacked with an arbitrary twisting angle will not form a periodic lattice (except at particular discrete angles when they form a periodic, or commensurate structure), it is found that incommensurate heterostructures can also form large-scale quasi-periodic moiré patterns and hence, the physical properties of heterostructures can be tuned by interfacial commensurability. These stacked layers are found to be exceptional in electrochemistry too. In most of the layered materials, basal plane (defect free) is inactive towards any electrochemical reactions (the reason is discussed in the later part of this chapter).[50] Hence, huge efforts have been undertaken with limited success on the vertical alignment of 2D layers to make sure that the entire exposed area is available for analytes (radicals), to make high sensitive sensors and catalysts.[51] However, it is found that the formation of staked layers and van der Waals solids can make the basal plane too electrochemically active due to the built-in electric field between the dissimilar layers.[3] This has been further established as substrate effects in 2D electrochemistry.

One of the methods to make stacked solids is the layer-by-layer transfer of different layers (schematic of multistep transfer process for graphene/$MoS_2$ hybrid, where graphene is grown on metal (Cu) and $MoS_2$ is grown on $Si/SiO_2$. Further the development of a graphite patterned polydimethyl siloxane (PDMS) substrate is also shown in Fig. 4).[52] This is a multi-step laboursome process, and one such scheme used for graphene/$MoS_2$ van der Waals solid is shown in Fig. 4. For example, $MoS_2$/$WS_2$ vertical heterostructure is achieved by sequentially transferring $MoS_2$ and $WS_2$ on $Si/SiO_2$ (similar to hBN/graphene structures where hBN is grown on $Si/SiO_2$ and graphene is grown on metals, followed by their sequential transfer).[54] However, these polymer-assisted layer-by-layer transfer routes can introduce organic and inorganic impurities at the interfaces, along with random relative orientations of two layers, leading to ill-defined electrical and optical properties (such a transferred microscopic image of triangle shaped $MoS_2$/$WSe_2$ hybrid is shown in Fig. 4. Its Raman imaging is also shown).[53] Layer by layer multi-step growth of different atomic layers is also attempted to get van der Waals vertical structures. However, the cross-contamination of different elements in the as-grown $MX_2$ heterostructures has frequently been observed as an obstacle in this method, since all the precursors coexist in the vapor phase throughout the growth process.[53]

Self-assembly of graphene and hBN in solution is also found to be an effective way to develop graphene/hBN van der Waals solids.[55,56] Though the mechanism leading to the stacked graphene/hBN solids with the mixing of different weight ratios of graphene and hBN in low surface tension solvents is not clear, a large amount of van der Waals solids with tunable opto-electronic properties are found to be formed by this method.[56] Another important feature observed in 2D electrochemistry is the role of another degree of freedom called stacking sequence. In the van der Waals 2D electrochemistry, not only the stacked materials, but their stacking sequence is also found to be important. This effect has been observed in graphene/$MoS_2$ and $WS_2$/$MoS_2$ van der Waals stacks in HER process.[52] Hence, the growth process also needs to be optimized for the best geometry.

## Electrochemistry of 2D Materials

### *Role of inherent defects*

As it has been discussed before, electrochemistry of 2D materials highly depends on the method of preparation of the layers. For example, when CDGs are prepared from graphene oxide by different reduction routes, like chemical, thermal, or electrochemical, it will end up with different number/types of oxygen functionalities. It should be noted that the heterogeneous electron transfer (HET) rates of various reactions are influenced by the C/O ratio. This effect is more prominent in the case of surface sensitive inner sphere redox probes like $[Fe(CN)_6]^{3-/4-}$, but less effective or ineffective in the case of outer sphere redox probes, like $[Ru(NH_3)_6]^{3+/2+}$.[57,58]

Another influential parameter in 2D electrochemistry is the adsorbed/intercalated molecules such as water. One of the recent studies demonstrated the role of intercalated water in electrocatalytically active nitrogen-doped graphitic systems.[59] In that study, graphite is oxidized to form GO sheets. These GO sheets have intercalated water molecules while exposed to the air due to their hydroscopic nature. To make these sheets effective catalyst (GO is an insulator and not an effective electrocatalyst), oxygen functionalities are reduced and doped with the heteroatom (nitrogen) to form nitrogen-doped graphene sheets, and further their ORR activity in acidic medium is tested using rotating disc electrode measurements. It is found that the removal of water molecules affects the nanosheets' morphology and macroscopic structure, hence affecting the catalytic activity of layers. Further, it can be concluded that the replacement of trapped water with ether is giving higher electrocatalytic activity towards the ORR compared to other solvents.

Electronic structures of perfect and defective graphene determine its resultant electrocatalytic activities. Another factor in graphene synthesis is the presence of inherent defects, such as point and line defect (Fig. 5 shows different types of defects which exist in graphene). Some of these inherent defects can introduce low energy fluctuations, such as spin and charge density in graphene (carbon atoms with high spin or charge density are most likely to be catalytically active sites, and there is no spin density on pure graphene), which in turn will affect the HET rate. It is found that many of these point and line defects will not contribute to electrochemical reactions such as ORR (due to the absence of spin density like pure/perfect graphene), but defects such as pentagon rings at the zigzag edge, or line defects (grain boundaries) consisting of pentagon–pentagon–octagon, or pentagon–heptagon chains also at the edges, show the electrocatalytic capability towards ORR.[60] Spin and charge density distributions determine the adsorption centers for radicals and further electron transfer (HET) to/from these radicals to the working electrode. It is also found that both four-electrons and two-electrons ORR pathways can occur on these defective graphenes simultaneously, and the reaction energy barriers of the four-electron transfer pathway on defective graphene clusters are comparable to that of platinum, a benchmarked catalyst for ORR.[60] This in turn motivates scientists towards the development of defect controlled metal free graphene systems, which can replace platinum in ORR, though it can be quite challenging.

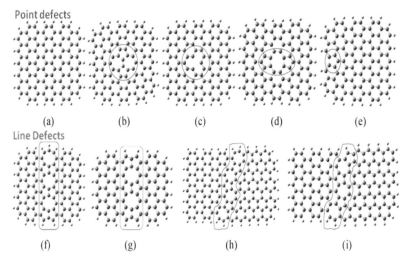

**Fig. 5.** The possible defects (point and line) in graphene.[60]

Unlike graphene, TMDs are found to possess distinctive inherent electroactivities, but their catalytic effects towards various reactions, say HER, are found to be strongly dependent on the chemical exfoliation route and metal-to-chalcogen composition.[61] Despite its inherent activity, large variations in activeness are observed depending on the exfoliation procedure. The TMDs with metal edge or chalcogen edge also vary in their activities.[62] These variations in the activities are dependent on the electrochemical reactions concerned and the radicals of interests, since the metal edge/chalcogen edge will act like adsorption centers for radicals, and in many of the electrochemical reactions adsorption is the reaction rate determining step.[61] Further, some of the exfoliation agents such as n-butyllithium are found to be aiding the phase transition to more catalytically active 1T polymorph (though this phase is less stable), due to electron transfer from the C-Li bond in n-butyllithium[63]. However, with methyllithium as an exfoliating agent, inherent electrocatalytic activities of TMDs are found to be deteriorated due to the coverage of methyl radicals after exfoliation.[64]

The catalytic activities of TMDs can also be affected by the valence chalcogen impurities or metal oxide presence.[65] For example, the presence of valence sulfide impurities, such as $MoS_3$ and $WS_3$, and their oxide counterparts, such as $MoO_2$, $MoO_3$ and $WO_2$, $WO_3$ can contribute to the enhanced catalytic activities towards hydronium reduction to hydrogen in case of $MoS_2$ and $WS_2$.[65] Therefore, it is highly possible that the differences in the reported reaction onset potentials and current densities can be due to the presence of above mentioned inherent catalytic impurities due to the synthesis procedure.

Another factor affecting the catalytic performance of 2D layers is the number of layers present (though there is a study which states that atomic thickness and structure are insignificant in electron transfer process and electrochemical response will not vary much from single to a few layered graphene,[66] the layer thickness will determine the active surface area, edge exposure, easiness of charge transfer with the electrode, and hence for all practical electrochemistry it can affect the electrochemical response). For example, the HET rates of single and bi-layer graphene in potassium ferricyanide redox probe is studied, and it is found that the standard HET rate constant estimated is higher for monolayer graphene flakes.[67] In another study, the layer dependent electrocatalysis of $MoS_2$ towards HER is found to be decreased (in terms of current density) by a factor of ~ 4.47 (current density) for the addition of every layer.[68]

### Role of induced defects

#### Dopants

In the case of graphene, it is well known that it has abundant free $\pi$ electrons, and they hold promise to act as catalysts for reactions that require electrons, such as the ORR. However, these $\pi$ electrons are too inert to be directly employed for ORR, and activation is essential prior to their use. The most straightforward method to activate the $sp^2$ carbon is to put more electrons into the $\pi$ cloud, and this has been achieved by doping $sp^2$ carbon materials with electron-rich or electron deficient atoms.[69] In the case of electron-rich atoms like nitrogen (N), the extra electrons contributed by N increase the electron density and raise the highest occupied molecular orbital (HOMO) energy level of the $sp^2$ carbons, facilitating the ORR. Nitrogen can be accommodated in three different geometries within the graphene lattice, namely, graphitic, pyiridinic, and pyrrolic. The electrochemical performance of all these three geometries are not same. For example, the electrocatalytic activity of nitrogen doped graphene is found to be dependent on the graphitic N content, which determines the limiting current density, while the pyiridinic N contributes to the major part of capacitance increment in N-doped graphene. It is important to note that it is quite challenging to keep one type of nitrogen geometry in N-doped graphene system during growth (various types of nitrogen environment can be identified from the X-ray photoelectron spectroscopy (XPS) analysis).[70] Further, it is found that the delocalized $\pi$ electrons can also be perturbed by conjugation with vacant orbitals, i.e., doping with electron-deficient elements such as boron (B).[71] The vacant $2p_z$ orbital of B conjugates with the carbon $\pi$ system to extract electrons. These electrons become active because of the low electronegativity of B, and therefore can be used in the ORR. The dopants can significantly bring spin and charge density to the graphene than the inherent defects, and the effect of dopants also depends on their distribution within the graphene.[72] It is also found that doping with multiple elements can enhance most of the electrochemical activities, since large scale spin and charge density can be brought in to graphene by multiple dopants of different characters.

Modification/doping with elements like Pt, Co, Fe, Ni, Cu, etc., is possible in the case of TMDs too, and some of these doping favor the catalytic activity towards certain reactions.[47,73] The edge states of TMDs determine their catalytic activity. For example, in the case of $MoS_2$, Mo-edge has been found to be active for HER, while the S-edge is inert. However, it is found that the incorporation of transition metal atoms (Fe, Co, Ni, or Cu) at the S-edge site can make it become HER active.[73] The transition metals doped $MoS_2$ show an increase in exchange current densities by at least two-fold higher than their undoped/pristine ones. Hence, it has been established that improving the electrocatalytic efficiency by incorporating promoters into particular atomic sites is possible in TMDCs. The hydrogen adsorption free energy, $\Delta G_H$, is found to be an accurate descriptor for determining the HER activity of a material, where the optimal HER catalyst will be thermo-neutral, $\Delta G_H \sim 0$ eV. Density functional theory (DFT) calculations can accurately predict the $\Delta G_H$ of a given structure,[47] and the calculated $\Delta G_H$ values for various doped $MoS_2$ systems are shown in Fig. 6 (hydrogen atoms bound at the most thermoneutral site are rounded in red, the other hydrogen atoms are bound too strongly (far away from thermoneutral) and can be considered poisoned at the edge). Further, it is theoretically established that the preferential doping of the S-edge is a general phenomenon arising from the relative stabilities of the Mo-edge and the S-edge $MoS_2$, where the Mo-edge is generally more stable. Moreover, Co can only be doped at the S-edge, whereas Ni can be doped both edges. It can be seen from Fig. 6 that the doped S-edges have $|\Delta G_H|$ values closer to thermo-neutral than the pristine $MoS_2$ S edge, indicating improved activity of the S-edge. In the cases of $MoS_2$ doped with Fe and Co, there is even improvement in $|\Delta G_H|$ values over the pristine Mo-edge by at least 0.02 eV. Hence it can be concluded that doping with Co, Fe, Ni, and Cu increases the S-edge activity to make it comparable or better than the pristine Mo-edge $MoS_2$. The $\Delta G_H$ of the doped Mo-edges are all further rising from thermoneutral than the pristine Mo-edge, indicating that Mo-edged $MoS_2$ are deactivated while doping. But in the case of Ni doped $MoS_2$, the Mo-edges are found to be still active. In Fe, Co, and Cu doping, the Mo-edge is expected to be unaffected while the

| $MoS_2$ catalyst | Edge | Structure | Active sites | $\Delta G_H$ (eV) |
|---|---|---|---|---|
| Pristine | Mo-edge | 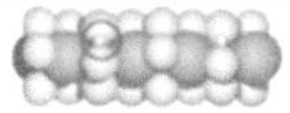 | 1 | 0.06 |
| | S-edge | 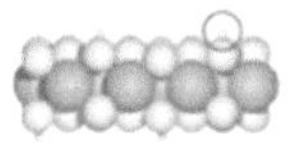 | 0 | −0.45 |
| Fe-doped | Mo-edge | 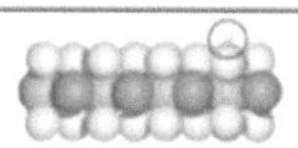 | 0 | 0.48 |
| | S-edge |  | 1 | 0.04 |
| Co-doped | Mo-edge | 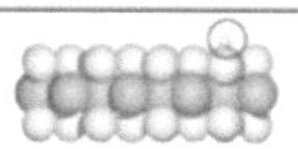 | 0 | 0.30 |
| | S-edge | 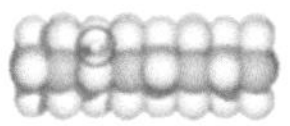 | 1 | 0.01 |
| Ni-doped | Mo-edge | 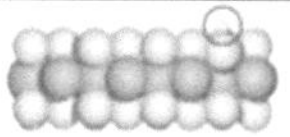 | 1 | 0.15 |
| | S-edge | 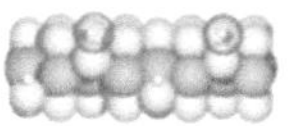 | 2 | −0.18 / −0.15 |
| Cu-doped | Mo-edge | 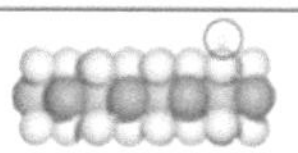 | 0 | 0.47 |
| | S-edge | 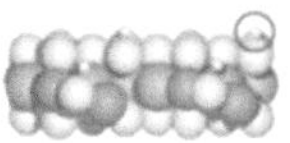 | 1 | 0.05 |

**Fig. 6.** Relevant edge structures, hydrogen adsorption free energies, and hydrogen coverage for pristine $MoS_2$ and TMs doped $MoS_2$ catalysts.[47]

results from the $\Delta G_H$ indicate that HER should be activated on the S-edge, making it comparable to pristine Mo-edged $MoS_2$ (which is inherently active).

It is also found that the HER activity of inert $MoS_2$ surface can be triggered via Pt-single-atom metals replacing 'Mo' atoms in $MoS_2$.[73] Here, the doped metal atoms can tune the adsorption behavior of H atoms on the neighboring S sites, resulting in a significantly enhanced HER activity on the $MoS_2$ surface. More importantly, in such a case, the reactants or harsh environments do not directly contact the metal dopants due to protection from the tri-layered sandwich structure of $MoS_2$, which thereby increases the catalytic stability and anti-poison ability. Hence, such an approach opens the possibility for developing 2D layered catalysts with long term performance, and also avoiding issues such as poisoning, those normally observed in the benchmarked metal based catalysts.[74]

However, electrochemical properties of 2D materials can often be misinterpreted due to the presence of impurities too. These impurities often originate from the starting materials used to fabricate or from the substrates which they grow. These impurities are wide-ranging, from electrochemically inactive $SiO_2$ to electrocatalytic transition metals, such as Ni, Fe, and Co, and most of the time it is very hard to completely get rid of them from the system. For example, metal impurities can remain in graphene despite strong oxidative treatment of graphite with concentrated sulfuric and nitric acids, as well as subsequent thermal or ultra-sonication/chemical exfoliation/reduction treatments. The CVD grown graphene transfer procedure also introduces metallic impurities which can alter the graphene electrochemical properties. In particular, Fe metal impurities could be introduced by the usage of $FeCl_3$ or $Fe(NO_3)_3$ as etching solutions to dissolve the metal substrate prior to the graphene transfer.[75] Hence, one should be extremely careful in interpreting the data from 2D electrochemistry, and also 2D materials need to be characterized for residual impurities in parts per million level (using techniques such as Inductively Coupled Plasma—Atomic Emission Spectroscopy (ICP-AES) or atomic resolution electron microscopy) and care must be taken to avoid the electrochemical interference effects from these impurities (for example, metal sites can be blocked by cyanide poisoning).[74]

### Role of substrates in 2D electrochemistry

The substrate used to support the working electrode/active catalyst is often considered to be idle (than as a electron source or sink) in conventional electrochemistry.[76] However, it is found that the substrate has a crucial role in 2D electrochemistry. For example, it is found that the stronger binding between $MoS_2$ and substrate can make the hydrogen adsorption weaker, hence affecting the HER activity of $MoS_2$.[68] The hydrogen adsorption properties of $MoS_2$ supported on various substrates, such as gold, graphene, and another $MoS_2$, are studied. $\Delta G_H$ for the Mo edged $MoS_2$ on graphene or gold is significantly higher than that of the freestanding Mo-edged $MoS_2$, suggesting that further improvements in designing appropriate substrates are required to decrease $\Delta G_H$ to make it close to thermo-neutral value.[77] The synergistic effects observed from the substrate include that from the plasmon-enhanced strong visible light photocatalysis from defect engineered CVD grown graphene and gold substrate, and electrochemical activities of hBN when deposited on gold substrate, otherwise inactive hBN.[78] It has also been established that the underlying substrate and its defects can influence the spin and charge density distributions in 2D layers, similar to the effect of heteroatom doping in graphene.[52]

The success story of in-plane engineering of atomic layers with dopants motivates the development of out of plane stacks of these in-plane bond saturated layers resulting in van der Waals solids. This has been identified as a strategic route for the development of new functional atomic layers/solids with controllable properties. For example, DFT studies show that the existence of graphene layer below the $MoS_2$ has a noticeable influence on the charge density distribution of $MoS_2$.[50] The built-in electric field in $MoS_2$-graphene hybrid forms an excessive negative charge density in the structure, which eventually enhances the HER process. Moreover, this sandwich structure can also make the basal plane of $MoS_2$ close to the thermo-neutral, enabling the basal plane of $MoS_2$ too to be active towards the HER. Further, the high optical transparency (visible region) and electronic conductivity of graphene can provide efficient current injection into the $MoS_2$ layers—augmenting the photo-electrocatalysis.[52] The structural defects occurring during the transfer of layers such as ripples, wrinkles, etc., can also help the catalysis process due to effects from graphene plasmonics and strain induced catalysis. The wrinkling and rippling are prominent while transferring graphene (other layers) to soft substrates/templates. The top graphene structure on TMDs (graphene/TMD stack) can protect them from further corroding, and hence this will also help to get a long term stable activities from these catalysts.[52]

Another factor, recently emerged as an important parameter in the designing of photo-electrocatalytic surfaces, is the stacking sequence.[52,54] According to the theoretical predictions which are correlating with experimental data, the photoexcited electrons and holes in such vertical heterostructures prefer to stay at different layers.[54] This means that stacking order of various layers with respect to the current collecting electrode is important in catalysis. For example, in graphene/$MoS_2$ vertical heterostructures in HER, graphene on top of $MoS_2$, where $MoS_2$ in direct contact with the current collector (electrode), is found to have high exchange current density in HER reaction than $MoS_2$ on top of graphene, where graphene is in contact with the electrode.[52] A similar observation has been made in $WS_2$/$MoS_2$/Au stacks, where $MoS_2$/$WS_2$/Au (in dark) is found to be more active in HER than $WS_2$/$MoS_2$/Au.[54] The conduction band minimum position of $WS_2$ is higher than that of $MoS_2$, which impedes the electron transfer from $MoS_2$ to $WS_2$ in $WS_2$/$MoS_2$/Au, and hence reduces the $H_2$ evolution. In the reverse structure, electron transfer from $WS_2$ to $MoS_2$ facilitates the $H_2$ reaction. However, upon illumination, in $MoS_2$/$WS_2$/Au, the photoexcited electrons from $WS_2$ will be injected into $MoS_2$ via the stepwise band alignment and contribute to the reduction of $H^+$ ions, while the holes at the $WS_2$ can be neutralized by the electrons from the electrode. While in $WS_2$/$MoS_2$/Au structure, the photoexcited electrons in $WS_2$ are transferred to $MoS_2$, resulting in a decrease in the electrons at the $WS_2$, hence impeding the evolution of $H_2$.[54]

Similar work function difference can be explained the reason for the stacking sequence dependent HER performance of graphene/$MoS_2$ van der Waals solids too. However, it is found that more than work function difference is happening in these solids where $\Delta G_H$ is also getting modified with the stacking sequencing (these studies are not complete, and initial studies show very promising aspects of stacking sequence in electrocatalysis). One thing that needs to be emphasized is that the interface induced effects of staking are limited to adjacent layers, and hence many layered stacked solids may not observe promising effects like single layered stacks (say single layered graphene-$MoS_2$ will have higher effect than tri-layered $MoS_2$—graphene stacks).

Lack of highly crystalline TMDs based heterostructures and the elastic deformation of layers during the layer-by-layer transfer and post transfer treatment processes hinder the development of van der Waals solids based high performance electrocatalysts and optoelectronic devices. Further, high quality crystals are also important in avoiding the recombination of photocarriers generated during the light exposure, leading to a high current and low over potential for the photo-electrochemical reactions of interest.

Four important electrochemical reactions, namely ORR, OER, HER, and HOR, which are relevant in electrochemical energy conversion, but are kinetically sluggish (need catalyst to undergo the reaction except HOR, where kinetics is faster) are intensely pursued with the 2D electrochemistry.[79] Graphene and its functional derivatives are extensively studied for all the four reactions mentioned above.[80] However, TMDs are mostly studied for electrocatalytic and photo-electrocatalytic HER, and though a few reports exist on their other electrocatalytic activities too, their activities are much inferior to the benchmarked systems in the concerned reaction.[81] The next section briefly accounts important reports on these four reactions with metal free 2D electrochemistry.

Being a very important reaction for fuel cells and metal-air batteries, ORR is intensively studied with graphene systems. N-doped graphene (N-graphene) is the most widely studied dopant in graphene engineering.[82] N is next to C in the periodic table of elements, and its electronegativity (3.04) is higher than that of C (2.55). The introduction of N into graphene sheets can modify its local electronic structures. The N heteroatom can be introduced directly during the graphene growth, by a CVD method using $NH_3$ and $CH_4$ as the N and C sources[83] or via a solvothermal process using lithium nitride and tetrachloromethane.[83] Alternatively, N can be doped through the post treatment of graphene or GO, such as hydrazine reduction,[84] thermal annealing in ammonia,[85] and the arc-discharge method.[86] It has been well established that the incorporation of N atoms into the graphene matrix can lead to three main types of N formats, including graphitic N with direct substitution structure, and pyiridinic N and pyrrolic N structures.[87] Graphitic N corresponds to the doping of N atom into a hexagonal ring. Pyiridinic and pyrrolic N donate one and two π electrons to the π system, forming $sp^2$ and $sp^3$ hybridized bonds, respectively.[88] N-graphene was first reported as a valuable catalyst for ORR in 2010[89] and then showed that N-doped graphene can act as a metal free catalyst in ORR, competing with benchmarked Pt. Later various attempts were made to incorporate nitrogen in graphene and a very high ORR activity through a four electron transfer process in oxygen saturated 0.1 M KOH electrolyte, comparable or even better than commercial Pt/C (Pt loading: 4.85 mg cm$^{-2}$) electrodes is also obtained. However, in another work it is found that N-graphene developed via a solvothermal method showed four electron oxygen reduction pathway in alkaline solution, and two electron reduction in acidic solution.[88] Nitrogen doped graphene with varying nitrogen content is also developed and studied for the ORR activity.[90] It is found that up to certain levels of nitrogen doping, the activity is enhanced successively, but after that it decreases (Fig. 7A). Another interesting factor observed was the stability of the nitrogen doped graphene (NG) over potential cycling, where it is found to be giving a constant performance even after 10000 cycles (Fig. 7B). However, the activity of Pt/C is found to be sluggish after 5000 cycles (Fig. 7C).

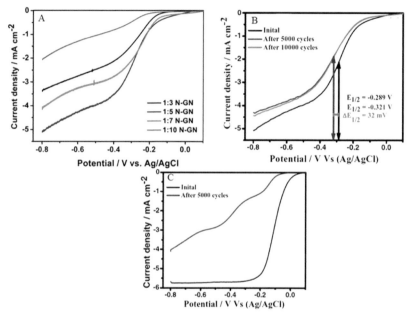

**Fig. 7.** (A) ORR performance of nitrogen doped graphene with varying nitrogen amount, in alkaline medium, (B) cyclic stability of optimum amount nitrogen containing graphene, and (C) ORR cycling stability of Pt/C (benchmarked catalysts).[90]

After the introduction of N-doped graphene as an electrocatalyst for ORR, other dopants like B, P, S, F, etc., are also tested towards ORR.[91] Although, the electronegativity of B is less (electronegativity-2.04) than that of C (electronegativity-2.55), B-graphene also shows good ORR activity. DFT calculations show that B doping can induce local high spin density on the basal plane owing to its strong electron withdrawing capability, which plays a key role in facilitating the adsorption of oxygen and OOH, thereby facilitating the ORR compared to pristine graphene.[82] P-doped graphene also showed outstanding ORR activity as well as excellent selectivity and stability, which endow its great potential as an efficient metal-free electrocatalyst.[92] Further, S could be introduced into graphene sheets in the form of thiophene-like S.[93] In this case too, excellent catalytic activities towards ORR are observed, demonstrating S-graphene to be an excellent metal-free catalyst with high stability and selectivity.

Halogenated graphene is another rapidly growing graphene derivative. It is well known that halogens possess higher reactivity than group IIIa–VIa elements. As halogen-doping transforms sp² carbon bonding to the sp³ state, it results in the drastic distortions in the geometric and electronic structures of graphene. For example, facile and scalable synthesis of edge-halogenated graphene nanoplatelets (XGnPs) and its efficacy as a metal-free electrocatalysts for oxygen reduction reaction is reported. The resultant XGnPs are also tested as cathode electrodes of fuel cells, and revealed remarkable electrocatalytic activities towards ORR with high tolerance to methanol crossover/CO poisoning effects and longer-term stability than those of the pristine graphene and commercial Pt/C catalysts.[94]

Graphene doped with more than one type of dopant elements having different electronegativity can create a unique electronic structure and it might bring synergistic effects. The electronegativity values of B and N are 2.04 and 3.4, which are lower and higher than that of C element (2.55), respectively. First principle calculations suggest that B and N co-doped graphene (BN-graphene) can be a good catalyst for the ORR.[95] The catalytic performance of BN-graphene (the introduction of both boron and nitrogen precursors during the graphene growth often end up with the patches (domains) of hBN and graphene) is also studied both experimentally and theoretically. The results show higher cathode current with better potential for BN-graphene system compared to the individually (B and N) doped graphene, indicating a more efficient and facile ORR process on the co-doped graphene. The S and N dual-doped graphene has also shown increased ORR activity in NaOH electrolyte.[96] In another work, boric acid and phosphoric acid are used to introduce B and P into N-graphene, obtaining BN-graphene and P and N dual-doped graphene (PN-graphene),[97] and both of them exhibited very good ORR activity. In general it can be concluded that with the optimum amount of dopant, doped graphene can compete in ORR performance with benchmarked systems such as Pt (Pt/C) (Fig. 8 showing the ORR performance of boron doped graphene (BG), nitrogen doped graphene (NG), fluorine doped graphene (FG)). However, the co-doping can even enhance the ORR performance where it can even beat benchmarked metals in performance (Fig. 8B) showing nitrogen and fluorine doped graphene has better current density than Pt/C).

The type of dopants, amount of dopants, and their distribution are important in deciding the catalytic activities (ORR), since they determine the spin and charge density of the doped graphene. It has also been observed that residual oxygen functionalities have little role in deciding the spin and charge density of heteroatoms doped graphene, and hence even doped CDGs having residual oxygen functionalities can perform well in ORR.

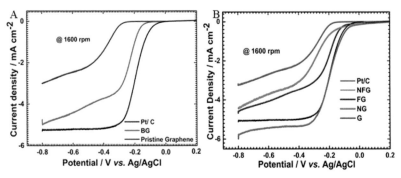

**Fig. 8.** Linear sweep voltamograms of various active catalysts towards ORR reaction. The effect of doping and co-doping of graphene is exemplified.[49,72]

Water splitting is a promising route for hydrogen production without the formation of greenhouse gases along with oxygen.[98] The cathodic half-cell reaction in water electrolysis process is HER, while the anodic reaction is OER. As promising metal-free catalysts, almost all types of carbon materials show certain activity toward HER including activated CNT,[99] graphene doped with heteroatoms,[100] and functionalized fullerenes.[101] One featured example is the chemically coupled $C_3N_4$–N doped graphene hybrid, which showed a comparable activity with most popular $MoS_2$ material in acid solutions, in agreement with both calculated $\Delta G_H$ and measured exchange current density ($j_0$).[102] It is found that through the chemical coupling of g-$C_3N_4$ and N-graphene sheet, there is a synergistic effect between $C_3N_4$ (too strong adsorption) and N-graphene (too weak adsorption) to create a proper adsorption of hydrogen on hybrid's surface close to that of Pt, which results in favorable HER activity. The studies of metal-free catalysts for OER have just started to be investigated and shown to be competitive. The currently available OER 2D metal-free catalysts are analogous to those developed for ORR, namely single/double doped heteroatoms with graphene[103] and $C_3N_4$–graphene.[104] Most of these materials can be obtained with high surface area to achieve apparently very high anodic current densities.

HOR is another important reaction, particularly important in fuel cells.[105] Different types of carbon materials have been used as catalyst supports for HOR,[106] but only a few studies have been reported about the graphene-supported catalysts for HOR. In HOR, the CO tolerance is the important parameter determining the efficiency of the catalyst. Graphene nanosheets showed a high CO tolerance performance in HOR. Pt clusters supported on graphene nanosheets (Pt/GNs) showed much higher CO tolerance performance in the HOR than the Pt clusters dispersed on carbon black (Pt/CB). It was found that in pure $H_2$ atmosphere, Pt/GNs, Pt/CB, and PtRu/CB showed similar electrocatalytic activities for the HOR despite the difference in the types of carbon material used.[107]

## Atomic Layers in Sensing

There are two critical aspects of 2D materials based sensor systems: (i) surface to volume ratio of the 2D substrate, and (ii) the electronic transfer characteristics of the material.[51] For sensors having similar overall size, the surface to volume ratio determines the number of analytes that are in contact with the electrode and the efficiency of the electronic interactions between analyte and the substrate directly affect the measurement. Hence the large surface area coupled with high mobility and conductivity are beneficial advantages of graphene like 2D materials for electronic and electrochemical biosensors. Various types of small molecule sensors are investigated with 2D materials as active sensing platform, namely, glucose sensors,[108,109] neurotransmitter sensors,[110] pathogenic cell detection sensors,[111] protein markers,[112] etc., are a few names apart from various nucleic acid sensors.[113] The electrochemical sensing platforms are realized based on any of the following properties, say electrochemical activity of the analyte (oxidation/reduction current of the analyte with 2D materials as working electrodes using standard two/three electrode electrochemistry) or impedance variation upon successive addition of the analytes (using electrochemical impedance spectroscopy and variation in the charge transfer resistance) or by enzymatic sensing where the enzymes are mobilized over the 2D electrode and electron transfer from or to the 2D materials is monitored or amperometric sensing or solution gated—field effect transistor (FET) based sensors (which operates at low voltages (it is important in bioanalyte sensing) and the changes in the drain-source current with gate voltage (in enzymatic sensing gate will be modified with the enzyme and the electrochemical reactions happening at the gate will be reflected in drain-source current) can be used for sensing) or electronic (FET)/electrochemical sensing integrated to a microfluidic sensing platforms.[51] In all the above mentioned electrochemical biosensing platforms, carbon based materials are considered to be ideal electrode materials due to their wide anodic potential range, low residual current, chemical inertness, easy availability, and reduced cost.[108] They are mechanically robust, exhibit fast response and can be easily fabricated in different configurations and sizes. Further, carbon materials such as graphene based electrodes can overcome the issues related with CNTs based benchmarked sensors (residual metal impurities are inherent in CNTs due to their role in CNT growth) where signal interference due to the presence of metals can be avoided by providing graphitic metal free edges with high surface area.[114]

Large surface area and ease of functionalization bring the opportunity to bind biomolecules to 2D layer surfaces via covalent and non-covalent bonding.[115] This property of 2D layers has been utilized in biomedical applications, such as controlled drug delivery, targeted imaging, biosensors, immunosensors, etc. The fluctuations in the electronic structure of the atomic layers caused by this binding can be transformed (role of a transducer) to electronic or electrochemical signal. However, understanding the bonding between the basic biomolecule, say protein, and the atomic layer will play a significant role in the development of sensors and biomarkers using 2D materials.[116] In this context, it is found that graphene interacts with the hydrophobic components of biomolecules and it alters their conformation and disrupts the biological activities.[117] In proteins, these conformational changes can completely denature them upon adsorption on graphene surface, and hence graphene-biomolecule interface should be taken seriously (the loss of specific function of Concanavalin, a protein, by a nonspecific binding with singlelayer graphene and its functioning after a specific binding are schematized in Fig. 9).

Studying their interactions in presence of other molecules such as water will help to unravel the interface phenomena, where these molecules can largely affect the binding and structure of biomolecules over atomic layers.[116] The conformational changes of the adsorbed protein may depend on various other factors, such as orientation of graphene, number of layers, and energetics of the equilibrium between folded and denatured states. Further, water molecules can play a crucial role in the assembly of proteins on a solid substrate by regulating the interactions between binding partners and conformations of proteins. Hence, graphene-protein assembly can be affected by the non-covalent interactions and heterogeneous and dynamic hydration environments. However, the properties and structure of water regulated graphene/other 2D materials-protein interfaces are not well characterized or explored till date. A direct contact of graphene with protein may be potentially hazardous and lead to unfolding of protein. However, a close contact between protein and graphene is necessary for obtaining a direct electron transfer from the enzyme active site.

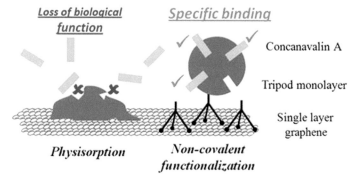

**Fig. 9.** The loss of specific function of Concanavalin, a protein, by a nonspecific binding with single layer graphene and its functioning after a specific binding. Reproduced with permission from American Chemical Society.[117]

For example, the molecular dynamic (MD) studies of interaction between graphene and an important enzyme glucose oxidase (GOx) were recently conducted, due to its importance in the development of glucometers and point of care devices for high sensitive glucose detection.[51,118] The GOx catalyzes the oxidation of glucose to D-glucono-d-lactone and hydrogen peroxide. It has one very tightly, non-covalently bound, flavine adenine dinucleotide cofactor per monomer, and a homodimer with molecular mass of 130–320 kDa (depending on the extent of glycosylation).[51] The MD simulations revealed that a very tight wrapping of the enzyme is possible, but it generates structural changes in the protein, including partial impairment of its catalytic function because of relocation of catalytic residues and cofactor. The role of different shapes of graphene in graphene-enzyme interaction is also studied, and quantified their ability to wrap the enzyme (glucose oxidase dimer). Such an encapsulation leads to the achievement of a high surface contact with graphene enabling direct electron transfer. However, the preservation of structure and function of enzyme are challenging and they depends on the enzyme shape and location of the substrate binding pocket.

This is true for other layered materials too. Though $MoS_2$ has been proven as a potential material for various bio-sensing applications,[52] interactions between $MoS_2$ and bio-molecules are complicated. In one of the theoretical studies, $MoS_2$ is found to be exhibiting van der Waals interactions with a model protein, villin headpiece (HP35), a commonly used protein for folding studies.[116] These interactions resulted in to a robust denaturation of HP35, with protein secondary structure severely destroyed within a few 100 nm simulations. Here also it is found that the water molecules play a key role in the adsorption and denaturation process. However, these issues can be addressed by suitably modifying the interactions between $MoS_2$ and biomolecules. It is predicted that one such change can be brought by surface modification of $MoS_2$ via oxygen functionalities.

TMD atomic layers such as $MoS_2$ have advantages over graphene in sensing due to the following two reasons: pristine surfaces lead to excellent electrostatic control, and the presence of a finite band gap helps in suppressing the leakage currents (for example, it has shown that in a typical $MoS_2$-FET biosensor can lead to ultra-high sensitivity, which can surpass that of FET biosensors based on graphene by more than

74-fold).[116] These factors in combination are influential in increasing the sensitivity and reducing the minimum detection limit of the TMD based FET biosensors, as it is shown for $MoS_2$-FET biosensors. In $MoS_2$-FET, the excellent electrostatics permit ultra-low voltage and low off state-leakage (and hence low power) operation that is highly desirable for hand-held, battery operated biosensors.[116] Further, it is established through theoretical analysis that $MoS_2$ is greatly advantageous for biosensor device scaling without compromising its sensitivity.[116] Most of the TMDs such as $MoS_2$ are also mechanically robust. Hence these features make TMDs highly promising materials for next generation label-free FET biosensors, where electrochemical reactions along with field effect principle of FET can be utilized for high sensitivity. Apart from proteins, 2D surfaces are studied for the electrochemical detection of various electrochemically active neurotransmitters such as dopamine (DA), serotonin (ST), etc., too.[120] In these sensors, the sensitivity and reliability of graphene based electrodes in comparison to single-wall CNTs (SWCNTs) are found to be better, and graphene exhibits better signal-to-noise ratio and stability than SWCNTs for electrochemical detection.[121] Further, simultaneous multi-analyte detection, say, DA, ascorbic (AA) acid and ST, is also reported with graphene based electrochemical sensors making them ideal for '*in vivo*' applications. DFT based studies on the effects of external electric field, heteroatoms and defects on graphene in the interactions with DA show that their interactions can be tuned externally, leading to the development of biosensors based on doped graphene (Fig. 10).[119] Selective binding of amino acids to heteroatoms doped graphene is also theoretically proven via *ab initio* calculations.[119]

It is found that DA is attached on the surface of graphene via physisorption, and the cohesive strength varies in various graphenes' as the cohesive energy pursues the following order: B-doped > N-doped > vacancy defect > Thrower−Stone−Wales defect.[119] It is found that the boron-doped graphene exhibits valence states with DA

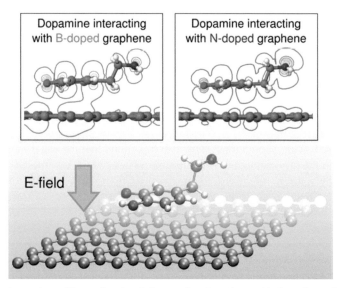

**Fig. 10.** The interactions of boron doped and nitrogen doped graphene with dopamine and the effect of electric field on this interactions. Reproduced with permission from American Chemical Society.[119]

molecules and this system shows the strongest cohesive energy. Further, when an electric field is applied, there is a shift observed in the valence states near the Fermi level producing a decrease in the molecule−layer interactions (this is prominent in B-doped graphene-DA, while the effect of electric field is insensitive in N-doped graphene-DA and apparently less in defective graphene-DA interactions).[119] All these studies show that defective or selectively doped graphene can act as potential candidates in the development of high sensitive neurotransmitter sensors. However, it is also important to emphasize that the nature and amount of defects, number of atomic layers, etc., matter while using 2D layers in sensing, similar to that in electrocatalysis.

Introducing selectivity to a sensor is the bottleneck in atomic layers based sensing. Apart from the selective covalent functionalization of 2D backbones for selectivity towards an analyte, bringing electrostatic interactions via doping is found to be an effective route towards selectivity. For example, high sensitive ammonia (both unionized (breathe ammonia) and ionized (blood)) detection is important in healthcare and medicine.[15] DFT studies show that ammonia (unionized) has little interactions with graphene, while doped graphene such as fluorographene, can interact with ammonia through hydrogen bonding interactions (fluorine-hydrogen interactions, Fig. 11).[15]

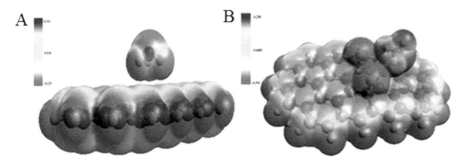

**Fig. 11.** Electrostatic potential plots for (A) graphene-ammonia and (B) FG-ammonia.[15]

This high sensitivity of fluorographene towards ammonia has been experimentally also proven via impedance measurements. However, the amount of fluorine in graphene also matters in the designing of fluorographene based ammonia sensor, since a large amount of fluorine can make graphene an insulator ($C_4F$ has a bandgap ~ 3.4 eV).[14] Theoretically also it is found that ammonia cannot be stabilized on large fluorine containing graphene surface.[15] Hence, the amount of dopants is also a critical parameter in designing high sensitive sensor platforms using graphene. Effect of van der Waals solids in 2D electrochemical sensing is not explored much, but will have plenty of opportunity if it can bring selectivity and sensitivity to the sensing platform. The use of atomic layers in biosensing, particularly for *in vivo* applications, demands their bio-compatibility too—a field still nascent in 2D materials based biosensors.

## Important Notes for Experimental 2D Electrochemistry

- The immobilization/transfer of 2D materials on different conventional carbon substrates can give different electrochemical activities, depending on their surface roughness. Hence, surface roughness of the current collector also matters in the electrochemical performance of 2D materials.

- It is important to note that the current density should be optimized by electrochemical surface area instead of geometrical surface area to avoid (minimize) the contribution from the non-faradaic currents, which can lead to wrong interpretation in the onset potential of a reaction.

- The substrate (underlying supporting electrode/current collector) matters in 2D electrochemistry, and hence the synergistic effects from the current collector (substrate) should be taken care.[122]

- Surface coverage of a 2D crystals is another important factor, since the area of a 2D crystal is limited to the size of the temperature zone (CVD growth) or size of the bulk crystal (in chemical/liquid exfoliation method).[123] Hence, the number of edge sites exposed can be varied with the same 2D layer having different lateral sizes. Further, in van der Waals heterostructures, the stacking sequence of different layers is also found to be playing a significant role in 2D electrochemistry.[52,124]

- Effects from residual impurities (metallic/non metallic) and surfactants should be taken care of. Similar to metallic/non-metallic impurities in 2D crystals (as discussed before—they mostly come during the synthesis or transfer processes of atomic layers), surfactants can also play a significant role in 2D electrochemistry. Surfactants mediated exfoliation of defect-free 2D materials is becoming a benchmarked method, and it has been recently found that surfactants can have detrimental effects on the electrocatalytic activities compared to that of surfactant free 2D crystals.[125]

# References

1. P. Ajayan, P. Kim and K. Banerjee, *Phys. Today*, 2016, **69**, 38–44.
2. K. S. Novoselov, A. K. Geim, S. V. Morozov, D. Jiang, Y. Zhang, S. V. Dubonos, I. V. Grigorieva and A. A. Firsov, *Science*, 2004, **306**, 666–669.
3. P. T. Araujo, M. Terrones and M. S. Dresselhaus, *Materials Today*, 2012, **15**, 98–109.
4. A. Kuc, T. Heine and A. Kis, *MRS Bulletin*, 2015, **40**, 577–584.
5. A. Ramasubramaniam, D. Naveh and E. Towe, *Phys. Rev. B*, 2011, **84**, 205325.
6. K. Watanabe, T. Taniguchi and H. Kanda, *Nat. Mater.*, 2004, **3**, 404–409.
7. M. Z. Hasan and C. L. Kane, *Reviews of Modern Physics*, 2010, **82**, 3045–3067.
8. N. E. Staley, J. Wu, P. Eklund, Y. Liu, L. Li and Z. Xu, *Phys. Rev. B*, 2009, **80**, 184505.
9. R. K. Biroju, P. K. Giri, S. Dhara, K. Imakita and M. Fujii, *ACS Appl. Mater. Interfaces*, 2013, **6**, 377–387.
10. K. B. Ravi, T. Nikhil, R. Gone, S. Dhara and P. K. Giri, *Nanotechnology*, 2015, **26**, 145601.
11. A. C. Ferrari, F. Bonaccorso, V. Fal'ko, K. S. Novoselov, S. Roche, P. Boggild et al., *Nanoscale*, 2015, **7**, 4598–4810.
12. A. K. Geim and I. V. Grigorieva, *Nature*, 2013, **499**, 419–425.
13. S. Hu, M. Lozada-Hidalgo, F. C. Wang, A. Mishchenko, F. Schedin, R. R. Nair, E. W. Hill, D. W. Boukhvalov, M. I. Katsnelson, R. A. W. Dryfe, I. V. Grigorieva, H. A. Wu and A. K. Geim, *Nature*, 2014, **516**, 227–230.
14. K. K. Tadi et al., *Particle & Particle System Characterization* 2016, DOI:10.1002/ppsc.201600346.
15. K. K. Tadi, S. Pal and T. N. Narayanan, *Sci. Rep.*, 2016, **6**, 25221.
16. C. Mattevi, H. Kim and M. Chhowalla, *J. Mater. Chem.*, 2011, **21**, 3324–3334.
17. W. Yang, G. Chen, Z. Shi, C.-C. Liu, L. Zhang, G. Xie, M. Cheng, D. Wang, R. Yang, D. Shi, K. Watanabe, T. Taniguchi, Y. Yao, Y. Zhang and G. Zhang, *Nat. Mater.*, 2013, **12**, 792–797.
18. R. K. Biroju, G. Rajender and P. K. Giri, *Carbon.*, 2015, **95**, 228–238.

19. A. Ambrosi, C. K. Chua, A. Bonanni and M. Pumera, *Chem. Rev. (Washington, DC, U. S.)*, 2014, **114**, 7150–7188.
20. A. H. R. Palser, *Phys. Chem. Chem. Phys.*, 1999, **1**, 4459–4464.
21. M. J. Jaison, T. N. Narayanan, T. Prem Kumar and V. K. Pillai, *J. Mater. Chem. A.*, 2015, **3**, 18222–18228.
22. G. Eda and M. Chhowalla, *Adv. Mater. (Weinheim, Ger.)*, 2010, **22**, 2392–2415.
23. W. Gao, L. B. Alemany, L. Ci and P. M. Ajayan, *Nat. Chem.*, 2009, **1**, 403–408.
24. M. Baleeswaraiah, T. N. Narayanan, B. Kaushik, M. A. Pulickel and T. Saikat, *2D Materials*, 2014, **1**, 011008.
25. P. M. Sudeep, S. Vinod, S. Ozden, A. Sruthi, A. Kukovecz, Z. Konya, R. Vajtai, M. R. Anantharaman, P. M. Ajayan and T. N. Narayanan, *RSC Advances*, 2015, **5**, 93964–93968.
26. L. Zhou, B. He, Y. Yang and Y. He, *RSC Advances*, 2014, **4**, 32570–32578.
27. J. N. Coleman, M. Lotya, A. O'Neill, S. D. Bergin, P. J. King, U. Khan, K. Young, A. Gaucher, S. De, R. J. Smith, I. V. Shvets, S. K. Arora, G. Stanton, H.-Y. Kim, K. Lee, G. T. Kim, G. S. Duesberg, T. Hallam, J. J. Boland, J. J. Wang, J. F. Donegan, J. C. Grunlan, G. Moriarty, A. Shmeliov, R. J. Nicholls, J. M. Perkins, E. M. Grieveson, K. Theuwissen, D. W. McComb, P. D. Nellist and V. Nicolosi, *Science*, 2011, **331**, 568–571.
28. J. M. Tour, *Nat. Mater.*, 2014, **13**, 545–546.
29. K. R. Paton, E. Varrla, C. Backes, R. J. Smith, U. Khan, A. O'Neill, C. Boland, M. Lotya, O. M. Istrate, P. King, T. Higgins, S. Barwich, P. May, P. Puczkarski, I. Ahmed, M. Moebius, H. Pettersson, E. Long, J. Coelho, S. E. O'Brien, E. K. McGuire, B. M. Sanchez, G. S. Duesberg, N. McEvoy, T. J. Pennycook, C. Downing, A. Crossley, V. Nicolosi and J. N. Coleman, *Nat. Mater.*, 2014, **13**, 624–630.
30. S. Pal, K. K. Tadi, P. M. Sudeep, S. Radhakrishnan and T. N. Narayanan, *Mater. Chem. Front.*, 2017, DOI:10.1039/C6QM00081A.
31. J. Zheng, H. Zhang, S. Dong, Y. Liu, C. Tai Nai, H. Suk Shin, H. Young Jeong, B. Liu and K. Ping Loh, *Nature Communications*, 2014, **5**, 2995.
32. D. Deng, X. Pan, L. Yu, Y. Cui, Y. Jiang, J. Qi, W.-X. Li, Q. Fu, X. Ma, Q. Xue, G. Sun and X. Bao, *Chem. Mater.*, 2011, **23**, 1188–1193.
33. A. Ambrosi, Z. Sofer and M. Pumera, *Small*, 2015, **11**, 604–604.
34. M. Acerce, D. Voiry and M. Chhowalla, *Nat. Nano*, 2015, **10**, 313–318.
35. C. Z. Congqin Miao, Owen Liang and Ya-Hong Xie, *Physics and Applications of Graphene-Experiments*, Intech, 2011.
36. H. O. Pierson, *Handbook of Chemical Vapor Deposition*, William Andrew Publishing, LLC, Norwich, New York, U.S.A., 1999.
37. D. M. Mattox, *Handbook of Physical Vapor Deposition (PVD) Processing*, Noyes Publications, 369 Fairview Avenue, Westwood, New Jersey, U.S.A., 1998.
38. W. C. Xuesong Li, Jinho An, Seyoung Kim, Jinghyo Nah, Dongxing Yang, Richard Piner, Aruna Velamakanni, Inchwa Jung and Rodney S. Ruoff, *Science*, 2009, **324**, 1312–1314.
39. S. B. Young Bin Lee, Houk Jang, Sukjae Jang, Jong-Hyun ahn et al., *Nano Lett.*, 2010, **10**, 490–493.
40. C. S. R.-V. Mark, P. Levendorf, Shicank Garg and Jiwong Park, *Nano Lett.*, 2009, **9**, 4479–4483.
41. F. Lili, Z. Jie, L. Zhen, L. Xiao, W. Kunlin, W. Jinquan, Z. Minlin, W. Dehai, X. Zhiping and Z. Hongwei, *Nanotechnology*, 2012, **23**, 115605.
42. L. Song, Z. Liu, A. L. M. Reddy, N. T. Narayanan, J. Taha-Tijerina, J. Peng, G. Gao, J. Lou, R. Vajtai and P. M. Ajayan, *Adv. Mater. (Weinheim, Ger.)*, 2012, **24**, 4878–4895.
43. K.-K. Liu, W. Zhang, Y.-H. Lee, Y.-C. Lin, M.-T. Chang, C.-Y. Su, C.-S. Chang, H. Li, Y. Shi, H. Zhang, C.-S. Lai and L.-J. Li, *Nano Lett.*, 2012, **12**, 1538–1544.
44. A. Govind Rajan, J. H. Warner, D. Blankschtein and M. S. Strano, *ACS Nano*, 2016, **10**, 4330–4344.
45. G. Babu, N. Masurkar, H. Al Salem and L. M. R. Arava, *J. Am. Chem. Soc.*, 2017, **139**, 171–178.
46. T. Kiran Kumar, M. P. Anil, P. Shubhadeep, P. M. Sudeep and N. N. Tharangattu, *Nanotechnology*, 2016, **27**, 275402.
47. H. Wang, C. Tsai, D. Kong, K. Chan, F. Abild-Pedersen, J. K. Nørskov and Y. Cui, *Nano Research*, 2015, **8**, 566–575.
48. Z. Kou, T. Meng, B. Guo, I. S. Amiinu, W. Li, J. Zhang and S. Mu, *Adv. Funct. Mater.*, 2017, DOI:10.1002/adfm.201604904, 1604904-n/a.
49. T. V. Vineesh, M. P. Kumar, C. Takahashi, G. Kalita, S. Alwarappan, D. K. Pattanayak and T. N. Narayanan, *Advanced Energy Materials*, 2015, **5**, 1500658.

50. K. Y. Honglin Li, Chao Li, Zheng Tang, Bangjun Guo, Xiang Lei, Hao Fu and Ziqiang Zhu, *Scientific Reports*, 2015, **5**, 18730.
51. S. Viswanathan, T. N. Narayanan, K. Aran, K. D. Fink, J. Paredes, P. M. Ajayan, S. Filipek, P. Miszta, H. C. Tekin, F. Inci, U. Demirci, P. Li, K. I. Bolotin, D. Liepmann and V. Renugopalakrishanan, *Materials Today*, 2015, **18**, 513–522.
52. K. B. Ravi, P. Shubhadeep, S. Rahul, P. K. Giri and N. N. Tharangattu, *Nanotechnology*, 2017, **28**, 085101.
53. Y. Shi, H. Zhang, W.-H. Chang, H. S. Shin and L.-J. Li, *MRS Bulletin*, 2015, **40**, 566–576.
54. J. Shi, R. Tong, X. Zhou, Y. Gong, Z. Zhang, Q. Ji, Y. Zhang, Q. Fang, L. Gu, X. Wang, Z. Liu and Y. Zhang, *Adv. Mater. (Weinheim, Ger.)*, 2016, **28**, 10664–10672.
55. G. Gao, W. Gao, E. Cannuccia, J. Taha-Tijerina, L. Balicas, A. Mathkar, T. N. Narayanan, Z. Liu, B. K. Gupta, J. Peng, Y. Yin, A. Rubio and P. M. Ajayan, *Nano Lett.*, 2012, **12**, 3518–3525.
56. M. B. M. Krishna, M. K. L. Man, S. Vinod, C. Chin, T. Harada, J. Taha-Tijerina, C. S. Tiwary, P. Nguyen, P. Chang, T. N. Narayanan, A. Rubio, P. M. Ajayan, S. Talapatra and K. M. Dani, *Advanced Optical Materials*, 2015, **3**, 1551–1556.
57. M. Pumera, *Chem. Soc. Rev.*, 2010, **39**, 4146–4157.
58. P. Chen, M. A. Fryling and R. L. McCreery, *Anal. Chem.*, 1995, **67**, 3115–3122.
59. U. Martinez, J. H. Dumont, E. F. Holby, K. Artyushkova, G. M. Purdy, A. Singh, N. H. Mack, P. Atanassov, D. A. Cullen, K. L. More, M. Chhowalla, P. Zelenay, A. M. Dattelbaum, A. D. Mohite and G. Gupta, *Science Advances*, 2016, **2**.
60. L. Zhang, Q. Xu, J. Niu and Z. Xia, *Phys. Chem. Chem. Phys.*, 2015, **17**, 16733–16743.
61. X. Yu, M. S. Prévot, N. Guijarro and K. Sivula, *Nature Communications*, 2015, **6**, 7596.
62. H. Zhu, M. Du, M. Zhang, M. Zou, T. Yang, S. Wang, J. Yao and B. Guo, *Chem. Commun. (Cambridge, U. K.)*, 2014, **50**, 15435–15438.
63. M. Kertesz and R. Hoffmann, *J. Am. Chem. Soc.*, 1984, **106**, 3453–3460.
64. D. Voiry, H. Yamaguchi, J. Li, R. Silva, D. C. B. Alves, T. Fujita, M. Chen, T. Asefa, V. B. Shenoy, G. Eda and M. Chhowalla, *Nat. Mater.*, 2013, **12**, 850–855.
65. N. Mohamad Latiff, L. Wang, C. C. Mayorga-Martinez, Z. Sofer, A. C. Fisher and M. Pumera, *Nanoscale*, 2016, **8**, 16752–16760.
66. M. S. Goh and M. Pumera, *Anal. Chem.*, 2010, **82**, 8367–8370.
67. A. T. Valota, I. A. Kinloch, K. S. Novoselov, C. Casiraghi, A. Eckmann, E. W. Hill and R. A. W. Dryfe, *ACS Nano*, 2011, **5**, 8809–8815.
68. Y. Yu, S.-Y. Huang, Y. Li, S. N. Steinmann, W. Yang and L. Cao, *Nano Lett.*, 2014, **14**, 553–558.
69. K. Gong, F. Du, Z. Xia, M. Durstock and L. Dai, *Science*, 2009, **323**, 760–764.
70. L. Lai, J. R. Potts, D. Zhan, L. Wang, C. K. Poh, C. Tang, H. Gong, Z. Shen, J. Lin and R. S. Ruoff, *Energy & Environmental Science*, 2012, **5**, 7936–7942.
71. L. Yang, S. Jiang, Y. Zhao, L. Zhu, S. Chen, X. Wang, Q. Wu, J. Ma, Y. Ma and Z. Hu, *Angew. Chem. Int. Ed.*, 2011, **50**, 7132–7135.
72. T. V. Vineesh, M. A. Nazrulla, S. Krishnamoorthy, T. N. Narayanan and S. Alwarappan, *Applied Materials Today*, 2015, **1**, 74–79.
73. J. Deng, H. Li, J. Xiao, Y. Tu, D. Deng, H. Yang, H. Tian, J. Li, P. Ren and X. Bao, *Energy & Environmental Science*, 2015, **8**, 1594–1601.
74. N. Wagner and M. Schulze, *Electrochim. Acta*, 2003, **48**, 3899–3907.
75. R. K. Biroju and P. K. Giri, *J. Phys. Chem. C.*, 2014, **118**, 13833–13843.
76. W. Chen, E. J. G. Santos, W. Zhu, E. Kaxiras and Z. Zhang, *Nano Lett.*, 2013, **13**, 509–514.
77. D. Voiry, J. Yang and M. Chhowalla, *Adv. Mater. (Weinheim, Ger.)*, 2016, DOI:10.1002/adma.201505597, n/a-n/a.
78. W. Chen, J. Kim, S. Sun and S. Chen, *J. Phys. Chem. C.*, 2008, **112**, 3891–3898.
79. D. Deng, K. S. Novoselov, Q. Fu, N. Zheng, Z. Tian and X. Bao, *Nat. Nano*, 2016, **11**, 218–230.
80. Y. Jiao, Y. Zheng, M. Jaroniec and S. Z. Qiao, *J. Am. Chem. Soc.*, 2014, **136**, 4394–4403.
81. M. Chhowalla, H. S. Shin, G. Eda, L.-J. Li, K. P. Loh and H. Zhang, *Nat. Chem.*, 2013, **5**, 263–275.
82. X. Kong, Q. Chen and Z. Sun, *ChemPhysChem.*, 2013, **14**, 514–519.
83. D. Wei, Y. Liu, Y. Wang, H. Zhang, L. Huang and G. Yu, *Nano Lett.*, 2009, **9**, 1752–1758.
84. C. Huang, C. Li and G. Shi, *Energy & Environmental Science*, 2012, **5**, 8848–8868.
85. X. Li, H. Wang, J. T. Robinson, H. Sanchez, G. Diankov and H. Dai, *J. Am. Chem. Soc.*, 2009, **131**, 15939–15944.
86. N. Li, Z. Wang, K. Zhao, Z. Shi, Z. Gu and S. Xu, *Carbon.*, 2010, **48**, 255–259.

87. D. Usachov, O. Vilkov, A. Grüneis, D. Haberer, A. Fedorov, V. K. Adamchuk, A. B. Preobrajenski, P. Dudin, A. Barinov, M. Oehzelt, C. Laubschat and D. V. Vyalikh, *Nano Lett.*, 2011, **11**, 5401–5407.
88. J. Bai, Q. Zhu, Z. Lv, H. Dong, J. Yu and L. Dong, *Int. J. Hydrogen Energy*, 2013, **38**, 1413–1418.
89. L. Qu, Y. Liu, J.-B. Baek and L. Dai, *ACS Nano*, 2010, **4**, 1321–1326.
90. M. P. Kumar, M. M. Raju, A. Arunchander, S. Selvaraj, G. Kalita, T. N. Narayanan, A. K. Sahu and D. K. Pattanayak, *J. Electrochem. Soc.*, 2016, **163**, F848–F855.
91. Z.-H. Sheng, H.-L. Gao, W.-J. Bao, F.-B. Wang and X.-H. Xia, *J. Mater. Chem.*, 2012, **22**, 390–395.
92. C. Zhang, N. Mahmood, H. Yin, F. Liu and Y. Hou, *Adv. Mater. (Weinheim, Ger.)*, 2013, **25**, 4932–4937.
93. S. Yang, L. Zhi, K. Tang, X. Feng, J. Maier and K. Müllen, *Adv. Funct. Mater.*, 2012, **22**, 3634–3640.
94. I.-Y. Jeon, H.-J. Choi, M. Choi, J.-M. Seo, S.-M. Jung, M.-J. Kim, S. Zhang, L. Zhang, Z. Xia, L. Dai, N. Park and J.-B. Baek, *Scientific Reports*, 2013, **3**, 1810.
95. Y. Zheng, Y. Jiao, L. Ge, M. Jaroniec and S. Z. Qiao, *Angew. Chem. Int. Ed.*, 2013, **52**, 3110–3116.
96. J. Xu, G. Dong, C. Jin, M. Huang and L. Guan, *ChemSusChem.*, 2013, **6**, 493–499.
97. C. H. Choi, M. W. Chung, H. C. Kwon, S. H. Park and S. I. Woo, *Journal of Materials Chemistry A*, 2013, **1**, 3694–3699.
98. X. Li, X. Hao, A. Abudula and G. Guan, *Journal of Materials Chemistry A*, 2016, **4**, 11973–12000.
99. W. Cui, Q. Liu, N. Cheng, A. M. Asiri and X. Sun, *Chem. Commun. (Cambridge, U. K.)*, 2014, **50**, 9340–9342.
100. B. R. Sathe, X. Zou and T. Asefa, *Catalysis Science & Technology*, 2014, **4**, 2023–2030.
101. J. Zhuo, T. Wang, G. Zhang, L. Liu, L. Gan and M. Li, *Angew. Chem. Int. Ed.*, 2013, **52**, 10867–10870.
102. Y. Zheng, Y. Jiao, Y. Zhu, L. H. Li, Y. Han, Y. Chen, A. Du, M. Jaroniec and S. Z. Qiao, *Nature Communications*, 2014, **5**, 3783.
103. S. Chen, J. Duan, M. Jaroniec and S.-Z. Qiao, *Adv. Mater. (Weinheim, Ger.)*, 2014, **26**, 2925–2930.
104. J. Tian, Q. Liu, A. M. Asiri, K. A. Alamry and X. Sun, *ChemSusChem.*, 2014, **7**, 2125–2130.
105. M. Carmo, V. A. Paganin, J. M. Rosolen and E. R. Gonzalez, *J. Power Sources*, 2005, **142**, 169–176.
106. B. Genorio, D. Strmcnik, R. Subbaraman, D. Tripkovic, G. Karapetrov, V. R. Stamenkovic, S. Pejovnik and N. M. Marković, *Nat. Mater.*, 2010, **9**, 998–1003.
107. E. Yoo, T. Okada, T. Akita, M. Kohyama, I. Honma and J. Nakamura, *J. Power Sources*, 2011, **196**, 110–115.
108. M. O. Valappil, S. Alwarappan and T. N. Narayanan, *Curr. Org. Chem.*, 2015, **19**, 1163–1175.
109. Z. Zhelev, R. Bakalova, H. Ohba, R. Jose, Y. Imai and Y. Baba, *Anal. Chem.*, 2006, **78**, 321–330.
110. M. Mallesha, R. Manjunatha, C. Nethravathi, G. S. Suresh, M. Rajamathi, J. S. Melo and T. V. Venkatesha, *Bioelectrochemistry (Amsterdam, Netherlands)*, 2011, **81**, 104–108.
111. L. Feng, Y. Chen, J. Ren and X. Qu, *Biomaterials*, 2011, **32**, 2930–2937.
112. W. Zhao, L. Wang and W. Tan, in *Bio-Applications of Nanoparticles*, ed. W. C. W. Chan, Springer New York, New York, NY, 2007, DOI:10.1007/978-0-387-76713-0_10, pp. 129–135.
113. Y. Bo, H. Yang, Y. Hu, T. Yao and S. Huang, *Electrochim. Acta*, 2011, **56**, 2676–2681.
114. S. Mao, K. Yu, J. Chang, D. A. Steeber, L. E. Ocola and J. Chen, *Scientific Reports*, 2013, **3**, 1696.
115. V. Georgakilas, M. Otyepka, A. B. Bourlinos, V. Chandra, N. Kim, K. C. Kemp, P. Hobza, R. Zboril and K. S. Kim, *Chem. Rev. (Washington, DC, U. S.)*, 2012, **112**, 6156–6214.
116. K. K. Tadi, T. N. Narayanan, S. Arepalli, K. Banerjee, S. Viswanathan, D. Liepmann, P. M. Ajayan and V. Renugopalakrishnan, *J. Mater. Res.*, 2015, **30**, 3565–3574.
117. T. Alava, J. A. Mann, C. Théodore, J. J. Benitez, W. R. Dichtel, J. M. Parpia and H. G. Craighead, *Anal. Chem.*, 2013, **85**, 2754–2759.
118. U. Ghoshdastider, R. Wu, B. Trzaskowski, K. Mlynarczyk, P. Miszta, M. Gurusaran, S. Viswanathan, V. Renugopalakrishnan and S. Filipek, *RSC Advances*, 2015, **5**, 13570–13578.
119. O.-Medina, F. Lopez-Uŕías, H. Terrones, F. J. Rodríguez-Macías, M. Endo and M. Terrones, *J. Phys. Chem. C*, 2015, **119**, 13972–13978.
120. N. N. Tharangattu, S. R. V. Chiranjeevi and A. Subbiah, *Nanotechnology*, 2014, **25**, 335702.
121. S. Alwarappan, A. Erdem, C. Liu and C.-Z. Li, *J. Phys. Chem. C.*, 2009, **113**, 8853–8857.
122. S. J. Rowley-Neale, J. M. Fearn, D. A. C. Brownson, G. C. Smith, Xiaobo Ji and C. E. Banks, *Nanoscale*, 2016, **8**, 14767–14777.
123. M. Velický and P. S. Toth, *Appl. Mater. Today*, 2017, **8**, 68–103.
124. R. K. Biroju, D. Das, R. Sharma, S. Pal, L. P. L. Mawlong, K. Bhorkar, P. K. Giri, A. K. Singh and T. N. Narayanan, *ACS Energy Lett.*, 2017, **2**, 1355–1361.
125. G. B. de-Mello, L. Smith, S. J. Rowley-Neale, J. Gruber, S. J. Hutton and C. E. Banks, *RSC Adv.*, 2017, **7**, 36208–36213.

# CHAPTER 8

# 2D Metal-free Photocatalyst for Hydrogen Generation from Water Splitting

*Xue Jiang,*[a] *Peng Wang*[b] and *Jijun Zhao**

## Introduction

The depletion of fossil fuels and the serious environmental problems associated with their combustion motivate us to search for alternative sources of sustainable, cheap, and clean energy. One such promising energy carrier is hydrogen. As an ultimate source of clean energy, hydrogen production via photocatalytic water splitting under visible-light radiation offers a viable strategy for solving energy and environmental problems simultaneously.

$$H_2O(l) \xrightarrow{h\nu \geq 1.23eV} H_2(g) + \frac{1}{2}O_2(g) \ (\Delta G = 237kJ/mol) \tag{1.1}$$

Water splitting photocatalysts are materials that can photocatalyze the water splitting reaction (Eqn. 1.1) under certain condition. According to Eqn. 1.1, the only production of water oxidation and reduction are oxygen and hydrogen. The latter one promises clean energy carrier with high energy density. However, this reaction is an "uphill" type reaction, which needs the standard Gibbs free energy change of 237 kJ/mol or 1.23 eV. Figure 1 shows a schematic diagram of water splitting into $H_2$ and $O_2$ over photocatalytic reaction.[1] They include light absorption of the semiconductor photocatalyst, generation of the excited charges, recombination of the excited charges,

Key Laboratory of Materials Modification by Laser, Ion and Electron Beams (Dalian University of Technology), Ministry of Education, Dalian 116024, China.

[a] E-mail: jiangx@dlut.edu.cn
[b] E-mail: ppwppy@mail.dlut.edu.cn
* Corresponding author: zhaojj@dlut.edu.cn

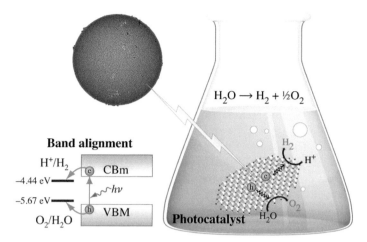

**Fig. 1.** Schematic energy diagram and mechanism of photocatalytic water splitting.[1]

separation of excited charges, migration of the charges, trapping of excited charges, and transfer of excited charges to water or other molecules. All of these processes will directly affect the total amount of hydrogen generation from the semiconductor photocatalyst system.

Taking into account the basic mechanism and processes of photocatalytic water splitting, there are many keys to developing a suitable high-efficiency semiconductor for the visible light driven photocatalytic splitting of water into $H_2$ and $O_2$: (1) the band gap must exceed the free energy of water splitting of 1.23 eV, and be smaller than about 3 eV for effective utilization of the maximum portion of the solar visible light; (2) the water oxidation ($H_2O/O_2$) and reduction ($H^+/H_2$) potentials must lie between conduction band minimum (CBM) and valence band maximum (VBM); (3) a small exciton binding energy facilitates the splitting of excitons into free charge carriers; (4) the surface area, porosity or reactive facets should be large enough for high reactivities;[2-5] (5) good motilities of charge carriers; (6) Recombination of the photogenerated electron and hole waste energy in the form of heat or light.[6-7] In addition to all these issues, low cost, abundant and environmentally-friendly materials are necessary for sustainable energy production.

Since the discovery of the first photocatalytic water splitting system on $TiO_2$ electrodes by Fujishima and Honda in 1972,[8] a wide range of semiconducting materials have been explored to function as photocatalysts for hydrogen production, which have be reported in the previous reviews.[9-13] Generally, there are three overall deficiencies of current reported water splitting photocatalytic materials: (1) Most of them are based on transition metal oxides, nitrides, and carbides with $d^0$, $d^{10}$, or $f^0$ configurations, along with group VA or VIA ions as counter-anion components.[9] High cost and dissolved metal ions of the photocatalyst materials actually hinder the wide applications of photocatalysis. In the drive towards green and sustainable chemistry, photocatalysts without metal elements have been highly anticipated for hydrogen production; (2) Most of water splitting photocatalytic materials have a band gap larger than 2.7 eV (450 nm), which is too high for achieving the viable electrochemical level under solar illumination. (3) Most of them suffer from low quantum efficiency in the visible range, with solar-to-hydrogen conversion efficiencies less than 0.1%. Therefore, the key issue of hydrogen production is to develop new photocatalysts with nontoxic or precious metals, lower band gap, and higher efficiency.

Due to the limited thickness, tunable electronic properties, high mobility, and large surface area, two dimensional (2D) metal-free materials are very promising candidates for developing appropriate photocatalysts mentioned above. They stand out with many advantages: (1) Metal free overcome the limitation of high cost and low stability of metal elements; (2) Limited thickness minimize the migration distance for the generated electrons and holes, thus enhancing the photocatalytic performance by reducing the possibility of electron-hole recombination;[1] (3) Tunable electronic properties determine the vital points in semiconductor photocatalysts: ideal band gap and the energy level of the conduction and valence bands; (4) High mobility of charge carriers improved e-h pairs separation; (5) Photocatalysts with large specific surface area will be good for the adsorption and dissociation of the water molecular. These systems are prominently represented by light-weight elementary semiconductors and carbon-based materials, including graphite-like carbon nitride (g-$C_3N_4$), graphene oxide (GO), organic materials, phosphorene, and so on. They have shown good performance in both the reduction and oxidation reaction. For example, Liu et al.[14] very recently reported a quantum efficiency of 16% for the carbon nanodot-2D carbon nitride nanocomposite, which meets the requirement of the DOE price target for $H_2$ generation.

This chapter summarizes the recent progress on the design and fabrication of 2D metal-free photocatalysts for hydrogen production from water splitting under visible light. As shown in Fig. 2, a number of representative metal-free photocatalytic materials will be addressed, including elemental semiconductor phosphorene, g-$C_3N_4$, graphene oxide, B-C-N hybrid sheets, 2D organic materials, and so on. On the other hand, the universal origin of photocatalytic mechanism and the modification of existing 2D metal-free photocatalysts to improve their photocatalytic activity by various strategies will be also discussed. We aim to give comprehensive references for the scientists

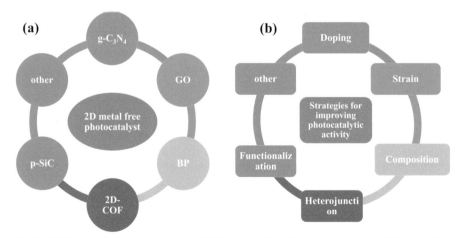

**Fig. 2.** (a) 2D metal free photocatalysts and (b) Strategies for tailoring the properties of 2D metal free photocatalysts for $H_2$ production from photocatalytic water splitting.

in the related fields or the graduate students who want to know photocatalysts for hydrogen generation.

## Two Dimensional Metal-free Photocatalytic Materials

### *Graphite carbon nitride as a photocatalyst*

Polymer graphitic carbon nitride is a layered material similar to graphene, which is composed of only C, N, and H. The discovery of graphitic carbon nitride could be traced back to 1834, when polymeric melon was reported by Liebig.[15] Subsequently, there are several phases of carbon nitride that have been found by scientists. As the most stable allotrope, graphite carbon nitride (g-$C_3N_4$) has attracted much attention. Generally, C, N, and H can form two configurations of g-$C_3N_4$ reported in the previous literatures,[16] that is s-triazine-based gt-$C_3N_4$ and s-heptazine-based gh-$C_3N_4$ (Fig. 3a). Both of them are formed by N-bridged triazine or heptazine with strong C-N bonds. However, the difference building blocks of gt-$C_3N_4$ and gh-$C_3N_4$ will consequently bring out diverse electronic and optical properties. The heptazine-based gh-$C_3N_4$ was demonstrated to be a semiconductor with a direct medium band gap of 2.7 eV from the ultraviolet–visible spectrum and density-functional-theory band structure in 2009 (Fig. 3b and Fig. 3c).[5] Besides the large band gap, the reduction and oxidation levels for $H_2O$ just locate inside the valence and conduction band of g-$C_3N_4$, which are suitable for hydrogen evolution. For the triazine-based g-$C_3N_4$, the band gap is deduced to be 1.6–2.0 eV from UV/Vis measurements and DFT results (Fig. 3d and Fig. 3e) later in 2014.[17] Due to the better stability and easier processibility, heptazine-based gh-$C_3N_4$ is investigated in more detail in the previous work. Hence, g-$C_3N_4$ usually refers to the heptazine-based graphite carbon nitride.

Due to its special geometry configuration and electronic structure, Wang et al. pioneer investigated the photocatalytic activity of native g-$C_3N_4$ under visible light in 2009.[5,18] Under different thermal condensation conditions, they first prepared a series of g-$C_3N_4$ by heating cyanimide to temperatures between 673 and 873 K. They have diverse geometry, electronic, and optical properties. For the material condensed at 823 K, an in-planar repeat period of 0.681 nm in the crystal is evidenced from the X-ray powder diffraction (XRD) pattern. Then, using the powder formed g-$C_3N_4$ prepared, they were surprised to find that steady $H_2$ evolution from water under visible light could be achieved with triethanolamine as a sacrifice agent, even without noble metal as co catalyst. Moreover, they also observed Pt loading on g-$C_3N_4$ was very helpful in enhancing the photocatalytic activity, which would result in the quantum efficiency (QE) of approximately 0.1% at 420–460 nm. Although the hydrogen production activity of g-$C_3N_4$ was low in this paper, this finding opened a new pathway to searching for promising photocatalytic water splitting materials among countless 2D metal free materials family.

Since being disclosed as a promising metal free photocatalyst, a tremendous amount of research and strategies were devoted to improve the hydrogen evolution rate of g-$C_3N_4$. So far, the efficiency of native g-$C_3N_4$ was improved step by step from 0.1% to 50.7%,[19] which is closely meeting the requirement of the DOE price target for $H_2$ generation. For example, sulfur-doped g-$C_3N_4$ with a unique electronic

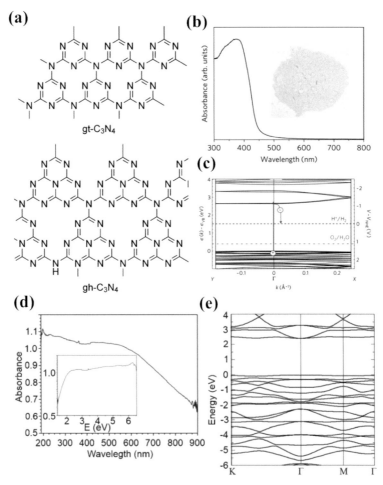

**Fig. 3.** (a) Two graphitic carbon nitrides discussed in literature based on s-triazine and s-heptazine building units.[16] (b, c) Ultraviolet–visible diffuse reflectance spectrum of the polymeric carbon nitride and DFT calculated band structure for gh-$C_3N_4$.[5] (d, e) UV/Vis diffuse-reflectance spectrum and DFT calculated band structure for gt-$C_3N_4$.[17]

structure showed a photoreactivity of $H_2$ production about 8 times higher than pure $C_3N_4$ under incident wavelength from 300 to 420 nm.[4] By introducing mesoporosity (67 nm–373 nm) into pristine g-$C_3N_4$, Wang et al.[20] (2012) observed that the efficiency of hydrogen evolution from the photochemical reduction of water could be improved by 8.3 times. After that, Schwinghammer et al. (2013) reported poly(triazine imide) (PTI/Li⁺Cl), a g-$C_3N_4$ structure based on triazine building blocks.[21] Using 4-amino-2,6-dihydroxypyrimidine as a dopant, a carbon- and oxygen-enriched PTI was obtained, and the QE for hydrogen generation reached as high as 3.4%. Doping with aromatic motifs was also used to enhance the catalytic activity of heptazine-based graphite carbon nitride. Selecting four typical organic agents, barbituric acid (BA), 2-aminobenzonitrile (ABN), 2-aminothiophene-3-carbonitrile (ATCN), and diaminomaleonitrile (DAMN) as co-monomer, Zheng et al. (2014) prepared g-$C_3N_4$ modified by aromatic hetero-structures by a simple one-pot chemical condensation of urea.[22] It is found that hetero-molecules can act as the chromophore center to harvest photons and efficiently reduce the charge transfer resistance. Owing to the red-shifted optical absorption and improved charge transporting, the photocatalytic $H_2$ evolution rate is significantly improved. The apparent quantum efficiency (AQE) of an optimized Pt and ATCN modified g-$C_3N_4$ is determined to be about 8.8% at 420 nm.

Recently, the quantum efficiency was further elevated to a much higher level. Martin et al.[23] (2014) synthesized a structure-controlled g-$C_3N_4$ with low-cost urea as a precursor. The material showed reproducible and excellently stable catalytic activity for $H_2$ evolution, resulting in a QE of 26.5% at 400 nm. From XPS analysis and time dependent DFT simulations, a protonation mechanism was suggested to explain the extraordinary photocatalytic activity. Liu et al.[14] reported that a combination of carbon nanodot and carbon nitride could constitute a high performance composite photocatalyst for water splitting via a stepwise two-electron/two-electron process, and the QE could reach up to 16% under light irradiation of wavelength $\lambda = 420 \pm 20$ nm. Similarly, by implanting metal cobalt hydroxide nanoparticles on g-$C_3N_4$ as a co-catalyst, Li et al.[24] (2016) found that the strong interaction could enhance the transfer rate of the photogenerated electron to reduce the carrier recombination. The light harvesting can be effectively improved after co-sensitized by Eosin Y and Rose Bengal dyes. And the achieved AQE for $H_2$ evolution reached 29.6% and 27.3% at 520 nm and 550 nm, respectively. Very recently, Lin et al.[19] (2016) reported a record-breaking AQE of 50.7% at 405 nm, which is the highest reported value of QE for hydrogen production by graphite carbon nitride-based photocatalysts. Preheated melamine was used as a precursor to synthesize crystalline tri-s-triazine based (heptazine-based) semiconductors. The characterization indicated that this synthetic method could improve the crystallinity of the sample and enhance the photo generated charge carrier mobility, as well as increase the surface area. When triethanolamine was used as a sacrificial agent, with addition of phosphate to mimic natural photosynthetic environment, the AQE of the biomimetic photocatalytic hydrogen evolution system was determined to be 50.7% at 405 nm.

## *Graphene oxide as a photocatalyst*

Graphene oxide (GO) is a monolayer of graphite oxide, which can be obtained by exfoliating graphite oxide into layered sheets through sonicating, thermal, and mechanical stirring.[25] Even though GO has been known for over 150 years, it has recently attracted resurgent interests since it is an important material to massively produce graphene with low cost, and has easy processibility and compatibility with various substrates.[26] A number of spectroscopic techniques and theoretical simulation have been carried out to determine the atomic structure of GO.[25,27] GO are composed of two primary regions: (1) the hydrophobic $\pi$ conjugated $sp^2$ domains, and (2) the $sp^3$ domains with hydrophilic oxygen-containing functional group. A structural model with carbonxylic acid at the edges, and epoxides and hydroxyls orderly arranged in a chained manner is widely accepted to be thermodynamically stable for GO.[28]

The fundamental physical properties of GO are closely related to the coverage, ratio, and arrangement of the functional groups. Typically, GO is insulating due to the large portion of sp[3] hybridized carbon atoms bonded with the oxygen-containing groups, which results in a sheet resistance of ~ $10^{12}$ $\Omega$sq$^{-1}$ or higher.[29] However, after reduction, the sheet resistance of reduced GO can be degraded by several orders of magnitude, hence transforming the material into a semiconductor, or even into a graphene-like semimetal.[30] For example, Guo et al.[31] (2012) demonstrated that oxygen contents of GO in the reduced region could be tailored by changing the laser power. Thus, the band gap of reduced GO was precisely modulated from 0.9 eV to 2.4 eV. On the other hand, Ito et al.[32] (2008) studied the GO with only epoxide groups using first-principles calculations. They found that as O/C ratio increase from 0 to 50%, the band gap increases significantly from 0 to 3.39 eV. A large range of 0–4.0 eV by varying the coverage of oxygen-containing groups is also supported by Liu et al.'s (2012) density functional theory simulations.[33] As shown in Fig. 4, the band gaps between the conduction band minimum (CBM) and valence band maximum (VBM) of the ordered GO structures with coverage of 20, 40, and 50% are 1.22, 1.92, and 2.06 eV, respectively, while for the disordered GO structures are 1.77, 0.80, and 0.53 eV, showing a clear increasing with coverage. In addition, the electronic property of GO can be affected by ratio of the OH and O functional groups. As the OH:O ratio increases, the band gap increases accordingly due to enhanced degree of sp[3] hybridization.[33]

The tunable band gap also leads to the tunable optical property of graphene oxide. Li et al.[34] (2008) found the absorption peak of the GO dispersion at 231 nm gradually red shifts to 270 nm, and the absorption region increases with reaction time. Moreover,

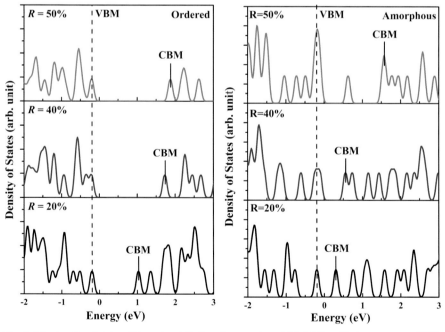

**Fig. 4.** The density of states of the ordered (left) and armouphous (right) GOs with the coverage of 20%, 40%, and 50%, respectively.[33]

Johari and Shenoy[35] (2011) investigated the effects of different functional groups on the optical properties of GO structures by performing DFT calculations. A series of GO structures with different coverage and functional groups (epoxy, hydroxyl, and carbonyl) are constructed and studied. According to the calculated electron energy loss spectra (EELS), the concentration of epoxy and hydroxyl functional groups varies from 25 to 75%, and the $\pi + \sigma$ plasmon peak shows a clear blue shift of about 1.0−3.0 eV. Mathkar et al.[36] (2012) suggested that the optical gap of GO can be tuned from 3.5 eV down to 1.0 eV with a concurrent increase of C/O ratio.

The tunable band gap and optical properties make GO materials are highly active at visible spectrum. Moreover, GO has mobility of charge carriers of 10 cm$^2$/Vs and conductivity of 10000 S/m, which are also sufficiently large to migrate the electrons fast toward surface for efficient photocatalytic application. As a result, GO materials could be potential candidates for photocatalytic applications. Indeed, such application have been supported by many experimental and theory research groups. Yeh et al.[37] (2010) proved the photocatalytic H$_2$ evolution activity of GO with a band gap of 2.4–4.3 eV for the first time. Electro-chemical analysis along with the Mott-Schottky equation illustrated that GO exhibited stable H$_2$ generation from an aqueous methanol solution or pure water, even in the absence of Pt cocatalyst under mercury light irradiation (Fig. 5). The quantum efficiencies of hydrogen production were calculated to be 2.7% and 0.010% for the mercury lamp and visible-light irradiations from the methanol solution.

After that, Yeh et al.[38] (2011) have further investigated the photocatalytic activity of GO with various oxidation levels. An inverse relationship between the amount of H$_2$ evolution and the population of the oxygen-containing groups on the GO sheets have been build. They concluded that GO with higher oxidation degree had a larger band gap and limited absorption of light, thus exhibiting a lower photocatalytic activity than the GO with lower oxidation degree. In addition, Matsumoto et al.[39] (2011) have reported the photoreactions to generate H$_2$ from an aqueous suspension of GO nanosheets under UV irradiation. Their findings have also shown that the GO with an appropriate reduction level can serve as a photocatalyst for H$_2$ production.

Considering the p-type conductivity hinders hole transfer for water oxidation and suppresses O$_2$ evolution, Yeh et al.[40] (2013) have further introduced amino and amide groups to GO surface. Their results demonstrated that the ammonia-modified GO exhibits n-type conductivity, and simultaneously catalyzes H$_2$ and O$_2$ evolution.

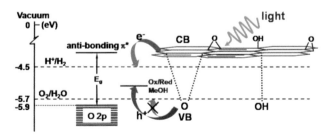

**Fig. 5.** Schematic energy-level diagram of GO relative to the levels for H$_2$ and O$_2$ generation from water. The top-of valence energy level was obtained using density functional theory for GO with 12.5% O coverage.[37]

For further study of the overall photocatalyst of GO, Yeh et al.[41] (2014) synthesized nitrogen-doped graphene oxide quantum dots (NGO-QDs) as the catalyst to fulfill the evolution of $H_2$ and $O_2$ at a molar ratio of approximately 2:1. This could be explained that p-n diodes configuration of NGO-QDs resulted in an internal Z-Scheme charge transfer for effective reaction at the QD interface. For this reason, visible light irradiation on the NGO-QDs bring about simultaneous $H_2$ and $O_2$ evolution from pure water.

To explain these experimental observations about GO as promising visible light driven photocatalyst materials and to clarify the optimal composition of GO for higher photocatalytic activity, Jiang et al.[42] (2013) have studied the electron properties responsible for photocatalytic water splitting using density functional theory calculations, especially on the effect of epoxy and hydroxyl functionalization on the work function, band gap, CBM/VBM position, and optical absorption spectra. The structures adopted in their research were followed by the stable GO structures with different OH:O ratio and coverage, which is confirmed by previous theoretical studies.[43–44] The epoxy and hydroxyl groups aggregate along armchair direction and form stable one-dimensional chain configurations on the basal plane.

With the coverage increasing, they have observed band gap of epoxy and/or hydroxyl functionalized graphene sheets are indeed continuously tunable from metallic to insulating. Meanwhile, the work function of GO increases with coverage rate, which is primarily due to hybridization between carbon atoms and functional species, as well as formation of hydrogen bonds. By varying the coverage and relative ratio of the epoxy and hydroxyl group, both the band gap and work function of the GO can meet the requirements of the photocatalyst. More importantly, they have aligned the redox energy levels of GOs with respect to the water oxidation/reduction potential levels, and simulated the optical absorption spectra, which are displayed in Fig. 6a and Fig. 6b. Their results show that the electronic structures of GO materials with 40–50% (33–67%) coverage and OH:O ratio of 2:1 (1:1) are suitable for both reduction and oxidation reactions for water splitting. Among these systems, the GO composition

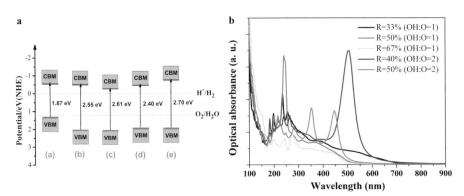

**Fig. 6.** (a) Site levels of VBM and CBM for OH:O = 1 and OH:O = 2 graphene oxide with different coverage rate. The dot lines are standard water redox potentials. The reference potential is the vacuum level. (b) Total calculated optical absorption curves for GO with OH:O = 1:1 and OH:O = 2:1 under different coverage rates, respectively.[42]

with 50% coverage and OH:O (1:1) ratio can be a very promising material for visible light driven photocatalysts.[42]

## 2D covalent organic frameworks as photocatalyst

2D covalent organic frameworks (2D-COF) are a class of porous crystalline materials constructed by linking organic secondary building blocks with covalent bonds to produce predetermined structures, which feature high surface areas, excellent thermal stability, and extremely low density. 2D-COF differ from 2D polymers with intrinsic microporosities due to their crystalline nature. Generally, the approaches to obtain 2D-COF can be summarized as solvothermal, ionothermal, and microwave methods.[45] Methods for growing COFs onto crystalline metal surfaces via the sublimation of building units have also been developed.[46]

Based on those synthesized methods described above, lots of high qualitied 2D-COF materials have been successfully prepared under optimal reaction conditions.[45] Generally, they can be classified into three categories: boron-containing, triazine-based, and imine-based COFs. For example, Côté et al.[47] ingeniously constructed the first 2D COF materials, namely 2D COF-1, which belongs to one type of boron-containing COFs and was synthesized through the self-condensation reaction of 1,4-benzenediboronic acid. It exhibits a layered staggered structure with surface area of 711 $m^2g^{-1}$ and an average pore size of 0.7 nm. The other example of boron-containing COFs is 2D HHTP-DPB COF, which has also been successfully obtained by the co-condensation of 2,3,6,7,10,11-hexahydroxytriphenylene (HHTP) and 4,40-diphenylbutadiynebis (boronic acid).[48] It possesses the pore size of 4.7 nm. Triazine-based COFs (CTFs) were first developed by Thomas' group,[49] which are based on the cyclotrimerisation of nitride building units in the presence of $ZnCl_2$. Cyclotrimerisation of 1,4-dicyanobenzene (DCB) affords the representative CTF-1 material with a BET surface area of 791 $m^2g^{-1}$ and a pore size of 1.2 nm. The imine-based 2D COFs have been synthesized by Wang et al.[50] via the co-condensation of 1,3,5-triformylbenzene and 1,4-diaminobenzene in a pioneering manner. COF-LZU1 possesses a 2D eclipsed structure with the layer distance of 3.7 Å.[50]

Motivated by synthesizing the high crystallized 2D COFs, the optical and electrical properties have been further investigated. Generally, 2D COFs have been predicted to be insulating or semiconducting. For example, Jiang et al.[51] found their synthesized 2D TP-COF is highly luminescent with a capacity of harvesting photons from ultraviolet to visible regions, while PPy-COF is electrically conductive, revealing photoconductivity with a quick response to light irradiation. In addition, Lotsch's group[52] have developed the hydrazone-based COF (TFPT-COF), which is featuring mesopores of 3.8 nm in diameter. They estimated its optical band gap of roughly 2.8 eV from the absorption edge based on the Kubelka-munk function. Theoretically, Lukose et al.[53] have investigated the electronic properties of a set of reported and hypothetical 2D COFs using density functional based tight-binding (DFTB) method, including COF-1, COF-2, COF-1M, COF-2M, COF-3M, PPy-COF, COF-5, COF-10, COF-8, COF-6, TP-COF, COF-4M, COF-5M, COF-6M, COF-7M, TP COF-1M, COF-8M, COF-9M, COF-10M, COF-11M, TP, and COF-2M. Their band gaps are not only confirmed by the previous experimental results, but also found to be in the range of 1.7–4.0 eV.

Interestingly, the electronic and optical properties might be tailored by the chemical properties of the building blocks, the unit of linkers, and the topology of networks. The monomer dependent band gaps originate from the concept of reticular chemistry.[54–55] That is, the band gap of 2D-COFs is determined by the smallest HOMO-LUMO gap of the constituting molecules. For example, phthalocyanine molecules have strong absorption capacities of visible light. The planar $\pi$-electronic macrocycles make them fascinating building units for constructing functional 2D COF materials. Using a Lewis acid-catalysed protocol, Spitler and Dichtel[56] (2010) synthesized an eclipsed 2D COF (Pc-PBBA-COF) containing metal-free phthalocyanine groups. Phthalocyanine-based Pc-PBBA-COF only shows a slight blue shift with the absorption maxima comparable to the phthalocyanine tetra (acetonide) powder, and shows an absorption band of most of the solar spectrum.

The optical gap could be also determined by the choice of monomers. Porphyrins play a central role in virtually all natural photosynthesis processes due to the large photoabsorption coefficient, and rather efficiently convert solar into electric energy under visible light.[57–59] Bearing this in mind, Jiang et al.[60] synthesized a series of 2D porphyrin COFs with different porphyrin derivatives using three component condensation systems, namely CuP-DHPh COF, H2P-DHPh COF, and NiP-DHPh COF. All those 2D porphyrin COFs are still semiconductors with a rather similar band gap of 1.36, 1.31, and 1.54 eV, respectively.

The linker unit size is another important factor that affects the electronic properties of 2D COFs. Due to the quantum confinement effect, the band gap of the unit becomes smaller as the size of the linked units increases. For example, the geometry of TP-COF, having the same building blocks as COF-5, is formed by incorporating a PDBA unit in place of PBBA linker into COF-5 structures. TP-COF and COF-5 have a relatively modest difference in the electronic character. The VBM of the PDBA unit in TP-COF shifts to higher energy by about 0.8 eV, in contrast to that of the PBBA in COF-5, and the CBM shifts to lower energy by about 0.2 eV.[61]

A recent study by Sakaushi and co-workers[62] used first principles calculations and electrochemistry to confirm the density of states of triazine-based frameworks, which proved the topology induced band gap tuning effects. In that work, the electronic structure of CTF-1 and the CTF synthesized using TCPB were measured by electrochemical methods and compared with the results of first-principles calculations. As CTFs are composed of triazine rings as electron acceptors (A) have an electron affinity of +0.46 eV, and benzene rings as electron donors (D) have an electron affinity of −1.15 eV, the electronic structures of CTFs depend on the ratio of the different aromatic rings. The band gap of CTF-1 and CTF-TCPB are 2.7 eV and 1.1 eV, corresponding to ratio of triazing and benzene rings of 1:4 and 1:5.[62] In addition, Shenoy et al.[63] have observed the interlayer coupling effect for 2D DA-COF based on density functional theory calculations. They found that the direct band gap (1.73 eV) in monolayer will shift to an indirect band gap (0.25 eV) in the multiple layers.

Those results suggest that one can start to control the electronic properties and optical properties of 2D-COFs in a wider range by judicial selection of composition structures, monomers, and functional groups. Besides the tunable optical and electronic properties, the effective charge separation have also enhanced due to $\pi$ stacking structure of 2D-COFs, which provide defined pathways for charge carrier transport. Yaghi et al.[50] have reported the integration of porphyrins into covalent organic frameworks (COFs), and have successively synthesized two porphyrin units based COFs (COF366 and COF66), which were determined to be hole conducting with motilities as high as 8.1 and 3.0 $cm^2V^{-1}s^{-1}$. They have also found that the lifetimes of the charged species for both COFs are ~ 80 us or even longer, in spite of the higher mobility of the charge carriers, which is the primary factor in the promotion of the effective charge carrier separation. Recently, a synthesized 2D DA-COF with donor and acceptor functional groups segregated in the basal plane has been experimentally[64] and theoretically[64] demonstrated to possess excellent carrier mobility. The mobility along vertical direction is 0.04 and 0.01 $cm^2 V^{-1}s^{-1}$ for electrons and holes, respectively. The effective mass of the carrier is also very high in plane of the DA-COF. The electron effective mass of monolayer DA-COF is 0.523 $m_e$, which is higher than the traditional semiconductors of Si (0.2 me) and GaAs (0.067 me), while the holes' effective mass of carriers in DA-COF is 21.225 $m_e$, which is two order magnetite higher than 0.16 $m_e$ of Si and 0.082 $m_e$ of GaAs.

All these advantages addressed above are implied by the 2D-COF to have the potential to be a new star metal free photocatalysts. Incorporating triazine units into 2D-COF systems successfully is the first step towards such photocatalyst applications, which correlate with efficient hydrogen evolution as the triazing nitrogen, with its free electron pair and electron poor character, which may act as the active site.[5] In addition, the triazine rings also offer the 2D-COF high chemical and thermal stability, which is preferred for photocatalysis.

Indeed, most 2D organocatalysts found so far are triazine-incorporated COF. The representative ones are the covalent triazine frameworks (CTF-0, CTF-1, and CTF-2), which have been synthesized experimentally[49,65,66] and firstly theorized to investigate as a new class of photocatalytically active 2D COF for light-driven water splitting by Jiang et al.[67] (2015). Some key physical properties that determine photocatalytic activity, including the electronic band structure, work function, CBM/VBM position, and optical absorption spectra, have been evaluated for 2D-CTFs using first-principles calculations. Taking CTF-0 as a starting point, the calculated band gap is 2.49 eV at PBE level and 3.32 eV with HSE06 hybrid functional (Fig. 7a). The partial density of states (PDOS) is shown in Fig. 7b. Almost all conducting states in the vicinity of the Fermi level originate from triazine nitrogen atoms. The band edge alignment of CTF-0 calculated with HSE06 functional is presented in Fig. 7c, which indicates that the CBM is 1.408 eV more negative than the reduction potential of $H^+/H_2$ and VBM is 0.682 eV more positive than the oxidation level of $O_2/H_2O$. The optical absorption spectra of CTF-0 further demonstrate that they meet the criteria for efficient visible-light absorption (Fig. 7d).

The band gap of 2D-CTFs can be enhanced via tailoring nitrogen concentration, which is 2.49 eV for CTF-0, 2.42 eV for CTF-1, and 2.07 eV for CTF-2, corresponding to the nitride content of 20%, 14.3%, and 10%, respectively. Moreover, the band gap reduces from 2.49 eV to 2.28 eV with increasing number of layers monotonically (N = 1–6). To evaluate the photocatalytic capability, the energy levels of CTF-0,

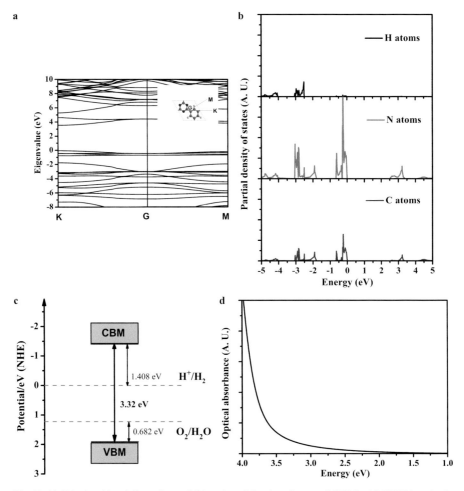

**Fig. 7.** (a) Calculated band dispersion and (b) projected density of states of CTF-0 with HSE06 methods, the valence band maximum is set to zero. (c) Band alignments of CTF-0 with respect to the standard water redox potentials. (d) Optical absorption of CTF-0.[67]

CTF-1, CTF-2, and multilayer CTF-0s with respect to the water oxidation/reduction potential levels are shown in Fig. 8a and Fig. 8b. Both band alignments indicate that tuning nitrogen concentration and stacking 2D-CTFs are effective means to improve the flow of the carrier and reduce the band gap, which reveal the great potential of CTF-0, CTF-1, CTF-2, which are a new generation of efficient visible light photocatalysts.

Inspired by this theoretical work, other 2D COFs containing triazine rings that act as photocatalysts for splitting of water have been studied experimentally. Bi et al.[68] (2015) have found the absorption edge of CTF-T1 located at about 422 nm, corresponding to a band gap of 2.94 eV, based on the Kubelka-Munk function (Fig. 9a). For comparison, the time course of $H_2$ evolution from water by CTF-T1 and g-$C_3N_4$ under visible light irradiation are shown in Fig. 9b. One can clearly see the amount of $H_2$ production of CTF-T1 are comparable to the g-$C_3N_4$.

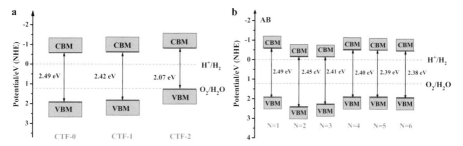

**Fig. 8.** Band edge alignments of CTF-0, CTF-1, and CTF-2 (a), and CTF-0 with AB stacking (b), respectively. The dot lines are standard water redox potentials. The reference potential is the vacuum level.[67]

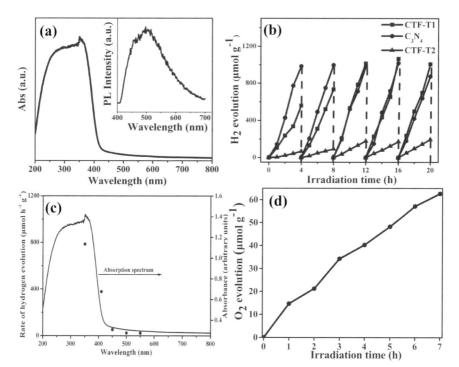

**Fig. 9.** (a) UV-vis diffuse reflectance spectrum and photoluminescence spectra with 360 nm light excitation at 298 K. (b) Time course of $H_2$ evolution from water by CTF-T1 and g-$C_3N_4$ under visible light irradiation ($\lambda \geq 420$ nm). (c) Wavelength dependent $H_2$ evolution from water by CTF-T1. (d) Time course of $O_2$ evolution from water by RuO$_2$ loaded CTF-T1 under visible light irradiation ($\lambda \geq 420$ nm).[68]

As shown in Fig. 9c, the $H_2$ evolution rate decreased along with the increasing wavelength of incident light, which matched well with light absorption in the optical spectrum (Fig. 9a). The quantum efficiency of $H_2$ evolving half reaction for CTF-T1 at 400–440 nm is about 2.4%. The other $O_2$ evolving half reaction for CTF-T1 also occurred using RuO$_2$ assistance (Fig. 9d). However, the activity of $H_2$ evolution is much smaller than that of $H_2$ evolving. Stegbauer et al.[52] (2014) also reported a hydrazine-based 2D-COF (TFPT) to produce hydrogen from water without signs of degradation when illuminated with visible light. Using aqueous triethanolamine solution as a sacrificial donor, the amount of hydrogen evolved in the first five hours was as high as 1970 umolh$^{-1}$g$^{-1}$, corresponding to a quantum efficiency of 2.2%, while maximum quantum efficiency of up to 3.9% was obtained for individual batches. However, no O$_2$ could be detected regrettably. In addition, Lotsch and coworker[69] found highly photoactive phenyl-triazine oligomers (PTO), which have a higher crystallinity as compared to CTF-1 obtained by lower temperature. The most active PTO samples efficiently and stably reduce water to hydrogen with an average rate of 1076 ± 278 umolh$^{-1}$g$^{-1}$ under simulated sunlight illumination, which is competitive with the best carbon nitride-based and purely organic photocatalysts.

2D COFs for photocatalytic hydrogen production beyond triazine rings have been firstly suggested by Jiang et al.'s theoretical work.[70] To design two dimensional (2D) organocatalysts, they constructed a series of covalent organic framework (COF) using bottom-up strategies, i.e., molecular selection, tunable linkage, and functionalization. First-principles calculations are performed to confirm their photocatalytic activity under visible light. Two of our constructed 2D COFs models (COF-B1 and COF-C3) are identified as sufficiently efficient organocatalyst for visible light water splitting. COF-B1 consists of boroxine rings and diazoniums linkage, while COF-C3 is constructed by the same diazoniums linkage and substitutes amine and cyano functionalized benzene ring. Their theoretical prediction provides essential insights into synthesizing more high-efficiency 2D-COF photocatalysts among the countless 2D polymers with other building blocks.

### *Other 2D metal free photocatalysts*

In addition to the above mentioned three types of 2D materials, other newly predicted 2D materials, such as C$_2$N,[71] BCN,[72] Si pentagonal monolayer p-SiX (X = B, C, and N),[73] elementary phosphorene,[74] as well as their derivatives, have also shown great potential for visible-light photocatalysis application. For example, the thinnest layered 2D crystal named C$_2$N, with uniform holes and nitrogen atoms, have been simply synthesized via bottom up wet chemical reaction[75] (Fig. 10a). Its optical band gap is about 1.96 eV. After that, Yang's group[71] performed first principles calculation to investigate monolayer and multilayer C$_2$N as a potential photocatalyst for water splitting. Both the monolayer and multilayer C$_2$N have a direct band gap (1.84–2.47 eV) and good visible light adsorption (Fig. 10b and 10c). The band alignments of monolayer and multilayer C$_2$N with respect to the water redox levels show that the layered C$_2$N are satisfied for the overall water splitting (Fig. 10d). More importantly, unlike g-C$_3$N$_4$ with many localized band edges in band structure which is considered to be the reason for its low quantum efficiency, the valence and conduction bands for monolayer C$_2$N are well dispersed, and no localized states are present to act as recombination centers for the photogenerated electron hole pairs. Hence, we can infer the lifetime of the photogenerated charge carriers in monolayer C$_2$N will be longer, which is very important to obtain a better photocatalytic efficiency than pristine g-C$_3$N$_4$.

Another example shown here is a new family of Si-based pentagonal monolayers p-SiX (X = B, C, and N), which is constructed on the basis of the okayamalite structure by means of first principle calculation.[73] Electronic structures by HSE06 hybrid

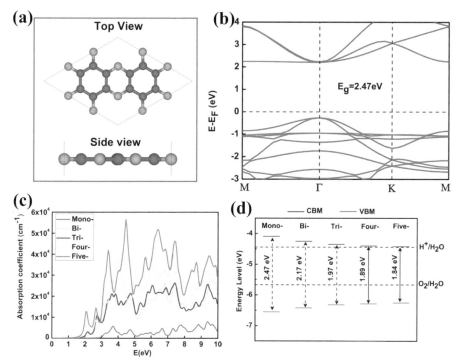

**Fig. 10.** (a) Top view and side view of the atomic structure of monolayer $C_2N$; (b) Band structures of monolayer $C_2N$ calculated by HSE06; (c) Optical spectra for few layer $C_2N$; (d) band edge position of monolayer and multilayer $C_2N$. The dashed lines are water redox potential.[71]

functional show that one of them, p-SiC, is a semiconductor with a band gap of 2.35 eV. For p-SiC, the electron mobility at the CBM along x direction is about $2.5 \times 10^3$ $cm^2V^{-1}s^{-1}$, while the hole mobility at the VBM is $1.9 \times 10^3$ $cm^2V^{-1}s^{-1}$ and $1.3 \times 10^3$ $cm^2V^{-1}s^{-1}$ for two different degenerate edges at the VBM. This carrier mobility can be compared to, or may even be higher than that of many other 2D semiconductors, such as $MoS_2$, phosphorene and 2D organic materials. The big difference between high carrier mobilities of electron and hole will decrease the recombination of photogenerated electron-hole pairs greatly. More meaningfully, their results indicated that both band gap and band edges of p-SiC can meet the requirement of the reduction and oxidation levels in water splitting. All those results reflected that p-SiC can be a potential photocatalyst for water splitting beyond g-$C_3N_4$ or other traditional photocatalysts with high carrier mobility and low recombination rate.

The last fascinating example shown here is monoelementary 2D phosphorene, which has also been observed as a class of efficient photocatalyst, due to its unexpected optoelectronic properties and high carrier mobility. Sa et al.[74] first unraveled it to be a photocatalyst in the application of water splitting hydrogen production in 2014. Its good stability in liquid water, suitable band gap (1.54–1.82 eV), and band edge alignments demonstrated its ability in hydrogen and oxygen production. At the same time, they have also found the water splitting process on phosphorene is energy favorable, and the water oxidation and reduction process will take place on different sides of phosphorene separately. Moreover, Zhou et al.[76] have also proposed its ability of photocatalytic solar water splitting are still retained even if it's cut to zero dimensional quantum dots. The quantum dots the consider yield band gap of 1.36–2.52 eV and 2.14–2.85 eV for black and blue phosphorus quantum dots.

# Strategies for Photocatalytic Performance Improvement

Based on the main process of photocatalytic water splitting, the $H_2$ or $O_2$ gas evolution rate of a photocatalyst depends on its ability of light harvesting, charge separation, and charge utilization. Recently, there have been many strategies proposed to improve the efficiency of light harvesting, charge separation, and charge utilization for the 2D metal free photocatalysts. For example, the band gap structure and band offset of some 2D semiconductors can be modified by elemental doping, chemical functionalization, and mechanical loading, which tune light adsorption and charge utilization property of the photocatalyst. Heterojunction and composition design of 2D metal free materials with other components have also been demonstrated as efficient ways to enhance charge carrier transportation and separation. Photocatalytic activity could also be improved by pH value of the solutions, which are beneficial for enhancing charge utilization by shifting the standard oxidation potential.

## *Doping of 2D metal free materials*

Doping refers to the introduction of foreign elements into the lattice of the host materials. Based on the roles of the foreign elements (donors or acceptors), it is classified into N type doping and P type doping. Experimentally, it can be fulfilled easily by a pre-treatment of precursors or post-treatment of the as-prepared 2D materials. For designing high performance 2D metal free photocatalysts for light harvesting, no metal doping with boron,[77] carbon,[78] sulfur,[4,79] nitride,[80] and phosphorus[81–82] have been investigated to tailor the chemical properties, band structure, and the charge carrier transport.[83] With suitable dopants, doping has been proved to be a powerful method to improve the activity of photocatalysis driven by visible light.

For instance, the element sulfur is known as a useful dopant which can efficiently improve the photocatalytic activity of both g-$C_3N_4$ and 2D COFs.[4,79] For solving the low quantity of hydrogen evolution of porous carbon nitride materials, Cheng et al.[4] have prepared the homogeneous sulfur-doped g-$C_3N_4$. The S-doping effect on the electronic structure of g-$C_3N_4$ has been characterized by the UV-visible absorption spectra, which is displayed in Fig. 11a. One can see that the intrinsic absorption edge of $C_3N_4$ has a blue shift upon sulfur doping. The band gaps of $C_3N_4$ and $C_3N_{4-x}S_x$ determined are 2.73 and 2.85 eV. At the same time, in contrast to $C_3N_4$, the doped $C_3N_{4-x}S_x$ demonstrated a VB maximum up-shifted from 2.12 to 1.79 eV, as well as the VB itself being widened (Fig. 11b). This suggested that the doping has indeed induced a simultaneous upward shift of the CB minimum by 0.45 eV (Fig. 11c). The elevated CB minimum and a widened VB resulted in sulfur doped graphitic $C_3N_4$ displaying a photorreactivity of $H_2$ evolution 7.2 and 8 times higher than pure $C_3N_4$ under 300 and 420 nm, respectively.

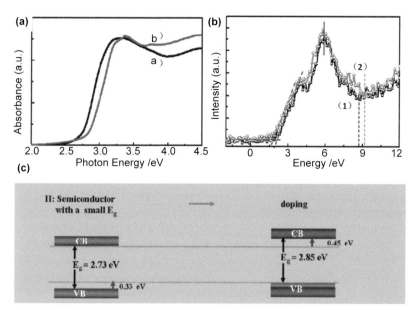

**Fig. 11.** (a) UV visible absorption spectra of $C_3N_4$ (1) and $C_3N_{4-x}S_x$ (2); (b) Total densities of states of XPS valence band spectra of $C_3N_4$ (1) and $C_3N_{4-x}S_x$ (2); (c) Schematic of the electronic structure of $C_3N_{4-x}S_x$ compared to that of $C_3N_4$.[4]

Li et al.[79] (2016) have also observed that the photocatalytic activity of 2D covalent triazine-based frameworks (CTFs) can be efficiently improved using such facile sulfur-doping approach. The sulfur-doped CTFs show superior photocatalytic activity and stability in hydrogen evolution from water under visible light irradiation, which is 5 times higher than that of the pristine CTFs and g-$C_3N_4$. The significantly improved catalytic efficiency was attributed to sulfur-doping in the frameworks, which results in enhanced adsorption of visible light, reduced recombination of free charge carriers, and rapid separation and transportation of photogenerated electron–holes.

In fact, both S-doped g-$C_3N_4$ and 2D COFs show n-type conductivity, which benefit the hydrogen and oxygen evolution. The n-type conductivity has also been observed in the other no metal element doped systems, including phosphorus-doped graphitic carbon nitride,[81–82] nitrogen doped graphene,[80] and so on. Those investigations further confirmed the n-type doping effect in promoting the photocatalytic activity. The p-type effect has been studied by Wang et al.[77] They found that the boron doping in g-CN can lead to the band gap changing from 2.88 to 2.83 eV. The obtained B-doped g-CN monolayer photocatalyze hydrogen evolution from a protic solution under visible light irradiation, and show a much higher activity than that of the pure g-CN catalyst.[77]

## Chemical functionalization

In addition to elemental doping, the chemical functionalization also plays an important role in improving the photocatalytic efficiency of 2D metal free materials. Functionalization is the addition of functional groups onto a material by chemical synthesis methods. In principle, the functionalization does not change the internal structures of materials. However, the introduction of organic functional groups may be n doped or p doped and band gap between the CBM and VBM can be tuned in a wide range. Therefore, it could be beneficial for the enhancement of photocatalytic $H_2$ production performance under visible light. Among countless functional groups, hydrogenation, fluorination, oxidation, hydroxylation, cyanidation, amination, and together on 2D metal free materials are feasible to obtain potential visible light driven photocatalysts, which have been confirmed by many experimental[37–41,84–87] and theoretical works.[42,70,88]

As a typical example, the effect of oxidation, hydroxylation, and amination have been widely observed on graphene oxide photocatalyst by tailoring its electronic structure. GO exhibits p-type conductivity or n-type conductivity when graphene covalently bonds to an electron accepting oxygen functionalities or electron donating nitrogen functionalities.[40] As oxygen bonds on graphene, the valence band changes from the π orbital of graphene to $O_{2p}$ orbital, leading to a larger band gap for a high oxygen coverage rate of GO, from semimetal to insulator, while the conduction band of GO is mainly formed by the anti-bonding π* orbital. Moreover, the effect of those functional groups can convert $sp^2$ hybridized carbons into $sp^3$ hybridizations and then tune the band gap from mid-ultraviolet to near infrared region (0.5–5.5 eV). Therefore, graphene oxide with OH, O, and $NH_2$ functional groups exhibited a suitable band gap for becoming an excellent photocatalytic $H_2$ production activity under visible light irradiation.[37–41,84–85] In fact, the other chemical functionalizations of graphene oxide, such as carboxylic acid functionalization with chloroacetic acid,[86] and triphenylamie functionalization with 4-benzaldehyde[87] was also applied to enhance photocatalytic $H_2$ production.

Hydrogenation is another common way to design new 2D metal-free photocatalysts. Based on the first principles calculations, Yang's group [88] predicted semihydrogenated BN is a potential metal-free visible light driven photocatalyst for water splitting. The band gap of bare graphitic BN sheet is calculated to be an insulator with a band gap of 5.56 eV. It is largely reduced to 2.24 eV after semihydrogenation (Fig. 12a), which corresponds to an absorption in the yellow-green region of the visible spectrum. At the same time, the position of the reduction level and oxidation potential revealed that the oxidation and reduction process is energetically favored with a relatively low driving force (Fig. 12a). Moreover, one can see from Fig. 12b, the probability distribution of VB and CB is well separated spatially: the former is localized in the left part of the supercell, whereas the latter is mainly in the right part. It leads to a low probability of recombination and avoids a big decrease in photocatalytic activity.

Cyanidation and amination on 2D COF could also act as a good candidate for water splitting. To design 2D organocatalysts, Jiang et al.[70] constructed a series of covalent organic framework (COF) with different functional groups. First-principles calculations were performed to confirm their photocatalytic activity under visible light. They found that one of the 2D COF with cyano/amine functional groups have a band gap of 2.09 eV at HSE06 functional level, which is suitable for visible light

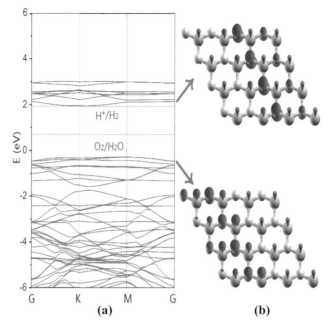

**Fig. 12.** (a) The band structure of sh-BN with HSE06 functional. The position of the reduction level for H⁺ to H₂ and the oxidation potential of H₂O to O₂ is indicated by the blue dashed line. The Fermi level is set to zero. (b) The charge distribution of VB (lower) and CB (upper) with isovalue 0.03 e/Å³.[88]

absorption. The redox potentials of $H_2O$ are both located in between the band gap, which indicates that it is a good candidate for water splitting.

## *Strain*

The general concept of strain is from the theory of elasticity. It is a tensorial quantity and defined by the amount of deformation under the application of an external force. Based on the deformation direction of 2D materials, the strain can be divided into uniaxial strain, biaxial, shear, and out plane strain. In experiments, in plane strain usually originates from surface stress or lattice mismatch with the substrate,[89] while out plane strain can be realized by pushing or compressing the single layer in a tailor made sandwich configuration.[90]

Recently, the effects of strain have been widely investigated, especially on 2D nanoscale materials. For example, by applying compressive or tensile biaxial strain on bilayer and monolayer $MoS_2$, the electronic properties are predicted to change from semiconducting to metallic.[91] The strain induced electronic structure modifications of bilayer $WSe_2$ has also been studied by Javey's group, [92] a indirect to direct band gap transition has been observed by DFT simulation, which explains the results of PL emission enhancement in experiments. By first principles calculation, Yang et al.[93] showed anisotropic free carrier mobility of phosphorene can be controlled by strain. With the appropriate biaxial or uniaxial strain (4–6%), they rotate the preferred conducting direction of phosphorene by 90°. The mechanism of such band gap reduction, indirect to direct band gap transition, and conducting pattern change was attributed to strain-induced symmetry breaking, which result in shifting in the energy levels of conduction and balance band or removal of degeneracy.

Generally, the 2D single layer materials can sustain high mechanical strain. Graphene and phosphorene can hold the uniaxial strain up to 25%[94] and 30%,[95] respectively. Such big strain provided enough space to modify the band gap, band offset, and charge separation path. Therefore, applying mechanical strain is also a useful method to improve the photocatalytic properties of 2D metal-free semiconductors.

For example, the potential application of phosphorene as a photocatalyst under strain have been assessed by density functional theory calculations.[74] The compression strains will reduce the band gap far from the visible light, showing no benefit to the photocatalysis properties. The band gap variation under tensile strain along a and b axes have been shown in the Fig. 13a. Both the band gaps first increase with the increasing strain and then decrease. Their peaks are 1.76 and 1.82 eV, respectively. The optical spectra of phosphorene having maximum band gap along the a and b axes have been plotted in the Fig. 13b. One can see that the absorption of phosphorene under strain in the visible light region is more noticeable than the strain free phosphorene.

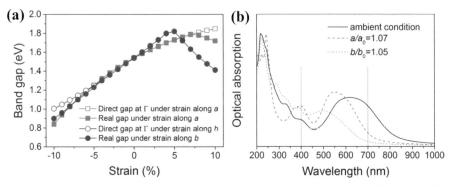

**Fig. 13.** (a) The band gap of phosphorene as a function of the uniaxial strain; (b) Optical spectra for phosphorene under strain free, 7% tensile strain along the a axis and under 5% tensile strain along the b axis. The solid vertical lines show the wavelength range of visible light.[74]

Moreover, the driving force of the splitting of water molecular on phosphorene have also been evaluated. Appropriate tensile strains further increase the driving force of the splitting of water.

## 2D metal free composite photocatalyts

Although the ability of visible light absorption have been improved by applying elemental doping, chemical functionalization, strain and stacking on photocatalysts, the high recombination rate between the photo-generated electrons and holes still restrict the hydrogen production activity. Cocatalysts, the light harvesting semiconductors loading with appropriate materials, have always been adopted to solve this problem. As famous candidates, the noble metals Pt, Pd, and Au are commonly used as cocatalysts due to their improved photocatalytic performance.[10] However, as Pt, Pd, and Au are rare and expensive, their use is restricted for practical applications. To tackle this noble metal related cocatalyst issue from water splitting, much attention has been focused on designing 2D metal free material based composites to replace noble metals due to the superior electrical mobility and large specific surface area. To date, large numbers of efficient 2D metal free composites for photocatalytic water splitting are synthesized, including 2D-0D composites and 2D-2D heterojuctions.

Kang et al.[14] from Soochow University have reported the design and fabrication of metal-free carbon nanodots and carbon nitride nanocomposite, and demonstrated their impressive performance for photocatalytic solar water splitting. First, the incorporation of carbon nanodots into the $C_3N_4$ matrix leads to an increase in the ultraviolet visible absorption over the entire wavelength range investigation. The optical band gap of carbon nanodots is 2.77 eV, which is almost identical to that of pure $C_3N_4$. Second, they show carbon nanodots and carbon nitride nanocomposite can split water into $H_2$ and $O_2$ with QEs of 16% for $\lambda = 420 \pm 20$ nm and 6.3% for $\lambda = 580 \pm 15$ nm. At the same time, overall solar energy conversion efficiency is 2.0, which is one order of magnitude larger than that of any previously reported stable water splitting photocatalyst. The other high quantum efficiency cases are GO and CdS quantum dots hybrid systems,[96] whose $H_2$ production rate and quantum efficiency is 1824 $\mu molh^{-1}g^{-1}$ and 23.4% at 420 nm, respectively, which is much better than Pt-CdS quantum dots photocatalysts. In addition, GO/3C-SiC QDs,[97] g-$C_3N_4$/CdS QDs [98] and g-$C_3N_4$/polypyrrol [99] binary composites have also been investigated. All these 2D-0D composites show a higher photocatalytic $H_2$ evolution as compared to their bare semiconductor. For GO/3C-SiC QDs system, the rate of $H_2$ production reaches to 95 $\mu L/h$ with 1 wt% GO content, which is 1.3 times larger compared to the case in pure SiC nanoparticle under visible light, while the $H_2$ evolution rate of the optimal g-$C_3N_4$ and CdS quantum dots composites was 17.27 $\mu molh^{-1}g^{-1}$ under visible light irradiation, which is 9 times that of pure g-$C_3N_4$. More promisingly, a very low mass organic molecule (polypyrrole) loading result in the hydrogen production was 50 times higher than pure g-$C_3N_4$.

Compared to 0D/2D heterostructures, the 2D/2D heterostructures possess relatively better coupling hetero-interfaces, which provide more abundant surface active sites and facilitate the transfer and separation of photogenerated electron and hole pairs, thus contributing to higher photocatalytic performance. Overall, there are three typical categories, including semiconductor-graphene heterojunction, semiconductor-semiconductor heterojunction, and isotype heterojunction.

For instance, Xiang et al.[100] (2011) hybridized the two star 2D materials, g-$C_3N_4$ and graphene through an impregnation chemical reduction route. The successful formation of 2D-2D layered junctions between them led to a very efficient interfacial charge separation, which enabled the spatial accumulation of photoinduced electrons and holes on the sides of graphene and g-$C_3N_4$, respectively (Fig. 14a). Consequently, with optimizing the interface effect at the amount of 1.0 wt% graphene, the photocatalytic $H_2$ production rate is 451 $\mu molh^{-1}g^{-1}$, which is 3.7 times more than that of pure g-$C_3N_4$.

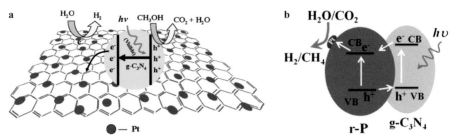

**Fig. 14.** (a) Schematic illustration of photocatalytic $H_2$ production over graphene and g-$C_3N_4$ composites.[100] (b) Energy band diagram of red phosphor and g-$C_3N_4$ heterojunction and schematic illustration of electron-hole separation process after excitation.[101]

The mechanism of such enhancement in g-$C_3N_4$/graphene hybrid structure was due to the ability of graphene sheets to act as conductive channels to efficiently separate the photogenerated charge carriers.

2D semiconductor and 2D semiconductor heterojunction also lead to considerable improvement on the photocatalytic activity for $H_2$ production. Yuan et al.[101] (2013) reported a composite photocatalyst by coupling red phosphor and g-$C_3N_4$ together. For the pure red phosphor and pure g-$C_3N_4$, the activity for $H_2$ production is 22 $\mu$molh$^{-1}$g$^{-1}$ and 340 $\mu$molh$^{-1}$g$^{-1}$, respectively. However, the $H_2$ evolution rate is about 310 $\mu$molh$^{-1}$g$^{-1}$ when loading of 5 wt% g-$C_3N_4$. The photocatalytic activity can be further enhanced to 1000 $\mu$molh$^{-1}$g$^{-1}$ for $H_2$ evolution by loading of 30 wt% g-$C_3N_4$. The mechanism of enhanced photocatalytic activity could be attributed to the effective interfacial charge transfer between red phosphor and g-$C_3N_4$ across the heterojunction, as illustrated in Fig. 14b. On the one hand, the photoinduced electrons on g-$C_3N_4$ surface could transfer to red phosphor through the interfaces since the CB edge of g-$C_3N_4$ is more negative than that of red phosphor. On the other hand, the photoexcited holes on the red phosphor surfaces move towards g-$C_3N_4$ due to the difference in valence band edge. Such effective separation of photogenerated charge carrier thus enhances the photocatalytic activity. Similar semiconductors and semiconductor heterostructures have also been observed in black-red phosphorus[102] and g-$C_9N_{10}$ and g-$C_{12}N_7H_3$ hybrid systems.[103]

Another novel isotype heterojunction was first found in g-$C_3N_4$ and sulfur mediated g-$C_3N_4$ binary system, which were synthesized using thiourea and urea as precursors.[104] The g-$C_3N_4$/g-$C_3N_4$ isotype heterojunction formation originated from g-$C_3N_4$, prepared using thiourea, show higher VB and CB positrons than urea. Such difference in band structure will cause the redistribution of the photoinduced electrons and holes, and reduce the energy wasteful electron and hole recombination. As a result, this g-$C_3N_4$/sulfur mediated g-$C_3N_4$ heterojunction shows an enhanced $H_2$ evolution activity over its corresponding host substrates.

## Chemical bias

While all the strategies for improving photocatalytic efficiency mentioned above are still focused on tuning electronic properties of the photocatalysts, electrolyte

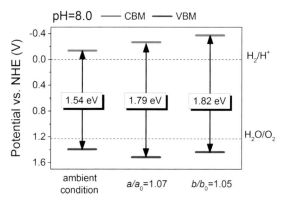

**Fig. 15.** Energy alignment of phosphorene at ambient condition, 5% and 7% strain, respectively. The black and green dashed lines are water redox potentials at vacuum level and pH = 8.0 solution.[74]

modification is also a feasible method. Adjusting the pH value of the electrolyte can tune the alignment of the energy levels between the 2D materials. The pH dependent reduction potential and oxidation potential are.[105]

$$E^{red}_{H^+/H_2} = -4.44eV + pH \times 0.059 \tag{1.2}$$

$$E^{oxd}_{O_2/H_2O} = -5.67eV + pH \times 0.059 \tag{1.3}$$

Here, −4.44 eV and −5.67 eV are the standard water reduction and oxidation potential at the vacuum level. According to the formula 1.2 and 1.3, the increasing pH shifts both the energy levels of $H^+/H_2$ and $O_2/H_2O$ potentials upward, which will make some unactive materials be possible photocatalysts or enhance the ability of charge carrier utilization. Phosphorene is an example of a 2D material benefitting from the pH modification.[74] For the phosphorene with strain and at ambient condition (Fig. 15), CBMs of all the phosphorene locate more negative than the redox potential of $H^+/H_2$, but the VBMs of all the phosphorene are not occupied by more positive than the redox potential of $O_2/H_2O$. That is to say, phosphorene is not suitable for water splitting under vacuum condition. However, the oxidation potential as well as the reduction potential will shift upward 0.472 eV in Fig. 15 in pH = 8.0 solutions. One can see phosphorene shows a favorable band position for water splitting for all the cases we have shown. Hence, the chemical bias engineering (pH) can also improve the photocatalysis properties by enhance the ability of charge carrier utilization.

## Summary and Perspectives

Two dimensional metal free photocatalytic materials are promising functional materials for the potential solution to the energy and environmental problems. In this chapter, 2D metal free materials such as elemental semiconductor phosphorene, g-$C_3N_4$, graphene oxide, B-C-N hybrid sheets, 2D organic materials, Si-based materials, etc. for $H_2$ production over photocatalytic water splitting were comprehensively overviewed. A systematic understanding of their electronic and optical properties in determining the ability of photocatalysts for $H_2$ production from photocatalytic water splitting was summarized. The strategies for tailoring the properties of 2D metal free photocatalytic materials for $H_2$ production from photocatalytic water splitting were also discussed in several aspects, including doping effect, chemical functionalization, mechanical loading, interface engineering, and external condition (pH value of solution).

Although considerable progress has been achieved, the studies in this field are still at the primary stage, and further developments are required. First, exploration of new metal free materials in the form of 2D nanosheets is greatly necessary. Although many semiconductors have been identified as suitable toward solar water splitting, not many of them are realized as 2D metal free materials. Based on the results of this chapter, we believe that there are more high-efficiency 2D metal free photocatalysts among countless unknown 2D materials. In addition, a better relationship between physical and chemical properties and photocatalytic activity should be correlated, which is beneficial for the purposeful design of high performance 2D metal free photocatalysts. This relies not only on advanced material characterization techniques, but also on rigorous theoretical simulation. New synthesis methods have to be developed for preparing high quality 2D metal free materials with high purity, compositions, and defects. High computation based on density functional theory have to be used for screening potential photocatalytic materials, which have been demonstrated as a fast, reliable, efficient and economical method. Second, developing new 2D metal free semiconductor and nanomaterials hybrid systems with synergic benefits from individuals is also of great significance. A rational design of combining 2D metal free semiconductor and nanomaterials will further optimize the performance of hydrogen production. Third, the new modification approach towards improvement of photocatalytic actively of 2D metal free photocatalysts is highly encouraged. For example, the influence of layer number, defects, morphology composition, electrical, and magnetic fields on the photocatalytic performance has rarely been investigated. Such modification will favor not only the adjustment of redox potential, but also the enhancement of light harvesting. Fourth, most previous findings about 2D metal free materials mainly observed the photocatalytic $H_2$ production from the reduction of water, while $O_2$ production from the oxidation of water is still an issue. 2D metal free photocatalysts with efficiency for simultaneous $H_2$ production and $O_2$ production is highly anticipated. Lastly, the mechanisms of photocatalytic enhancement of 2D metal free under many approaches are partly unclear. Therefore, more studies are still needed to improve the general understanding of enhancement mechanism.

Nevertheless, semiconductor photocatalysts research will continue to thrive because of the new opportunities provided by two dimensional metal free materials, which will require extensive efforts from experimental and theoretical scientists and researchers all over the world. It is expected in the future that the artificial photosynthesis devices can be well designed and materialize. We also hope that relevant exploration about photocatalytic water splitting will be expanded to solving other environmental and energy-related issues, such as photocatalytic $CO_2$ reduction and organic pollutant removal.

# Acknowledgement

This chapter is supported by the National Natural Science Foundation of China (11404050) and the Fundamental Research Funds for the Central Universities of China (DUT16RC(4)50, DUT16JJ(G)05, DUT16LAB01, DUT17LAB19). We also acknowledge the Supercomputing Center of Dalian University of Technology for providing the computing resource.

# References

1. A. K. Singh, K. Mathew, H. L. Zhuang and R. G. Hennig, *J. Phys. Chem. Lett.*, 2015, **6**, 1087–1098.
2. J. Yu, W. Wang, B. Cheng and B. L. Su, *The Journal of Physical Chemistry C*, 2009, **113**, 6743–6750.
3. N. Wu, J. Wang, D. N. Tafen, H. Wang, J. G. Zheng, J. P. Lewis, X. Liu, S. S. Leonard and A. Manivannan, *J. Am. Chem. Soc.*, 2010, **132**, 6679–6685.
4. G. Liu, P. Niu, C. Sun, S. C. Smith, Z. Chen, G. Q. Lu and H.- M. Cheng, *J. Am. Chem. Soc.*, 2010, **132**, 11642–11648.
5. X. Wang, K. Maeda, A. Thomas, K. Takanabe, G. Xin, J. M. Carlsson, K. Domen and M. Antonietti, *Nat Mater*, 2009, **8**, 76–80.
6. M. R. Hoffmann, S. T. Martin, W. Choi and D. W. Bahnemann, *Chem. Rev.*, 1995, **95**, 69–96.
7. D. Chen, L. Tang and J. Li, *Chem. Soc. Rev.*, 2010, **39**, 3157–3180.
8. A. Fujishima and K. Honda, *Nature*, 1972, **238**, 37–38.
9. X. Chen, S. Shen, L. Guo and S. S. Mao, *Chem. Rev.*, 2010, **110**, 6503–6570.
10. Q. Xiang, J. Yu and M. Jaroniec, *Chem. Soc. Rev.*, 2012, **41**, 782–796.
11. K. Maeda and K. Domen, *J. Phys. Chem. Lett.*, 2010, **1**, 2655–2661.
12. T. Hisatomi, J. Kubota and K. Domen, *Chem. Soc. Rev.*, 2014, **43**, 7520–7535.
13. S. J. A. Moniz, S. A. Shevlin, D. J. Martin, Z.-X. Guo and J. Tang, *Energy & Environmental Science*, 2015, **8**, 731–759.
14. J. Liu, Y. Liu, N. Liu, Y. Han, X. Zhang, H. Huang, Y. Lifshitz, S.- T. Lee, J. Zhong and Z. Kang, *Science*, 2015, **347**, 970–974.
15. J. V. Liebig, *Ann. Pharm*, 1834, **10**.
16. E. Kroke, *Angew. Chem. Int. Ed.*, 2014, **53**, 11134–11136.
17. G. Algara-Siller, N. Severin, S. Y. Chong, T. Björkman, R. G. Palgrave, A. Laybourn, M. Antonietti, Y. Z. Khimyak, A. V. Krasheninnikov and J. P. Rabe, *Angew. Chem.*, 2014, **126**, 7580–7585.
18. Y. Zheng, L. Lin, B. Wang and X. Wang, *Angew. Chem. Int. Ed.*, 2015, **54**, 12868–12884.
19. L. Lin, H. Ou, Y. Zhang and X. Wang, *ACS Catalysis*, 2016.
20. Y. Wang, X. Wang and M. Antonietti, *Angew. Chem. Int. Ed.*, 2012, **51**, 68–89.
21. K. Schwinghammer, B. Tuffy, M. B. Mesch, E. Wirnhier, C. Martineau, F. Taulelle, W. Schnick, J. Senker and B. V. Lotsch, *Angew. Chem. Int. Ed.*, 2013, **52**, 2435–2439.
22. Y. Zheng, L. Lin, X. Ye, F. Guo and X. Wang. *Angew. Chem. Int. Ed.*, 2014, **53**, 11926–11930.
23. D. J. Martin, K. Qiu, S. A. Shevlin, A. D. Handoko, X. Chen, Z. Guo and J. Tang, *Angew. Chem. Int. Ed.*, 2014, **53**, 9240–9245.
24. Z. Li, Y. Wu and G. Lu, *Applied Catalysis B: Environmental*, 2016, **188**, 56–64.
25. J. Zhao, L. Liu and F. Li, *Graphene Oxide: Physics and Application. Springer*, 2015.
26. D. Joung, A. Chunder, L. Zhai and S. I. Khondaker, *Appl. Phys. Lett.*, 2010, **97**, 093105.
27. L. K. Putri, L.-L. Tan, W.-J. Ong, W. S. Chang and S.-P. Chai, *Applied Materials Today*, 2016, **4**, 9–16.
28. D. R. Dreyer, S. Park, C. W. Bielawski and R. S. Ruoff, *Chem. Soc. Rev.*, 2010, **39**, 228–240.
29. H. A. Becerril, J. Mao, Z. Liu, R. M. Stoltenberg, Z. Bao and Y. Chen, *ACS Nano*, 2008, **2**, 463–470.
30. D. Chen, H. Feng and J. Li, *Chem. Rev.*, 2012, **112**, 6027–6053.
31. L. Guo, R.-Q. Shao, Y.-L. Zhang, H.-B. Jiang, X.-B. Li, S.-Y. Xie, B.-B. Xu, Q.-D. Chen, J.-F. Song and H.-B. Sun, *The Journal of Physical Chemistry C*, 2012, **116**, 3594–3599.
32. J. Ito, J. Nakamura and A. Natori, *J. Appl. Phys.*, 2008, **103**, 113712.

33. L. Liu, L. Wang, J. Gao, J. Zhao, X. Gao and Z. Chen, *Carbon*, 2012, **50**, 1690–1698.
34. D. Li, M. B. Muller, S. Gilje, R. B. Kaner and G. G. Wallace, *Nat. Nano*, 2008, **3**, 101–105.
35. P. Johari and V. B. Shenoy, *ACS Nano*, 2011, **5**, 7640–7647.
36. A. Mathkar, D. Tozier, P. Cox, P. Ong, C. Galande, K. Balakrishnan, A. Leela Mohana Reddy and P. M. Ajayan, *J. Phys. Chem. Lett.*, 2012, **3**, 986–991.
37. T. F. Yeh, J. M. Syu, C. Cheng, T. H. Chang and H. Teng, *Adv. Funct. Mater.*, 2010, **20**, 2255–2262.
38. T.-F. Yeh, F.-F. Chan, C.-T. Hsieh and H. Teng, *The Journal of Physical Chemistry C*, 2011, **115**, 22587–22597.
39. Y. Matsumoto, M. Koinuma, S. Ida, S. Hayami, T. Taniguchi, K. Hatakeyama, H. Tateishi, Y. Watanabe and S. Amano, *The Journal of Physical Chemistry C*, 2011, **115**, 19280–19286.
40. T.-F. Yeh, S.-J. Chen, C.-S. Yeh and H. Teng, *The Journal of Physical Chemistry C*, 2013, **117**, 6516–6524.
41. T.-F. Yeh, C.-Y. Teng, S.-J. Chen and H. Teng, *Adv. Mater.*, 2014.
42. X. Jiang, J. Nisar, B. Pathak, J. Zhao and R. Ahuja, *J. Catal.*, 2013, **299**, 204–209.
43. L. Wang, Y. Sun, K. Lee, D. West, Z. Chen, J. Zhao and S. Zhang, *Phys. Rev. B*, 2010, **82**, 161406.
44. L. Wang, K. Lee, Y.-Y. Sun, M. Lucking, Z. Chen, J. J. Zhao and S. Zhang, *Acs Nano*, 2009, **3**, 2995–3000.
45. S.-Y. Ding and W. Wang, *Chem. Soc. Rev.*, 2013, **42**, 548–568.
46. N. A. A. Zwaneveld, R. Pawlak, M. Abel, D. Catalin, D. Gigmes, D. Bertin and L. Porte, *J. Am. Chem. Soc.*, 2008, **130**, 6678–6679.
47. A. P. Côté, A. I. Benin, N. W. Ockwig, M. Keeffe, A. J. Matzger and O. M. Yaghi, *Science*, 2005, **310**, 1166.
48. E. L. Spitler, B. T. Koo, J. L. Novotney, J. W. Colson, F. J. Uribe-Romo, G. D. Gutierrez, *J. Am. Chem. Soc.*, 2011, **133**, 19416–19421.
49. P. Kuhn, M. Antonietti and A. Thomas, *Angew. Chem. Int. Ed.*, 2008, **47**, 3450–3453.
50. S. Wan, F. Gándara, A. Asano, H. Furukawa, A. Saeki, S. K. Dey, L. Liao, M. W. Ambrogio, Y. Y. Botros, X. Duan, S. Seki, J. F. Stoddart and O. M. Yaghi, *Chem. Mater.*, 2011, **23**, 4094–4097.
51. S. Wan, J. Guo, J. Kim, H. Ihee and D. Jiang, *Angew. Chem.*, 2008, **120**, 8958–8962.
52. L. Stegbauer, K. Schwinghammer and B. V. Lotsch, *Chemical Science*, 2014, **5**, 2789–2793.
53. B. Lukose, A. Kuc, J. Frenzel and T. Heine, *Beilstein Journal of Nanotechnology*, 2010, **1**, 60–70.
54. O. M. Yaghi, M. O'Keeffe, N. W. Ockwig, H. K. Chae, M. Eddaoudi and J. Kim, *Nature*, 2003, **423**, 705–714.
55. N. W. Ockwig, O. Delgado-Friedrichs, M. O'Keeffe and O. M. Yaghi, *Acc. Chem. Res.*, 2005, **38**, 176–182.
56. E. L. Spitler and W. R. Dichtel, *Nat. Chem.*, 2010, **2**, 672–677.
57. A. Kira, T. Umeyama, Y. Matano, K. Yoshida, S. Isoda, J. K. Park, D. Kim and H. Imahori, *J. Am. Chem. Soc.*, 2009, **131**, 3198–3200.
58. H. Hayashi, I. V. Lightcap, M. Tsujimoto, M. Takano, T. Umeyama, P. V. Kamat and H. Imahori, *J. Am. Chem. Soc.*, 2011, **133**, 7684–7687.
59. S. Mathew, A. Yella, P. Gao, R. Humphry-Baker, B. F. E. Curchod, N. Ashari-Astani, I. Tavernelli, U. Rothlisberger, M. K. Nazeeruddin and M. Grätzel, *Nat. Chem.*, 2014, **6**, 242–247.
60. Chen, X., M. Addicoat, E. Jin, L. Zhai, H. Xu, N. Huang, Z. Guo, L. Liu, S. Irle and D. Jiang, *J. Am. Chem. Soc.*, 2015, **137**, 3241–3247.
61. Y. Zhou, Z. Wang, P. Yang, X. Zu and F. Gao. *Journal of Materials Chemistry*, 2012, **22**, 16964–16970.
62. K. Sakaushi, G. Nickerl, H. C. Kandpal, L. Cano-Cortés, T. Gemming, J. Eckert, *J. Phys. Chem. Lett.*, 2013, **4**, 2977–2981.
63. D. Er, L. Dong and V. B. Shenoy, *The Journal of Physical Chemistry C*, 2016, **120**, 174–178.
64. X. Feng, L. Chen, Y. Honsho, O. Saengsawang, L. Liu, L. Wang, A. Saeki, S. Irle, S. Seki and Y. Dong, An, *Adv. Mater.*, 2012, **24**, 3026–3031.
65. M. J. Bojdys, J. Jeromenok, A. Thomas and M. Antonietti, *Adv. Mater.*, 2010, **22**, 2202–2205.
66. P. Katekomol, J. Roeser, M. Bojdys, J. Weber and A. Thomas, *Chem. Mater.*, 2013, **25**, 1542–1548.
67. X. Jiang, P. Wang and J. Zhao, *Journal of Materials Chemistry A*, 2015, **3**, 7750.
68. J. Bi, W. Fang, L. Li, J. Wang, S. Liang, Y. He, M. Liu and L. Wu, *Macromol. Rapid Commun.*, 2015, **36**, 1799–1805.
69. K. Schwinghammer, S. Hug, M. B. Mesch, J. Senker and B. V. Lotsch, *Energy & Environmental Science*, 2015, **8**, 3345–3353.

70. W. Peng, J. Xue and Z. Jijun, *J. Phys.: Condens. Matter*, 2016, **28**, 034004.
71. R. Zhang and J. Yang. 2015. Few-Layer C2N: A Promising Metal-free Photocatalyst for Water Splitting. arXiv:1505.02768.
72. R. Lu, F. Li, J. Salafranca, E. Kan, C. Xiao and K. Deng, *PCCP*, 2014, **16**, 4299–4304.
73. X. Li, Y. Dai, M. Li, W. Wei and B. Huang, *Journal of Materials Chemistry A*, 2015, **3**, 24055–24063.
74. B. Sa, Y.-L. Li, J. Qi, R. Ahuja and Z. Sun, *The Journal of Physical Chemistry C*, 2014, **118**, 26560–26568.
75. J. Mahmood, E. K. Lee, M. Jung, D. Shin, I.-Y. Jeon, S.-M. Jung, H.-J. Choi, J.-M. Seo, S.-Y. Bae, S.-D. Sohn, N. Park, J. H. Oh, H.-J. Shin and J.-B. Baek, *Nat Commun*, 2015, **6**.
76. S. Zhou, N. Liu and J. Zhao, *Comput. Mater. Sci.*, 2017, **130**, 56–63.
77. Z. Lin and X. Wang, *Angew. Chem. Int. Ed.,* 2013, **52**, 1735–1738.
78. C. Huang, C. Chen, M. Zhang, L. Lin, X. Ye, S. Lin, M. Antonietti and X. Wang, *Nat Commun*, 2015, **6**.
79. L. Li, W. Fang, P. Zhang, J. Bi, Y. He, J. Wang, *Journal of Materials Chemistry A*, 2016, **4**, 12402–12406.
80. D. W. Chang and J.-B. Baek, *Chemistry – An Asian Journal*, 2016, **11**, 1125–1137.
81. J. Ran, T. Y. Ma, G. Gao, X.-W. Du and S. Z. Qiao, *Energy & Environmental Science*, 2015, **8**, 3708–3717.
82. Y.-P. Zhu, T.-Z. Ren and Z.-Y. Yuan, *ACS Appl. Mater. Interfaces*, 2015, **7**, 16850–16856.
83. X. Wang, G. Sun, P. Routh, D.-H. Kim, W. Huang and P. Chen. *Chem. Soc. Rev.*, 2014, **43**, 7067–7098.
84. K. Krishnamoorthy, R. Mohan and S.-J. Kim, *Appl. Phys. Lett.*, 2011, **98**, 244101.
85. T.-F. Yeh and H. Teng, *ECS Transactions*, 2012, **41**, 7–26.
86. J. Liu, S. Xu, L. Liu and D. D. Sun, *Carbon*, 2013, **60**, 445–452.
87. Z. Li, Y. Chen, Y. Du, X. Wang, P. Yang and J. Zheng, *Int. J. Hydrogen Energy*, 2012, **37**, 4880–4888.
88. X. Li, J. Zhao and J. Yang, *Sci. Rep.*, 2013, **3**, 1858.
89. Z. H. Ni, T. Yu, Y. H. Lu, Y. Y. Wang, Y. P. Feng and Z. X. Shen, *ACS Nano*, 2008, **2**, 2301–2305.
90. S. Bertolazzi, J. Brivio and A. Kis, *ACS Nano*, 2011, **5**, 9703–9709.
91. E. Scalise, M. Houssa, G. Pourtois, V. Afanas'ev and A. Stesmans, *Nano Research*, 2012, **5**, 43–48.
92. S. B. Desai, G. Seol, J. S. Kang, H. Fang, C. Battaglia, R. Kapadia, J. W. Ager, J. Guo and A. Javey, *Nano Lett.*, 2014, **14**: 4592–4597.
93. R. Fei and L. Yang, *Nano Lett.*, 2014, **14**, 2884–2889.
94. C. Lee, X. Wei, J. W. Kysar and J. Hone, *Science*, 2008, **321**, 385–388.
95. Q. Wei and X. Peng. *Appl. Phys. Lett.*, 2014, **104**, 251915.
96. J. Zhang, J. Yu, M. Jaroniec and J. R. Gong. 2012. *Nano Lett.*, 12: 4584–4589.
97. J. Yang, X. Zeng, L. Chen and W. Yuan. *Appl. Phys. Lett.*, 2013, **102**, 083101.
98. L. Ge, F. Zuo, J. Liu, Q. Ma, C. Wang, D. Sun, L. Bartels and P. Feng, *The Journal of Physical Chemistry C*, 2012, **116**, 13708–13714.
99. Y. Sui, J. Liu, Y. Zhang, X. Tian and W. Chen. *Nanoscale*, 2013, **5**, 9150–9155.
100. Q. Xiang, J. Yu and M. Jaroniec, *The Journal of Physical Chemistry C*, 2011, **115**, 7355–7363.
101. Y.-P. Yuan, S.-W. Cao, Y.-S. Liao, L.-S. Yin and C. Xue, *Applied Catalysis B: Environmental*, 2013, **140-141**, 164–168.
102. Z. Shen, S. Sun, W. Wang, J. Liu, Z. Liu and J. C. Yu, *Journal of Materials Chemistry A*, 2015, **3**, 3285–3288.
103. H. Li, H. Hu, C. Bao, F. Guo, X. Zhang, X. Liu, J. Hua, J. Tan, A. Wang, H. Zhou, B. Yang, Y. Qu and X. Liu, *Sci. Rep.*, 2016, **6**, 29327.
104. J. Zhang, M. Zhang, R.-Q. Sun and X. Wang, *Angew. Chem.*, 2012, **124**, 10292–10296.
105. V. Chakrapani, J. C. Angus, A. B. Anderson, S. D. Wolter, B. R. Stoner and G. U. Sumanasekera, *Science*, 2007, **318**, 1424–1430.

# CHAPTER 9

# Transition Metal Dichalcogenides in Energy Applications

*Xinyu Cheng,*[1,‡] *Meng Sun,*[2,‡] *Haiming Xie*[3,*] and *Jinghong Li*[1,*]

## Introduction

With the explosion of global population and the huge consumption of the traditional fossil fuels, sustainable and eco-friendly energy sources are in urgent need in the new century. For the sake of the welfare of our offspring, people should spare no effort to harvest renewable energies by virtue of various high-efficient electrochemical energy applications. Recently, the promising energy devices such as supercapacitors, lithium-ion batteries, microbial fuel cells (MFCs), proton exchange membrane fuel cells (PEMFCs), direct methanol fuel cells (DMFCs), and solar cells, have made great strides in terms of developing active and durable electrode materials, improving the efficiency of electron transfer, and decreasing the energy barrier of complex electrocatalytic reactions.[1–5] For example, the progress on the high-density compacted energy storage devices, such as hybrid electric vehicles, large electrical equipment, and renewable energy power plants have played prominent roles in solving the inefficiency of energy storage. Besides that, the typical electrocatalytic reaction, including the hydrogen evolution reaction (HER), the oxygen evolution reaction (OER) originated from water splitting process, and the oxygen reduction reaction (ORR),

[1] Department of Chemistry, Tsinghua University, Beijing 100084, China.
[2] Department of Chemical and Environmental Engineering, Yale University, New Haven, Connecticut 06511, United States.
[3] National & Local United Engineering Laboratory for Power Battery, Department of Chemistry, Northeast Normal University, Changchun 130024, China.
* Corresponding authors: jhli@mail.tsinghua.edu.cn
‡ The authors contributed to this work equally.

naturally also exhibits important performance in maximizing energy density when they are utilized for fuel cell applications. Although noble metal Pt-based catalysts still occupy the largest market shares of large-scale production of electrode materials for fuel cell and battery industrializations, its exorbitant prices and scarcity limit its application. Therefore, the development of comparable candidates with earth-rich elemental component and preferable structural characteristics will exhibit desirable catalytic activity and architectural design, necessarily facilitating the replacement of novel materials and advanced structures.

Dimensionality of nanomaterials has great effect on their electrochemical behaviors, as well as the composition and atom arrangement. For example, zero-dimensional structures, such as nanoparticles and quantum dots, possess the largest surface area and lowest surface energy, exhibiting excellent activity and stability. Moreover, one-dimensional (1D) structures, like nanotubes and nanowires, pave the way in constructing micro-devices, such as fiber-shaped supercapacitors or microbial fuel cells.[6] Besides that, two-dimensional (2D) structures expose more surface area, while three-dimensional (3D) structures, as the combination of 1D and 2D construction, enable superior electron transfer efficiency via spatial construction. It is noted that the 2D materials provide superficial structures and advantages, promisingly facilitating effective active sites for catalytic reactions. They can also form net structures with larger active surface area and open pore structures through stacking, enhancing mass adsorption and transfer on the surface. Profiting from these superiorities, the 2D structure of nanomaterials have gotten extensive application, especially in the promising electrocatalytic energy conversion. For example, 2D layered hetero-doped graphene and transition metal dichalcogenide (TMD) nanosheets can both serve as active and sustainable electrocatalysts for ORR, HER, and OER.[7,8] Hence, it is of importance and significance to understand the structural characteristics, architectural features, and properties of 2D nanomaterials.

Graphene has aroused great interest all over the world due to its specific mono-layer of $sp^2$ bonding carbon atoms network structure.[9] Similarly, the graphene-like layered materials, such as TMDs, have recently jumped into the view of scientists due to their excellent electric and heat conductivity, strong mechanical strength, ultra-high carrier mobility, and large specific surface area.[9–11] The graphene-like TMDs, which can be represented as $MX_2$, attract most of the attention because of their 2D morphology, versatile physicochemical properties, and potential applications as appropriate semiconductors for its sizable bandgaps and large quantities.[12,13] The 2D layered graphene bears the brunt of the desirable electrode material with high conductivity and large surface area, which greatly facilitates the charge separation and transfer in photoelectrochemical conversion.[14,15] The layered TMDs have been proved to hold superior theoretical capacitance with the highest capacity of $MoS_2$ reaching 1131 mA h $g^{-1}$ at the current density of 50 mA $g^{-1}$ to store energy effectively through reversible redox reaction. Meanwhile, the various active sites on TMDs enhance the catalytic activity in water splitting and oxygen reduction reaction.[12,16] For instance, the monolayer $MoS_2$ reported by Zhang group[17] (2014) exhibited low Tafel slope of ~ 73 mV decade$^{-1}$ in HER, while Xie and co-workers[18] fabricated single-unit-cell orthorhombic $CoSe_2$ sheets as OER catalysts with a low overpotential of 0.27 V. Song group[19] applied oxygen-doped $MoS_2$ catalyst in ORR, which displayed onset and half-wave potentials of 0.94 and 0.80 V, respectively, overtaking the electrocatalytic performance of Pt/C. It can be seen that the morphologies of TMDs with multidimensional assemblies prominently enrich the patterns of charge storage and optimize the mechanism of electron transportation. Therefore, understanding the rational structure design of these TMDs is quite imperative.

A large number of books focus on introduction on graphene because of its leading role in ultrathin 2D materials. In this chapter, to avoid repetition, we largely limit our discussion and just focus on graphene-like 2D TMD materials. As known, structures, properties, and synthetic methods are the key factors in fabricating materials with excellent performance to be applied into target area. Therefore, the basic understanding and corresponding electrochemical energy applications are briefly introduced. In the first section of this chapter, the fundamental characterizations of those materials are described, including electronic structures and intrinsic properties. Then, the assortments of their synthetic methods are summarized. Moreover, we discuss different applications of those mentioned electrode materials in supercapacitors, lithium-ion batteries, and electrocatalytic reactions. Finally, the challenges and future perspectives facing TMDs are featured on the basis of its current development.

# Structures and Properties of Graphene-like 2D Layered Transition Metal Dichalcogenides

Since graphene was explored, its unique structure has attracted great attention from the public. Graphene is a single layer of graphite, so some compounds that have similar structures to graphite, such as laminar TMDs, could be exfoliated to form single-layered structures. In 2010, Rao group[20] defined those compounds with layered structures as "graphite-like" compounds, and the mono- or few-layered structure obtained from those compounds as "graphene-like" structures. Different structures contribute to various applications of TMDs. For example, $HfS_2$ is an absolute insulator, but $MoS_2$, $WS_2$, $WTe_2$, and $TiSe_2$ serve as semiconductor materials. $NbS_2$ and $VSe_2$ possess best conductivity, potentially substituting metal elements used in electronic devices. $NbS_2$ and $TaS_2$ exhibit unique properties such as superconductivity, charge density wave (a periodic distortion of the crystal lattice), and Mott transition (metal to non-metal transition).[21]

## Structures of Graphene-like 2D Layered Transition Metal Dichalcogenides

TMDs can be described with the formula $MX_2$, where M represents transition metal of group IV-X, and X is a chalcogenide (S, Se, or Te). The structure of layered TMDs is different from that of graphene, because the layered TMDs are composed of three layers of atoms (X-M-X), which means two hexagonal chalcogen planes are separated by one transition-metal plane. Mono-layered TMDs exhibit two combination modes of transition metal atoms and chalcogenides, which refer to trigonal prismatic and octahedral Mo coordination, respectively.[8,21] 1T $MoS_2$ represents octahedral coordination, while 2H and 3R $MoS_2$ represent trigonal prismatic coordination. In most cases, one of two modes is stable for $MX_2$, but in fact, in $MoS_2$, two phases exist at the same time due to the lattice match at the interface.[12] The 2H type is more stable in nature with two layers per unit (Fig. 1). When heating, the shift of one sulfur layer results in an ABC stacking sequence (1 T type). The *d*-orbital electrons of transition metal and its surrounding coordination have great influence on the electronic properties of graphene-like TMDs. If the orbitals are fully filled (e.g., M = Mo, W), the materials exhibit semiconductor behavior which corresponds to 2H type, while it shows metallic properties when the *d*-orbitals are partially occupied (e.g., M = Nb, Re), which refers to 1T type.[13,22,23]

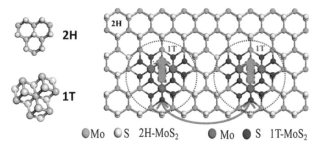

**Fig. 1.** The 2H and 1T phase of $MoS_2$, and the diagrammatic representation of the phase incorporation of the 1T phase of 2H-$MoS_2$ nanosheets. [Reprinted with permission from Ref. 8, Zhang et al. *Energy Environ. Sci.*, 2016, **9**, 1190, Copyright @ The Royal Society of Chemistry (2016)].

## Properties of Graphene-like 2D Layered Transition Metal Dichalcogenides

Layered TMDs have great electronic, mechanical, thermal, and optical properties. Taking $MoS_2$ for example, the room-temperature carrier motilies is of 100 cm$^2$ V$^{-1}$ S$^{-1}$ and Young's modulus was ~ 0.27 TPa.[13]

*Electronic Properties* TMDs can be categorized into absolute insulators, semiconductors, semi-metals, and metals. The bandgap structure of layered TMD materials is based on the number of the layers. When the bulk materials transform to multi- or single-layered materials, their electronic properties will be significantly changed. For example, $MoS_2$ is a typical kind of semiconductor material, bulk $MoS_2$ material has indirect bandgaps, while its single-layered form has direct ones.[24] Kuc group[25] (2011) reported that quantum confinement in layered *d*-electron dichalcogenides resulted in tuning the electronic structure. The bandgap structures of $WS_2$ and $MoS_2$ are shown in Fig. 2. When the number of the layers decrease, the indirect bandgap structures of $WS_2$ and $MoS_2$ achieve transformation to direct ones. However, $NbS_2$ and $ReS_2$ are metallic whose electronic performance has nothing to do with the number of the layers. Single-layered $MoSe_2$ and $MoTe_2$ also exhibited transformation of bandgap structures from indirect to direct style, with the bandgap energies of 1.44 and 1.07 eV, respectively.[26]

*Mechanical Properties* 2D TMD is a kind of flexible thin material, which has great potential to be utilized as mechanical materials. Kis and co-workers[27] measured the stiffness and breaking strength of single-layered $MoS_2$ exfoliated from bulk materials, with the in-plane stiffness and Young's modulus reaching 180 ± 60 N m$^{-1}$ and

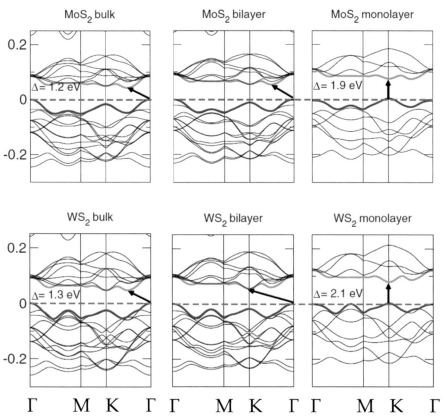

**Fig. 2.** Band structures of bulk $MoS_2$ and $WS_2$. The corresponding monolayer was calculated at the DFT/PBE level. The horizontal dashed lines indicate the Fermi level. The arrows indicate the fundamental band gap (direct or indirect) for a given system. The top of valence band (blue/dark gray) and bottom of conduction band (green/light gray) are highlighted. [Reprinted with permission from Ref. 26, Kuc et al., *Phys. Rev. B*, 2011 **83**, 245213, Copyright @ American Physical Society (2011)].

270 ± 60 GPa. Li[28] (2012) used first-principle calculation to predict ideal tensile strength of ultrathin $MoS_2$ sheets with Young's modulus reaching 250 GPa, which is in agreement with the previous experimental results. The excellent mechanical performance of layered TMDs promotes their application in flexible electronic devices or as reinforcing elements in composites.

*Thermal Properties* TMDs exhibit excellent thermal stability, which will not readily be decomposed at higher temperature. They facilitate outstanding thermal conductivity. Varshney and co-workers[29] (2010) used molecular dynamics simulations to investigate the anisotropic nature of thermal transport in $MoS_2$. The anisotropic factor of $MoS_2$ along two directions reached ~ 4 in the value of thermal conductivity. Liang group[30] used first-principles calculations to investigate the vibrational properties and intrinsic thermal electronic properties of layered $MoS_2$ and $WSe_2$. It was found that with the thickness of the thin-layered material increases, its thermal conductance will gradually approach to that of the bulk material.

*Optical Properties* Bandgap structures of the material influence absorption and emission of photons, which will eventually have great effect on their optical properties. When bulk TMDs with indirect bandgaps transfer to layered form with direct bandgaps, their optical properties change, especially in the change of photoluminescence, electroluminescence, absorption spectrum, vibrational spectrum, and photovoltaic effect. The 2D single-layered TMDs with direct bandgap structure have better performance in light utilization because photons with greater energy than the bandgap energy can be easily absorbed or emitted.[31] The photoluminescence quantum yield of 2D layered structures increases $1 \times 10^4$ fold than that of bulk ones. 2D layered TMDs also have excellent light sensitivity, for example, the light sensitivity of mono-layered $MoS_2$ reaches 7.5 mA $W^{-1}$. Based on the excellent optical properties, $MoS_2$ has been explored to be used as photoelectric detectors, photo sensors, and other advanced photochemical devices and holds great potential to be further utilized in large-scale practice. Besides, single-layered $MoS_2$ and $SnS_2$ have great advantages in flexible transparent LED devices.

# Synthesis of Graphene-like 2D Layered Transition Metal Dichalcogenides

Preparation of 2D layered TMDs with ideal structures and properties is as important as exploring potential applications of those materials. The synthetic methods for 2D TMDs can be categorized into two parts, including top-down methods and bottom-up approaches. Top-down methods refer to the exfoliation methods and laser thinning techniques, while bottom-up approaches cover chemical vapor deposition (CVD) and wet chemical synthesis (hydro/solvothermal methods). In this chapter, we just focus on the exfoliation method, CVD method, and hydro/solvothermal method in synthesizing 2D layered TMDs.

## Exfoliation Method

*Mechanical Exfoliation* Mechanical exfoliation is the most traditional method to produce 2D layered materials. The detailed process could be described as facilitating layered materials rubbed against $Si/SiO_2$ substrate, and then transforming them onto target substrate. In 2005, Novoselov group[32] applied this method to produce single-layered TMDs from their bulk material. Single-layered crystallites and atomic-resolution images of $NbSe_2$ and $MoS_2$ are respectively shown in Fig. 3. 2D layered TMDs obtained in this way have little lattice imperfections and remain stable for a long time, which is suitable for fundamental studies about physical properties of those materials.[33] Lee and co-workers[34] (2010) investigated the frictional characteristics of $MoS_2$ and $NbSe_2$ nanosheets exfoliated from their bulk material. However, this method takes a long time, and the output of 2D materials is too low to satisfy large-scale application.

*Lithium Intercalation and Exfoliation* Zhang group developed an electrochemical Li-intercalation and exfoliation method to produce multiple kinds of TMDs.[35] In a

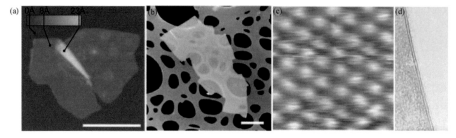

**Fig. 3.** Single-layered crystallites of (a) NbSe$_2$ and (b) MoS$_2$ characterized by atomic-resolution images. (c) Unfiltered scanning tunneling microscopy image of the crystal lattice in the NbSe$_2$ monolayer on top of an oxidized Si wafer. (d) HRTEM image of a double-layered MoS$_2$. [Reprinted with permission from Ref. 32, Novoselov et al., *PNAS.*, 2005, **102**, 10451. 2005 Copyright @ National Academy of Sciences (2005)].

lithium-ion battery configuration, the layered bulk material served as the cathode, such as MoS$_2$, while the anode was lithium foil. In the charging process, lithium ions were intercalated between layers of bulk materials to form LixMoS$_2$, while in the discharging process, lithium ions reacted with electrolyte (water or ethanol) to form hydrogen, resulting in the exfoliation of layered MoS$_2$ (Fig. 4). Apart from MoS$_2$, Zhang group also prepared other metal dichalcogenide nanosheets (NbSe$_2$, WSe$_2$, SbSe$_3$, BiTe$_2$, TiS$_2$, and TaS$_2$) by optimizing cut-off voltage and discharge current.[36,37] The reaction route could be detected by discharging curves. Although lithium-intercalation exfoliation takes little time, it might change the electronic properties of the layered TMDs, which leads to limitations in their applications.

*Liquid Exfoliation* The basic principle of liquid exfoliation is that target TMD bulk compounds are immersed in specific solvents, where TMDs could be efficiently dispersed and dissolved. Then bulk compounds can be transformed into individual flakes by sonication. It is indicated that the exfoliation enthalpy is minimized when the surface energy of the layered material matches that of the solvent, which facilitates

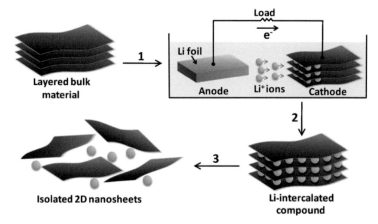

**Fig. 4.** Electrochemical lithiation process for the 2D layered bulk nanosheet. [Reprinted with permission from Ref. 35, Zeng et al., *Angew. Chem. Int.* Ed., 2011, **50**, 11093, Copyright @ Wiley-VCH Verlag GmbH & Co. KGaA, Weinheim (2011)].

the exfoliation process.[12] Coleman group[38] (2011) proved that 2D layered TMD compounds such as $MoS_2$, $WS_2$, $MoSe_2$, $MoTe_2$, $TaSe_2$, $NbSe_2$, $NiTe_2$, $Bi_2Te_3$ could be dispersed in common solvents and deposited as individual layers. For example, $MoS_2$ and $WS_2$ could be well dispersed in N-methyl-pyrrolidone (NMP) and exfoliated as single- or few-layered 2D flakes, as shown in Fig. 5. Liquid exfoliation could meet the requirement for mass production because of its convenience and low cost, however, it may be difficult to choose appropriate solvents. Common solvents, such as water and ethanol are considered poor solvents for the exfoliation of 2D materials. Therefore, finding unusual solvents or preparing mixed ones will be the solution,[39] which brings great trouble to scientists. Moreover, some factors such as solvent concentration, pH, and electrical conductivity might greatly affect the purity of the exfoliated samples.

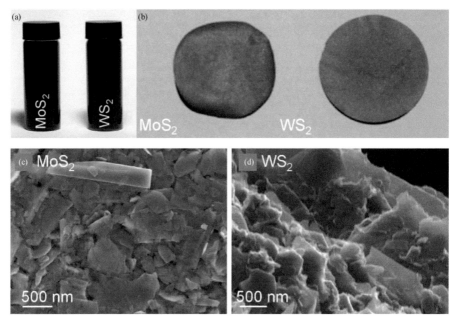

**Fig. 5.** (a) Photograph of the dispersion of $MoS_2$ (in N-methyl-pyrrolidone (NMP)) and $WS_2$ (in N-methyl-pyrrolidone (NMP)). (b) Photograph of freestanding films of $MoS_2$, and $WS_2$ (thickness ~ 50 mm). (c) The SEM image of the surface of the $MoS_2$ film. (d) He ion microscope image of the edge of the $WS_2$ film. [Reprinted with permission from Ref. 38, Coleman et al., *Science*, 2011, **331**, 568, Copyright @ 2011 American Association for the Advancement of Science (2011)].

## CVD Method

The CVD method, utilizing the reaction of metal and sulfur source, is more efficient in fabricating mono- or few-layered TMDs on certain substrates ($Si/SiO_2$).[65] Ultrathin 2D materials prepared in this way have high purity and crystal quality with controllable thickness and sizes. In the case of $MoS_2$, Li and co-workers[40] (2012) first proposed that annealing $(NH_4)_2MoS_4$ could result in mono-layered $MoS_2$ film directly. As is shown in Fig. 6a, the whole process included dip-coating $(NH_4)_2MoS_4$ on $Si/SiO_2$ foils, followed by annealing at 500°C, and then sulfurization under sulfur vapor at 1000°C to form MoS$_2$ nanosheets with large area and high crystallinity on the substrate, which could be readily transformed to other substrates. Besides, Lou and co-workers[41] put a SiO$_2$ substrate with molybdenum film deposited on it and pure sulfur powder on the

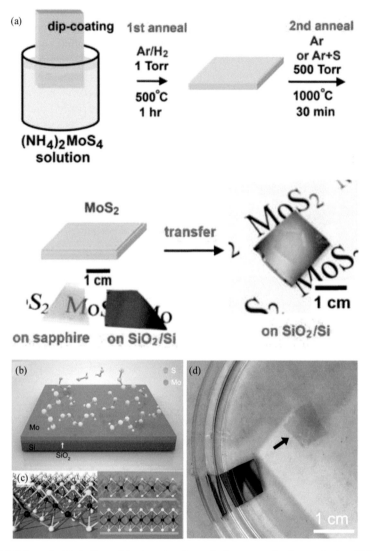

**Fig. 6.** (a) Schematic illustration of the two-step thermolysis process for the synthesis of MoS$_2$ thin layers on the insulating substrate. The precursor (NH$_4$)$_2$MoS$_4$ was dip-coated on SiO$_2$/Si or sapphire substrates followed by the two-step annealing process. The as-grown MoS$_2$ film can be transferred onto other arbitrary substrates. (b) Introducing sulfur on the Mo thin film that was pre-deposited on the SiO$_2$ substrate. (c) MoS$_2$ films that are directly grown on the SiO$_2$ substrate. (d) SiO$_2$ substrate (left) and peeled-off few-layered MoS$_2$ (right, indicated by the arrow). [Reprinted with permission from Ref. 40, Liu et al., *Nano Lett.*, 2012, **12**, 1538, Copyright @ American Chemical Society (2012)] [Reprinted with permission from Ref. 41, Lou et al. *Small*, 2012, **8**, 966, Copyright @ Wiley-VCH Verlag GmbH & Co. KGaA, Weinheim (2012)].

center and edge of the furnace, respectively. In the nitrogen atmosphere, the reactants were heated at 500°C and kept at 750°C for ten minutes, then cooled naturally to room temperature to obtain 2D $MoS_2$ (Fig. 6b–d). Except for molybdenum itself, heating molybdenum oxide and sulfur powder at high temperature could form 2D $MoS_2$.[42] As for other 2D TMDs, $MoSe_2$,[43,44] $WS_2$,[45,46] $WSe_2$,[47,48] and $ReS_2$[49,50] could be synthesized by changing the precursor, temperature, and substrates. In general, CVD method is the most effective way to produce massive and highly crystalline 2D TMDs because high-temperature annealing enhances the crystallinity of the material. Also, this method makes it easier to control the sizes and physical properties of the product.

## Hydro/Solvothermal Synthesis

Hydro/solvothermal synthesis indicates preparing 2D layered materials through chemical reactions in water or organic solvents in sealed vessels at high temperature and pressure. Xie group made great contribution in developing this method. For example, Xie and co-workers (2013) used hexaammonium heptamolybdate tetrahydrate ($(NH_4)_6Mo_7O_{24}\cdot 4H_2O$) and excess thiourea to fabricate defect-rich 2D layered $MoS_2$ through hydrothermal method.[51] They also adopted similar hydrothermal method at 200°C to synthesize oxygen-incorporated $MoS_2$ with enlarged interlayer spacing. Matte's group[20,52] reported many ways to synthesize layered $WS_2$, $WSe_2$, and $MoSe_2$ by hydrothermal process. For example, $MoSe_2$ and $WSe_2$ could be obtained by dissolving molybdic acid (or tungstic acid) and selenium metal in aqueous solution and enabling them to react under hydrothermal condition. Apart from hydrothermal method, solvothermal method is also largely utilized to synthesize 2D layered TMDs. For example, Gao and co-workers[53] (2015) prepared $MoS_2$-coated $CoSe_2$ in a sealed N, N-dimethylformamide (DMF)/hydrazine solvothermal system at 200°C for 10 hours (Fig. 7). The hydro/solvothermal method is simple with scalable yield and low cost. Products synthesized in this way have high qualities and controllable sizes, but it is difficult to ensure whether single-layered materials are obtained. Besides, the mechanism of the hydro/solvothermal is unsure, so it is hard to apply the same method in other synthesis processes.

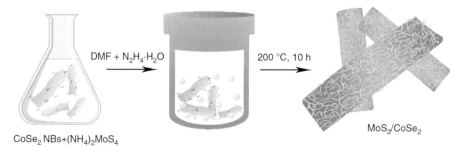

**Fig. 7.** Schematic illustration of the preparation of $MoS_2/CoSe_2$ hybrid: solvothermal synthesis with $CoSe_2$/DETA nanobelts as substrates for preparation of $MoS_2/CoSe_2$ hybrid. [Reprinted with permission from Ref. 53, Gao et al. *Nat Comm.*, 2015, **6**, 5982, Copyright @ Macmillan Publishers Limited (2015)].

# Graphene-like 2D Layered Transition Metal Dichalcogenides in Electrochemical Energy Utilization

## Supercapacitors

Supercapacitors play a crucial role among multiple energy storage devices with their excellent performance in energy and power densities, fast charge-discharge rate, and good stability. Supercapacitors can be categorized into three types, including electric double-layer supercapacitors (EDLCs), faradic pseudocapacitors, and hybrid supercapacitors. EDLCs store energy through charge accumulation at the interface of electrode and electrolyte, whose capacitance comes from physical processes, while the capacitance of faradic pseudocapacitors arises from fast and reversible redox reactions at the electrode/electrolyte interfaces. The hybrid supercapacitor is composed of a battery-type faradic cathode and a double-layer-type anode, which takes advantage of both two electrodes to widen working potential window of the whole device.[1]

TMDs are more likely to be applied in pseudocapacitors. As is already known, the emergence of pseudocapacitance comes from ion intercalation, such as the intercalation of lithium ions into $Nb_2O_5$.[54] The sandwich S-M-S structure of TMDs provides enough accommodation for ion intercalation. Moreover, transition metal atoms have multiple valance states, enabling them to achieve pseudocapacitance through fast and reversible redox reactions. Chhowalla's group[55] showed that single-layered 1T $MoS_2$ could be exfoliated and restacked to form electrode materials for supercapacitors without additives (Fig. 8). The restacked layered $MoS_2$ provided enough space for ion intercalation, which exhibited high volumetric capacitance of $\sim 400$ to $700$ F $cm^{-3}$ in different aqueous solutions, with the capacitance value of $650$ F $cm^{-3}$ in $H_2SO_4$ aqueous electrolyte. While the 1T $MoS_2$ was in organic electrolyte, it was suitable in high-voltage operation (3.5 V), and showed high volumetric energy and power densities with great cycling abilities for over 5000 cycles. The excellent performance of 1T $MoS_2$ is attributed to high electrical conductivity, hydrophilicity, and multiple channels in nanosheets for ions intercalation.[56] This pioneering work opens up the way for the utilization of 2D $MoS_2$ in supercapacitors. Afterwards, Yang and co-workers[57] developed a micro-supercapacitor based on $MoS_2$ films. The electrode was fabricated by spray painting of $MoS_2$ on $SiO_2$ chip and then laser patterning. The whole device manifested great energy storage abilities with the areal capacitance of $8$ mF $cm^{-2}$ and superior cycling performance. Zhu and co-workers[58] synthesized 2D multi-layered water-coupled metallic $MoS_2$ (M-$MoS_2$-$H_2O$) electrode with a large amount of nanotubes, which exhibited a high capacitance of $380$ F $g^{-1}$ at the scan rate of $5$ mV $s^{-1}$ and still retained $105$ F $g^{-1}$ at $10$ V $s^{-1}$. The symmetric supercapacitance based on this electrode reached high specific capacitance of $249$ F $g^{-1}$ at $50$ mV $s^{-1}$. Apart from $MoS_2$, other 2D TMDs also serve as efficient electrode material for supercapacitors. Xie and co-workers[59] presented an ammonia-assisted method to exfoliate $VS_2$ nanosheets with high conductivity. The 150 nm in-plane supercapacitor composed of $VS_2$ electrodes achieved high areal capacitance of $4760$ μF $cm^{-2}$ and exhibited great cycling performance for over 1000 cycles. Arul and co-workers[60] (2016) fabricated $NiSe_2$ electrode with the specific capacitance of $75$ F $g^{-1}$ and good cycling stability for 94% capacitance retention after 5000 cycles at $1$ mA $cm^{-2}$.

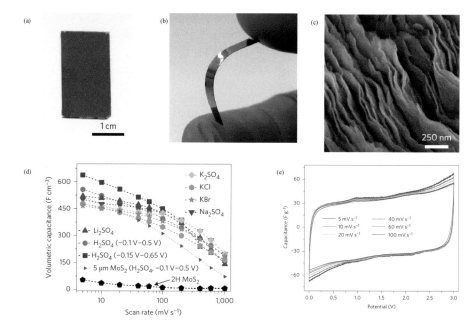

**Fig. 8.** (a, b) Photographs of electrodes consisting of a thick film of chemically exfoliated 1T MoS$_2$ prepared by vacuum filtration and transferred onto (a) rigid glass and (b) a flexible polyimide substrate. (c) High-magnification image of restacked MoS$_2$ nanosheets. (d) Evolution of the volumetric capacitance of the 1T phase MoS$_2$ electrodes with scan rate for different electrolytes and 1-μm- and 5-μm-thick films. The concentration of the cations in the electrolyte solutions was fixed at 1 M. (e) Electrochemical behavior of 1T phase MoS$_2$ electrodes in organic electrolytes. [Reprinted with permission from Ref. 55, Acerce et al. *Nat. Nano.*, 2015, **10**, 313, Copyright @ Macmillan Publishers Limited (2015)].

Except for layered TMDs themselves, their composite with carbon materials or conducting polymers all have wide applications in supercapacitors. Carbon materials such as activated carbon, carbon nanotubes, and graphene are widely applied in EDLCs because of their high conductivity and large surface area. Among them, graphene stands out for its unique properties and excellent performance to serve as the electrode material of supercapacitors. TMD/graphene composite takes advantage of both pseudocapacitance of TMDs and double-layer capacitance of graphene. Leite and co-workers[61] deposited layered MoS$_2$ on reduced graphene oxide by microwave with the first layer of MoS$_2$ bonded with oxygen of reduced graphene oxide. The MoS$_2$/graphene hybrid material exhibited specific capacitance of 128, 265, and 148 F g$^{-1}$ for materials with low, medium, and high concentrations of MoS$_2$ layers respectively, and the energy density of this hybrid material with low concentration of MoS$_2$ layers reached up to 64 Wh kg$^{-1}$. As for compositing with conductive polymers, polyaniline (PANI), polythiophene (PTH), and polypyrrole (PPy) are generously applied.[62] Yan and co-workers[63] reported a mild way to produce massive MoS$_2$@PANI architectures, enhancing the interaction between electrolyte and electrode materials, and reducing the diffusion distance of ions and electrons, enabling the composite to show excellent electrochemical properties. Besides, Tang and co-workers[64] (2015) synthesized 2D mono-layered MoS$_2$/PPy ultrathin films hybrids as the electrode for supercapacitors.

The high performance of the device was due to large surface area of PPy ultrathin films and strong interface interaction between PPy and $MoS_2$, facilitating efficient charge storage and transport of electrons and electrolytes.

## Lithium-Ion Batteries

Among the various energy storage devices, lithium-ion batteries play an important role because of their high specific energy and long cycle life, which can be used as commercial portable devices. Most rechargeable lithium-ion batteries are composed of a $LiCoO_2$ cathode and a graphite anode with a theoretical energy density of 387 Wh kg$^{-1}$.[65] In the charging process, lithium ions are embedded into the graphene layer to form $LiC_6$, while in the discharging process, they are desorbed from the electrode materials.[4] The key factor in improving the performance of lithium-ion battery is to design and fabricate novel electrode materials with high lithium-storage abilities.

TMDs, with their layered structures, have become strong candidates as anodes for lithium-ion batteries, since lithium ions could be inserted in and extracted from the space between layers of TMDs during the charge and discharge process without causing great structure deterioration. Cheon and co-workers[66] fabricated 2D $WS_2$ nanosheets by rolling out of 1D $W_{18}O_{49}$ nanorods by carbon disulfide in hot hexadecylamine solution (Fig. 9). The 2D morphology of $WS_2$ nanosheets resulted in a larger surface area and provided more space for ions to intercalate than that of the bulk material. The 2D $WS_2$ electrode for lithium-ion batteries exhibited the first reversible capacity of 377 mAh g$^{-1}$ and great cycling performance for 30 cycles, which was much higher than that of bulk $WS_2$. Guo and co-workers[67] reported $MoS_2$ with enlarged spacing between layers by exfoliation and restacking. The $MoS_2$ electrode exhibited large capacity of 800 mAh g$^{-1}$ and excellent cycling stability even at high charge current density, which could serve as promising anode materials for lithium ions batteries. Increasing the surface area and enlarging the space between layers in 2D materials would be widely utilized strategies for enhancing ions intercalation capability, which could largely improve the electrochemical performance of lithium ions batteries.

As for TMD composites, Zhao and co-workers[68] proposed a 2D $MoS_2$/2D single-layered mesoporous-carbon composite material for high-performance lithium-ion batteries. The sufficient mesopores on this composite could serve as an excellent pathway for lithium-ion transportation and accommodate its intercalation and deintercalation, contributing to the capacitive performance of the material. At the current density of 1000 mA g$^{-1}$, the electrode with sandwich-like structures (mesoporous carbon/$MoS_2$/mesoporous carbon) manifested high reversible capacity of over 1400 mAh g$^{-1}$. Furthermore, it still retained a reversible capacity of 400 mAh g$^{-1}$ at an ultrahigh current density of 10 A g$^{-1}$ for over 300 cycles. Hybridizing with graphene is also a good choice to construct high-performance electrodes for 2D $MoS_2$,[69–71] $ReS_2$[72] and other layered TMDs. For instance, Fu and co-workers[72] made $ReS_2$ nanosheets grow vertically on graphene. The uniform distribution of $ReS_2$ on graphene and the weak interaction between $ReS_2$ layers resulted in excellent performance for lithium-ion batteries. The novel structure offered more space for lithium-ion intercalation and the intimate contact between $ReS_2$ and graphene, which facilitated fast electron transfer. The $ReS_2$/graphene composite electrode showed excellent cycling stability,

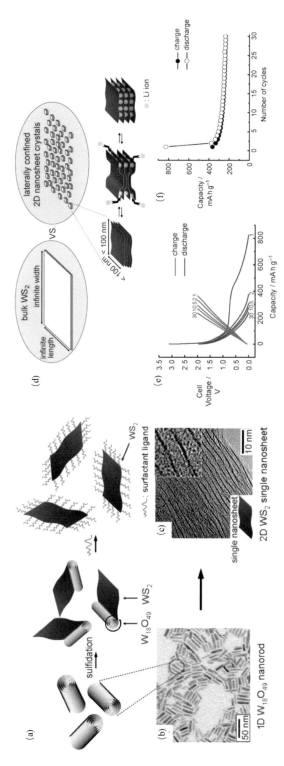

**Fig. 9.** (a) Schematic diagram for the synthesis of 2D WS$_2$ nanosheet crystals. (b) TEM image of W$_{18}$O$_{49}$ nanorod precursors. (c) TEM images of 2D WS$_2$ obtained with the reaction 10 min after the CS$_2$ injection. (d) Schematic representation of the lithium–ion intercalation process in 2D WS$_2$ nanosheet crystals. (e) Cell-voltage profiles of 2D WS$_2$ nanosheet crystals during the 1st, 2nd, 5th, 10th, and 30th cycles between 5 mV and 2.0 V. (f) Cycle-life performance of the 2D WS$_2$ nanosheet crystals cycled up to the 30th cycle under constant-current conditions (100 mA g$^{-1}$). [Reprinted with permission from Ref. 65, Seo et al., *Angew. Chem. Int. Ed.*, 2007, **46**, 8828, Copyright @ Wiley-VCH Verlag GmbH & Co. KGaA, Weinheim (2007)].

and at the low current density of 100 mA g$^{-1}$, the electrode exhibited a capacitance of 500 mAh g$^{-1}$ in the first eight cycles and the capacity was still over 200 mAh g$^{-1}$ after 500 cycles even at the high current density of 1000 mA g$^{-1}$.

## Electrocatalysis

To relieve the intense energy situation, electrocatalytic HER is regarded as a promising solution because the production of hydrogen could largely reduce the consumption of fossil fuels. Until now, Pt has outperformed as the most effective catalyst in HER. However, the high cost and small quantities of Pt make it urgent to find substitutions. 2D layered TMDs have been found to be the strongest candidates as substitutes for Pt in HER due to their low cost, large abundance, and excellent catalytic performance.

The mechanism of HER can be interpreted by Tafel analysis based on the Volmer-Heyrovsky and Volmer-Tafel mechanisms.

Volmer reaction indicates the adoption of H$^+$ on the interface of electrode,

In acidic condition:   $H_3O^+(aq) + e^- \rightleftarrows H* + H_2O(l)$
In alkaline condition:   $H_2O(l) + e^- \rightleftarrows H* + OH^-(aq)$

The volmer reaction can be followed by Heyrovsky reaction, which demonstrates the next-step reduction of H*,

In acidic condition:   $H* + H_3O^+(aq) + e^- \rightleftarrows H_2(g) + H_2O(l)$
In alkaline condition:  $H* + H_2O(l) + e^- \rightleftarrows H_2(g) + OH^-(aq)$

or Tafel reaction:

In both acidic and alkaline condition:  $H* + H* \rightleftarrows H_2(g)$

where H* refers to a hydrogen atom bounded to an active site. In all reactions, binding of the H* to the catalyst surface is paramount. The binding energy with extra low or high value limits the adsorption or desorption reaction during catalytic process. The rate-limiting step in the HER process can be identified as Volmer, Heyrovsky, or Tafel reactions with Tafel slopes of 120, 40, and 30 mV dec$^{-1}$, respectively. [73–78]

Tafel plots derived from polarization curves are utilized in the study of catalytic performance of HER. The Tafel equation is shown as follows:

$$j = -j_0 e^{-\frac{n}{b}} \tag{1.1}$$

$$\ln(-j) = \ln(j_0) - (\frac{1}{b})\eta \tag{1.2}$$

where $j$ is the current density, $j_0$ is the exchange current density, $\eta$ is the overpotential which indicates the increased potential beyond the reversible potential of the half-cell reaction, $b$ refers to Tafel slope. The smaller value of Tafel slope manifests faster electron transfer rate, indicating better electrocatalytic HER activity of the materials.

Theoretical studies based on DFT calculations indicate that the edges of 2D TMDs serve as active sites for HER. The atomic sites on edges of MoS$_2$ nanosheets have unsaturated coordination and dangling bonds, which are key factors in HER. Also, both computational and experimental studies prove that layered TMDs outperform bulk TMDs and other HER catalysts, because countless active edge sites with high catalytic activities can be found on layered TMDs.[8, 21] Therefore, increasing the number of edge sites is crucial in enhancing catalytic activities of layered TMDs. Xie group[51] (2013) fabricated $MoS_2$ with defect-rich structure by thiourea. On the one hand, excess thiourea served as reducing agent to reduce Mo(VI) to Mo(IV), on the other hand, it could stabilize $MoS_2$ nanosheets by separating them. Due to the defect-rich structure of ultrathin $MoS_2$ nanosheets, more edge sites were exposed on the surface, which resulted in excellent catalytic performance with the HER onset overpotential of just 120 mV and the Tafel slope value of 50 mV decade$^{-1}$ (Fig. 10). Chen and co-workers[79] developed 2D $TaS_2$ nanosheets with ultra-small pores by oxygen-plasma etching method. The exposed active sites along those pores largely increased the catalytic activities towards HER. Compared to pristine $TaS_2$, $TaS_2$ treated by oxygen plasma for 10 minutes showed better catalytic performance with the onset potential value of 200 mV and the Tafel slope value of 135 mV decade$^{-1}$. Gao and co-workers[53] (2015) fabricated $MoS_2/CoSe_2$ hybrid as HER catalyst by *in situ* growth of 2D layered $MoS_2$ nanosheets on the surface of $CoSe_2$. The hybrid catalyst exhibited onset potential of –11 mV, Tafel slope of 36 mV decade$^{-1}$ and high exchange current density in $H_2SO_4$ electrolyte. The good catalytic performance can be interpreted by increased catalytic sites between $MoS_2$ and $CoSe_2$ together with their electrocatalytic synergistic effects. Besides, a large number of studies of surfactant in 2D-$MoS_2$ nanomaterials have been conducted. Generally, surfactants are known to stabilize mono-layered 2D $MoS_2$ nanosheets and expose more active catalytic sites, which enhance the catalytic activity of the 2D $MoS_2$ nanomaterials. However, Craig E. Banks and co-workers[80] manifested that the surfactant (sodium cholate, SC) has detrimental effect on the HER activity of 2D-$MoS_2$. It can be attributed to a weaker catalytic performance of 2D-$MoS_2$-SC than the pristine 2D $MoS_2$ with higher onset potential of 0.19 V and the Tafel slope of 47 mV dec$^{-1}$. The study critically highlighted a significant role of surfactant for 2D-$MoS_2$ fabrication for HER.

Apart from the quantity of active sites, scientists are dedicated to exploring ways of enhancing the electric conductively of the materials because electric conductivity also plays a decisive role in catalytic performance. The electric conductivity of $MoS_2$ can be improved through compositing with conducting compounds. Dai and co-workers[81] reported $MoS_2$ particles grown on graphene. The catalytic performance of this hybrid material was considerably higher than that of $MoS_2$ alone with small onset potential and Tafel slope of about 100 mV and 41 mV decade$^{-1}$ respectively. Shin and co-workers[82] synthesized $WS_2$/graphene composite nanosheets through one-pot hydrothermal method at low temperature. Due to the strong interactions between $WS_2$ and graphene, the charge transfer rate was greatly enhanced with improved catalytic activities of HER. The $WS_2$/graphene catalysts exhibited onset potential ranging from –150 ~ –200 mV with Tafel slope of 58 mV decade$^{-1}$. It is also worth mentioning that Craig E. Banks and co-workers[76] fabricated 2D $MoS_2$ modified carbon-based electrodes, such as pyrolytic graphite (EPPG), glassy carbon (GC), boron-doped diamond (BDD)

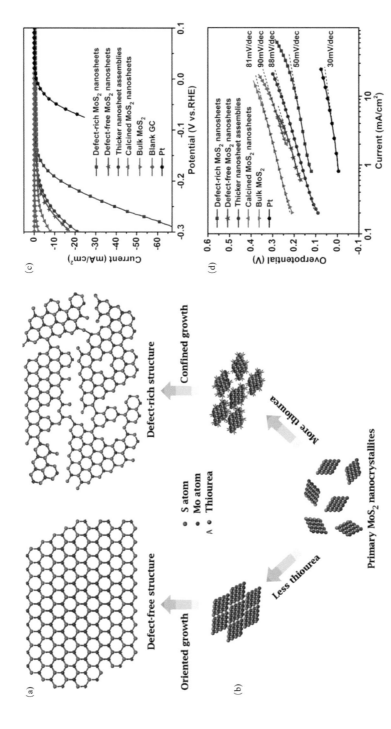

**Fig. 10.** (a) Structural models of defect-free and defect-rich structures. (b) As-designed synthetic pathways to obtain the above two structures. (c) Polarization curves and (d) corresponding Tafel plots of the defect-rich MoS$_2$ nanosheets, defect-free MoS$_2$ nanosheets, thicker nanosheet assemblies, calcined MoS$_2$ nanosheets, bulk MoS$_2$, blank glassy carbon and Pt electrodes. [Reprinted with permission from Ref. 51, Xie et al., *Adv. Mater.*, 2013, **25**, 5807, Copyright @ Wiley-VCH Verlag GmbH & Co. KGaA, Weinheim (2013)].

and screen-printed graphite electrode (SPE). The coverage of $MoS_2$ increased the catalytic activity of those supporting carbon-based electrodes within certain limits. The HER onset potentials of the modified EPPG, GC, SPE and BDD electrodes have reduced by 0.33, 0.57, 0.29 and 0.31 V compared with their unmodified electrodes, which revealed the increased catalytic activity of the hybrids over HER. Craig E. Banks and co-workers[77] also synthesized 2D-$MoSe_2$ incorporated screen-printed electrodes ($MoSe_2$-SPEs) for HER. The optimal mass ratio of the electrode should be 10% 2D-$MoSe_2$ to 90% carbon ink. The as-fabricated electrode exhibited low onset potential (–460 mV) and appropriate Tafel value (47 mV dec$^{-1}$), both of which contributed to a better catalytic performance than that of the unmodified SPEs (onset potential –880 mV and Tafel value 120 mV dec$^{-1}$). The modified electrode displayed excellent stability due to a highly overlapped polarization curve of HER with the initial scan after 1000 cycles. These 2D modified electrodes hold great potential to be used in real practice as ideal HER alternatives for commercial Pt/C.

# Conclusion and Prospective

In this chapter, we summarize the basic understanding of the layered 2D TMDs in terms of the structures, properties, synthetic methods, and applications in electrochemical energy. As for 2D layered TMDs, the multiple valence states and edge sites contribute a lot to the capacitive and catalytic performance. To better boost the performance of TMDs and address the problem of their poor conductivity and cycling performance, increasing the space between layers or creating more channels in the material could accommodate abundant ions intercalation, contributing to the capacitive capability of 2D layered TMDs. Exposing more edge sites on the surface of 2D layered TMDs could facilitate fast redox reactions for highly efficient charge conversion and transfer. The challenges and difficulties in fabricating 2D TMD materials and exemplified potential solutions are identified to improve their performance.

Facing the future, the promising 2D TMDs with unique structures and properties still have some space to develop for the potentially next-generation energy applications. The new strategies on improving the performance should be taken into further exploration, including constructing novel structures or finding new materials to form composites. Moreover, although there are increasing researches into applications of those materials, their functional mechanism is still unclear and needs to be further explored. In particular, we should pay more effort in investigating the working principle in those materials or at the interface between those materials and electrolyte. Combination between theoretical calculation and experimental studies will be a new trend in exploring deeper understanding and optimization. Our final goal is to produce high-quality materials in low cost and large scale, which could be used in real life to address energy crisis. To optimize these materials for deeper application, breaking the routine, inventing new synthetic methods for large scale and constructing novel structures may lead to surprising results. Therefore, there is still a long way to go. We believe that through combining innovation and hard work, 2D layered TMDs will have bright and shining future.

# Acknowledgments

We acknowledge the financial support of this work received from National Key Research and Development Program of China (No. 2016YFA0203101), National Natural Science Foundation of China (No. 51572139), and the Open Project of National & Local United Engineering Lab for Power Battery, Northeast Normal University (No. 130017501).

# References

1. X. Lu, M. Yu, G. Wang, Y. Tong and Y. Li, *Energy Environ. Sci.*, 2014, **7**, 2160–2181.
2. R. Raccichini, A. Varzi, D. Wei and S. Passerini, *Adv. Mater.*, 2016, **29**, 1603421.
3. M. Pumera, Z. Sofer and A. Ambrosi, *J. Mater. Chem. A*, 2014, **2**, 8981–8987.
4. M. Pumera, *Energy Environ. Sci.*, 2011, **4**, 668–674.
5. K. P. Loh, S. W. Tong and J. Wu, *J. Am. Chem. Soc.*, 2016, **138**, 1095–1102.
6. M. Yu, X. Cheng, Y. Zeng, Z. Wang, Y. Tong, X. Lu and S. Yang, *Angew. Chem. Int. Ed.*, 2016, **55**, 6762–6766.
7. S. Ratha and C. S. Rout, *ACS Appl. Mater. Interfaces*, 2013, **5**, 11427–11433.
8. G. Zhang, H. Liu, J. Qu and J. Li, *Energy Environ. Sci.*, 2016, **9**, 1190–1209.
9. A. K. Geim and K. S. Novoselov, *Nat. Mater.*, 2007, **6**, 183–191.
10. D. Chen, L. Tang and J. Li, *Chem. Soc. Rev.*, 2010, **39**, 3157–3180.
11. X. Huang, Z. Zeng, Z. Fan, J. Liu and H. Zhang, *Adv. Mater.*, 2012, **24**, 5979–6004.
12. X. Huang, Z. Zeng and H. Zhang, *Chem. Soc. Rev.*, 2013, **42**, 1934–1946.
13. M. Xu, T. Liang, M. Shi and H. Chen, *Chem. Rev.*, 2013, **113**, 3766–3798.
14. J. D. Roy-Mayhew and I. A. Aksay, *Chem. Rev.*, 2014, **114**, 6323–6348.
15. H. Wang and Y. H. Hu, *Energy Environ. Sci.*, 2012, **5**, 8182–8188.
16. Y. Sun, Q. Wu and G. Shi, *Energy Environ. Sci.*, 2011, **4**, 1113–1132.
17. Y. Zhang, Q. Ji, G.-F. Han, J. Ju, J. Shi, D. Ma, J. Sun, Y. Zhang, M. Li, X.-Y. Lang, Y. Zhang and Z. Liu, *ACS Nano*, 2014, **8**, 8617–8624.
18. L. Liang, H. Cheng, F. Lei, J. Han, S. Gao, C. Wang, Y. Sun, S. Qamar, S. Wei and Y. Xie, *Angew. Chem. Int. Ed.*, 2015, **54**, 12004–12008.
19. H. Huang, X. Feng, C. Du, S. Wu and W. Song, *J. Mater. Chem. A*, 2015, **3**, 16050–16056.
20. H. S. S. Ramakrishna Matte, A. Gomathi, A. K. Manna, D. J. Late, R. Datta, S. K. Pati and C. N. R. Rao, *Angew. Chem. Int. Ed.*, 2010, **49**, 4059–4062.
21. M. Chhowalla, H. S. Shin, G. Eda, L.-J. Li, K. P. Loh and H. Zhang, *Nat. Chem.*, 2013, **5**, 263–275.
22. D. Voiry, A. Goswami, R. Kappera, e. SilvaCecilia de Carvalho Castro, D. Kaplan, T. Fujita, M. Chen, T. Asefa and M. Chhowalla, *Nat. Chem.*, 2015, **7**, 45–49.
23. D. Voiry, A. Mohite and M. Chhowalla, *Chem. Soc. Rev.*, 2015, **44**, 2702–2712.
24. A. Splendiani, L. Sun, Y. Zhang, T. Li, J. Kim, C.-Y. Chim, G. Galli and F. Wang, *Nano Lett.*, 2010, **10**, 1271–1275.
25. A. Kuc, N. Zibouche and T. Heine, *Phys. Rev. B*, 2011, **83**, 245213.
26. Y. Ma, Y. Dai, M. Guo, C. Niu, J. Lu and B. Huang, *Phys. Chem. Chem. Phys.*, 2011, **13**, 15546–15553.
27. S. Bertolazzi, J. Brivio and A. Kis, *ACS Nano*, 2011, **5**, 9703–9709.
28. T. Li, *Phys. Rev. B*, 2012, **85**, 235407.
29. V. Varshney, S. S. Patnaik, C. Muratore, A. K. Roy, A. A. Voevodin and B. L. Farmer, *Comput. Mater. Sci.*, 2010, **48**, 101–108.
30. W. Huang, X. Luo, C. K. Gan, S. Y. Quek and G. Liang, *Phys. Chem. Chem. Phys.*, 2014, **16**, 10866–10874.
31. Q. H. Wang, K. Kalantar-Zadeh, A. Kis, J. N. Coleman and M. S. Strano, *Nat. Nanotechnol.*, 2012, **7**, 699–712.
32. K. S. Novoselov, D. Jiang, F. Schedin, T. J. Booth, V. V. Khotkevich, S. V. Morozov and A. K. Geim, *Proc. Natl. Acad. Sci. U. S. A.*, 2005, **102**, 10451–10453.
33. H. Li, G. Lu, Y. Wang, Z. Yin, C. Cong, Q. He, L. Wang, F. Ding, T. Yu and H. Zhang, *Small*, 2013, **9**, 1974–1981.

34. C. Lee, Q. Li, W. Kalb, X.-Z. Liu, H. Berger, R. W. Carpick and J. Hone, *Science*, 2010, **328**, 76–80.
35. Z. Zeng, Z. Yin, X. Huang, H. Li, Q. He, G. Lu, F. Boey and H. Zhang, *Angew. Chem. Int. Ed.*, 2011, **50**, 11093–11097.
36. Z. Zeng, T. Sun, J. Zhu, X. Huang, Z. Yin, G. Lu, Z. Fan, Q. Yan, H. H. Hng and H. Zhang, *Angew. Chem. Int. Ed.*, 2012, **51**, 9052–9056.
37. Z. Zeng, C. Tan, X. Huang, S. Bao and H. Zhang, *Energy Environ. Sci.*, 2014, **7**, 797–803.
38. J. N. Coleman, M. Lotya, A. O'Neill, S. D. Bergin, P. J. King, U. Khan, K. Young, A. Gaucher, S. De, R. J. Smith, I. V. Shvets, S. K. Arora, G. Stanton, H.-Y. Kim, K. Lee, G. T. Kim, G. S. Duesberg, T. Hallam, J. J. Boland, J. J. Wang, J. F. Donegan, J. C. Grunlan, G. Moriarty, A. Shmeliov, R. J. Nicholls, J. M. Perkins, E. M. Grieveson, K. Theuwissen, D. W. McComb, P. D. Nellist and V. Nicolosi, *Science*, 2011, **331**, 568–571.
39. K.-G. Zhou, N.-N. Mao, H.-X. Wang, Y. Peng and H.-L. Zhang, *Angew. Chem. Int. Ed.*, 2011, **123**, 11031–11034.
40. K.-K. Liu, W. Zhang, Y.-H. Lee, Y.-C. Lin, M.-T. Chang, C.-Y. Su, C.-S. Chang, H. Li, Y. Shi, H. Zhang, C.-S. Lai and L.-J. Li, *Nano Lett.*, 2012, **12**, 1538–1544.
41. Y. Zhan, Z. Liu, S. Najmaei, P. M. Ajayan and J. Lou, *Small*, 2012, **8**, 966–971.
42. Y.-H. Lee, X.-Q. Zhang, W. Zhang, M.-T. Chang, C.-T. Lin, K.-D. Chang, Y.-C. Yu, J. T.-W. Wang, C.-S. Chang, L.-J. Li and T.-W. Lin, *Adv. Mater.*, 2012, **24**, 2320–2325.
43. Y.-H. Chang, W. Zhang, Y. Zhu, Y. Han, J. Pu, J.-K. Chang, W.-T. Hsu, J.-K. Huang, C.-L. Hsu, M.-H. Chiu, T. Takenobu, H. Li, C.-I. Wu, W.-H. Chang, A. T. S. Wee and L.-J. Li, *ACS Nano*, 2014, **8**, 8582–8590.
44. J. C. Shaw, H. Zhou, Y. Chen, N. O. Weiss, Y. Liu, Y. Huang and X. Duan, *Nano Res.*, 2014, **7**, 511–517.
45. M. Okada, T. Sawazaki, K. Watanabe, T. Taniguch, H. Hibino, H. Shinohara and R. Kitaura, *ACS Nano*, 2014, **8**, 8273–8277.
46. Y. Zhang, Y. Zhang, Q. Ji, J. Ju, H. Yuan, J. Shi, T. Gao, D. Ma, M. Liu and Y. Chen, *ACS Nano*, 2013, **7**, 8963–8971.
47. S. M. Eichfeld, L. Hossain, Y.-C. Lin, A. F. Piasecki, B. Kupp, A. G. Birdwell, R. A. Burke, N. Lu, X. Peng, J. Li, A. Azcatl, S. McDonnell, R. M. Wallace, M. J. Kim, T. S. Mayer, J. M. Redwing and J. A. Robinson, *ACS Nano*, 2015, **9**, 2080–2087.
48. B. Liu, M. Fathi, L. Chen, A. Abbas, Y. Ma and C. Zhou, *ACS Nano*, 2015, **9**, 6119–6127.
49. X. He, F. Liu, P. Hu, W. Fu, X. Wang, Q. Zeng, W. Zhao and Z. Liu, *Small*, 2015, **11**, 5423–5429.
50. K. Keyshar, Y. Gong, G. Ye, G. Brunetto, W. Zhou, D. P. Cole, K. Hackenberg, Y. He, L. Machado, M. Kabbani, A. H. C. Hart, B. Li, D. S. Galvao, A. George, R. Vajtai, C. S. Tiwary and P. M. Ajayan, *Adv. Mater.*, 2015, **27**, 4640–4648.
51. J. Xie, H. Zhang, S. Li, R. Wang, X. Sun, M. Zhou, J. Zhou, X. W. Lou and Y. Xie, *Adv. Mater.*, 2013, **25**, 5807–5813.
52. H. S. S. R. Matte, B. Plowman, R. Datta and C. N. R. Rao, *Dalton Trans.*, 2011, **40**, 10322–10325.
53. M.-R. Gao, J.-X. Liang, Y.-R. Zheng, Y.-F. Xu, J. Jiang, Q. Gao, J. Li and S.-H. Yu, *Nat. Commun.*, 2015, **6**, 5982.
54. V. Augustyn, J. Come, M. A. Lowe, J. W. Kim, P.-L. Taberna, S. H. Tolbert, H. D. Abruña, P. Simon and B. Dunn, *Nat. Mater.*, 2013, **12**, 518–522.
55. M. Acerce, D. Voiry and M. Chhowalla, *Nat. Nanotechnol.*, 2015, **10**, 313–318.
56. C. Tan, X. Cao, X.-J. Wu, Q. He, J. Yang, X. Zhang, J. Chen, W. Zhao, S. Han, G.-H. Nam, M. Sindoro and H. Zhang, *Chem. Rev.*, 2017, **117**, 6225–6331.
57. L. Cao, S. Yang, W. Gao, Z. Liu, Y. Gong, L. Ma, G. Shi, S. Lei, Y. Zhang, S. Zhang, R. Vajtai and P. M. Ajayan, *Small*, 2013, **9**, 2905–2910.
58. X. Geng, Y. Zhang, Y. Han, J. Li, L. Yang, M. Benamara, L. Chen and H. Zhu, *Nano Lett.*, 2017, **17**, 1825–1832.
59. J. Feng, X. Sun, C. Wu, L. Peng, C. Lin, S. Hu, J. Yang and Y. Xie, *J. Am. Chem. Soc.*, 2011, **133**, 17832–17838.
60. N. S. Arul and J. I. Han, *Mater. Lett.*, 2016, **181**, 345–349.
61. E. G. da Silveira Firmiano, A. C. Rabelo, C. J. Dalmaschio, A. N. Pinheiro, E. C. Pereira, W. H. Schreiner and E. R. Leite, *Adv. Energy Mater.*, 2014, **4**, 1301380.
62. H. Wang, H. Feng and J. Li, *Small*, 2014, **10**, 2165–2181.
63. J. Zhu, W. Sun, D. Yang, Y. Zhang, H. H. Hoon, H. Zhang and Q. Yan, *Small*, 2015, **11**, 4123–4129.
64. H. Tang, J. Wang, H. Yin, H. Zhao, D. Wang and Z. Tang, *Adv. Mater.*, 2015, **27**, 1117–1123.

65. P. G. Bruce, S. A. Freunberger, L. J. Hardwick and J.-M. Tarascon, *Nat. Mater.*, 2012, **11**, 19–29.
66. J.-W. Seo, Y.-W. Jun, S.-W. Park, H. Nah, T. Moon, B. Park, J.-G. Kim, Y. J. Kim and J. Cheon, *Angew. Chem. Int. Ed.*, 2007, **46**, 8828–8831.
67. G. Du, Z. Guo, S. Wang, R. Zeng, Z. Chen and H. Liu, *Chem. Commun.*, 2010, **46**, 1106–1108.
68. Y. Fang, Y. Lv, F. Gong, A. A. Elzatahry, G. Zheng and D. Zhao, *Adv. Mater.*, 2016, **28**, 9385–9390.
69. X. Cao, Y. Shi, W. Shi, X. Rui, Q. Yan, J. Kong and H. Zhang, *Small*, 2013, **9**, 3433–3438.
70. K. Chang and W. Chen, *J. Mater. Chem.*, 2011, **21**, 17175–17184.
71. K. Chang and W. Chen, *Chem. Commun.*, 2011, **47**, 4252–4254.
72. Q. Zhang, S. Tan, R. G. Mendes, Z. Sun, Y. Chen, X. Kong, Y. Xue, M. H. Rümmeli, X. Wu, S. Chen and L. Fu, *Adv. Mater.*, 2016, **28**, 2616–2623.
73. A. B. Laursen, S. Kegnaes, S. Dahl and I. Chorkendorff, *Energy Environ. Sci.*, 2012, **5**, 5577–5591.
74. N. M. Marković, B. N. Grgur and P. N. Ross, *J. Phys. Chem. B*, 1997, **101**, 5405–5413.
75. N. M. Marković and P. N. Ross, *Surf. Sci. Rep.*, 2002, **45**, 117–229.
76. S. J. Rowley-Neale, D. A. C. Brownson, G. C. Smith, D. A. G. Sawtell, P. J. Kelly and C. E. Banks, *Nanoscale*, 2015, **7**, 18152–18168.
77. S. J. Rowley-Neale, C. W. Foster, G. C. Smith, D. A. C. Brownson and C. E. Banks, *Sustainable Energy Fuels*, 2017, **1**, 74–83.
78. X. Zhu, M. Liu, Y. Liu, R. Chen, Z. Nie, J. Li and S. Yao, *J. Mater. Chem. A*, 2016, **4**, 8974–8977.
79. H. Li, Y. Tan, P. Liu, C. Guo, M. Luo, J. Han, T. Lin, F. Huang and M. Chen, *Adv. Mater.*, 2016, **28**, 8945–8949.
80. G. B. de-Mello, L. Smith, S. J. Rowley-Neale, J. Gruber, S. J. Hutton and C. E. Banks, *RSC Adv.*, 2017, **7**, 36208–36213.
81. Y. Li, H. Wang, L. Xie, Y. Liang, G. Hong and H. Dai, *J. Am. Chem. Soc.*, 2011, **133**, 7296–7299.
82. J. Yang, D. Voiry, S. J. Ahn, D. Kang, A. Y. Kim, M. Chhowalla and H. S. Shin, *Angew. Chem. Int. Ed.*, 2013, **52**, 13751–13754.

# CHAPTER 10

# Surface Engineering with Chemically Modified Graphene

*Paul Sheehan,*[1,*] *D. R. Boris,*[2] *Pratibha Dev,*[3]
*S. C. Hernández,*[2] *Woo-Kyung Lee,*[1] *Shawn Mulvaney,*[1]
*T. L. Reinecke,*[4] *J. T. Robinson,*[4] *Stanislav Tsoi,*[1]
*S. G. Walton*[2] and *Keith Whitener*[1]

## Introduction

"God made the bulk; the surface was invented by the devil," quipped Wolfgang Pauli as a nod to the complex physics and chemistries that arise when a material ends, balancing the surface atoms between two material environments. Many fields have been established to understand and to control surfaces and so to attain desired interfacial properties whether it is the lubrication of moving parts, the wetting of surfaces, the corrosion of metal, or the biocompatibility of a hip implant. Achieving these desired properties has only grown more complicated as materials have been refined, structures have been miniaturized, and mechanical tolerances increased. Consequently, scientists and engineers now often receive a material or device that is highly optimized for an application and then are challenged to modify just one or two of its surface properties. With the many constraints that come with these devices, it is no longer feasible to bury the interface under a thick coating but, rather, one must carefully engineer the desired property using an appropriate thin film. In this chapter, the goal is to demonstrate that engineering surfaces with 2D materials could be the ultimate endpoint of this trend.

[1] Chemistry Division, US Naval Research Laboratory, 4555 Overlook Ave. SW Washington DC 20375.
[2] Department of Physics and Astronomy, Howard University, Washington, DC.
[3] Plasma Physics Division, 4555 Overlook Ave. SW Washington DC 20375.
[4] Electronic Sciences and Technology Division, 4555 Overlook Ave. SW Washington DC 20375.
* Corresponding author: paul.sheehan@nrl.navy.mil

While these thin films are only a few atoms thick—thinner than the adventitious carbon on most surfaces—they provide a range of surface properties, some not available with conventional treatments.

Before discussing the benefits of using a two-dimensional (2D) material to modify a surface, it would be useful to comment briefly on the other broad classes of surface modification. Clearly, the subject of coatings is immense, a subject beyond a single chapter and which is covered in many textbooks.[1,2] Even narrowing the subject to creating thin films is challenging since it has been a widespread endeavor for decades, providing many capable tools that rely on several different chemical or physical deposition strategies. As an example, one robust benchtop chemistry is the formation of self-assembled monolayers, self-terminating chemical reactions that form a single layer of organized molecules on the surface. These SAM molecules typically have bifunctional chemistry where one end binds well to the desired material, such as thiols for gold surfaces, and the opposite end has the desired functionality. By changing this exposed functionality one can dramatically change the properties of a material, even enabling a water drop to run uphill by generating a surface free energy gradient sufficient to drive the drop moving against gravity.[3] However, while highly effective for coating, SAMs require a chemical bond to the substrate to maintain their stability. Long lists of pairs of molecules and substrates have been assembled—alkanethiols for gold, silanes for silicon oxide, phosphonic acids for metal oxides—an illustration of the need to tailor the SAM to the substrate. Critically, high quality SAMs are not known for many surfaces, meaning that they are a useful but limited tool for surface functionalization.

Chemical strategies can be more complex such as atomic layer deposition (ALD), where a two-step, gas-phase chemical process leads to self-limiting growth of monatomic (e.g., metals) or diatomic (e.g., oxides) coatings on planar or textured surfaces. Similarly, chemical vapor deposition (CVD) relies on gases and vapors to react at a target surface, though it typically does not form a truly self-limiting film. The differentiating aspect of ALD and CVD as compared to SAM treatments is the temperature and pressure of the reactions; SAM chemistries proceed at roughly room temperature and atmospheric pressure, while ALD occurs at moderate temperatures ($\leq 300°C$) and sub-atmospheric pressures. CVD deposition can place even greater demands on the sample since the operating temperature can range up to and over 1000°C while the pressure can decrease to high vacuum. As such, substrates that are sensitive to heat or incompatible with vacuum may never receive the 'ideal' coating. Moreover, these techniques also require chemical reaction with the substrate to form high quality coatings. Besides chemical approaches, physical deposition strategies have been heavily developed since they are central to the microfabrication industry. Physical vapor deposition (PVD) techniques use direct heating, particle bombardment, or photon irradiation to vaporize material from a source or "target" in a low pressure environment ($< 10^{-2}$ Torr) and then transport it to a substrate. Such coatings are typically inorganic (e.g., metals, semiconductors, or oxides) and have a wide range of crystalline microstructures. Because film formation relies heavily on the source and not surface-driven reduction reactions, substrate temperatures will generally be lower than both ALD and CVD processes. However, because of the high energy of the evaporated metal as it hits the substrate, the continuity and the structure of the deposited film can again depend heavily on the substrate.

Using these chemical and physical techniques, researchers have accumulated an extensive toolbox of tips, tricks, and strategies for coating different materials. These capabilities are continually being challenged since, as the constraints tighten and the dimensions of the manufactured object shrink, the need for ever thinner films arises. For instance, in optical applications thinner films are needed since thicker protective films can mean a loss of incident light. In biomolecular sensors, the electronic signal decays rapidly with distance when attempting detection with a field-effect transistor. In heat exchangers, thicker protective coatings resist heat transfer. In all these cases, thinner coatings can substantially improve performance. However, thinning often brings a new limitation since for the thinnest conventional films to be stable they must be chemically bound or otherwise highly adherent to the surface. While one mechanism cannot describe why this is so for all systems, in general one can say that a mismatch of surface free energies between the film and surface can lead to dewetting if these two materials do not share a chemical bond. Dewetting leads to films that are pitted, that segregate into islands, or that otherwise do not provide continuous coverage across the surface. Consequently, many low surface energy materials cannot be effectively coated with high-quality thin films. Inhomogeneity in the wetting of a surface can also lead to inhomogeneity in the coating, hence the routine injunction to coat only clean, dry surfaces.

2D materials do not require chemical bonds to adhere to the surface nor highly favorable surface energetics to ensure continuity of the film. In principle, uniform coverage is assured since the 2D materials already have the bonding of all their surface atoms satisfied, while adhesion to the substrate is assured since they adhere to the substrate through van der Waals forces, which is relatively strong due to their large surface-to-volume ratio. Using van der Waals forces, which are present between all atoms, to adhere the film to the surface solves a second, more general problem. Currently, every time a new material/film combination is required, new deposition conditions must be developed which consumes significant time and money (although it does keep surface scientists gainfully employed). In general, developing reproducible and robust surface chemistries represents a major development cost for applications such as biosensors or medical implants, and not solving a surface chemistry property can often prematurely end a technology. Because van der Waals forces are universal, occurring between all materials, it should be possible to develop a surface coating just once and then apply it to a wide range of surfaces without the need for extensive priming of the surface. In this approach, the desired physical or chemical property would first be fully developed using either the intrinsic properties of a 2D sheet or through chemical modification. This engineered sheet would then be transferred *in toto* to the receiving substrate to which it would adhere through van der Waals forces, the equivalent of a chemical "sticky note" (Fig. 1). The development of a transferrable chemistry would greatly expedite research and development not only in reducing the resources required for developing a technology but in allowing new materials to be used that otherwise have poorly developed chemistries. Finally, it should be noted that

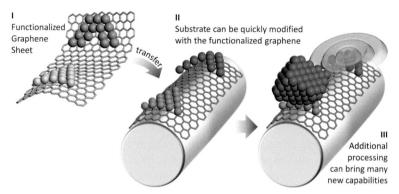

**Fig. 1.**  The ultimate incarnation of surface engineering would transfer prepared chemically functionalized graphene sheets onto the surface to be modified where it would adhere through generic van der Waals forces.

while adhesion through van der Waals forces is fairly robust—capable, for instance, of withstanding significant shear forces under fluid flow[4]—it is also the case that the 2D material can be covalently bound to the substrate if needed.[5]

Beyond solving the problem of difficult-to-coat surfaces, 2D materials bring several new properties that would be useful to apply to surfaces. Perhaps the pre-eminent such property is impermeability. Graphene, in particular, has been shown to be impermeable to virtually all gases, including helium.[6] Currently, this impermeability is not widely applicable since defects and grain boundaries can allow gas to bypass the graphene when used over large areas. However, extensive research on the growth and transfer of graphene continues with recent results showing single grains of graphene with widths approaching hundreds of microns, and some have even reported single crystals at truly wafer scales.[7] Secondly, 2D materials possess a wide range of chemical and physical properties. For instance, graphene can be an electrical conductor with one of the highest known carrier mobilities, approaching one million $cm^2$ $V^{-1}$ $s^{-1}$ at room temperature,[8] or after chemical reaction, an insulator with a band gap around 3.5 eV.[9] Chemical modification can also shift it from graphene fluoride's highly hydrophobic state to the highly hydrophilic properties of graphene oxide. Thirdly, graphene can be readily chemically patterned from the square meter level to the nanoscale. Finally, graphene and the wide assortment of 2D materials now in development promise new properties that are otherwise difficult to obtain, such as readily-patterned, room-temperature ferromagnetism.[10]

The use of graphene or any 2D materials to engineer a surface is now in its infancy. This chapter will discuss the state of the art in developing new properties in 2D materials, the transfer of these properties to the receiving surface, challenges to be addressed, theoretical considerations for how these materials interact with their surroundings, and applications that could benefit from these new coatings. The emphasis will be placed on graphene and its derivatives since they are the most heavily developed and available in large quantities.

## Chemical Strategies to Obtain New Properties from Graphene

A vast amount of chemistry has been performed on graphene, making it far too broad a subject to give a proper account here.[11–16] We will focus instead on common chemistries that find widespread use in graphene-based surface engineering: oxidation, fluorination, hydrogenation, and diazotization. This selection follows four criteria: (1) all four reactions can be performed on graphene directly, (2) the added functional groups impart interesting and useful surface properties in themselves, (3) the reactions are all starting points for further graphene functionalization, and (4) the reactions represent the overwhelming bulk of the chemistry of functional graphene found in the literature.

### *Oxidation*

Graphene oxide (GO) dominates the literature on functional graphene composites and coatings because of its chemical versatility and the ease and low cost of its production. Moreover, the epoxy and hydroxyl groups on the graphene basal plane, as well as edge-functionalized tears and pores bearing carbonyl, carboxyl, and ether groups are attractive starting points for further functionalization.[11] GO is highly resistive but can be chemically or thermally reduced to restore some of its conductivity.[17] It is also water-dispersible and films of GO are quite hydrophilic. This easy processability lends itself naturally to use in polymer composites and structural materials.

The most prevalent synthesis of graphene oxide proceeds from bulk graphite. In the popular Hummers method, graphite is exposed to sulfuric acid and a nitrate salt, generating a graphite bisulfate intercalation compound.[18,19] The material is further oxidized and functionalized using potassium permanganate, which adds the oxygen-bearing functional groups to the graphene. This method can be scaled up to multi-ton quantities, and graphene oxide suspensions are readily available commercially. One key point about GO prepared from bulk graphite is that coatings of GO do not retain the single-atom-layer thickness seen in graphene sheets prepared by other methods. Flakes of GO can be only as large as the grain size of the parent graphite (usually in the micron range).[20] Thus, a thin film coating of GO on a surface resembles less a single sheet and more a pile of leaves, a feature that could be an advantage or a detriment depending on the application.

For single-layer oxidized graphene, chemical processing such as the Hummers method is far too destructive. Instead, two main approaches exist. The first is plasma oxidation, which is covered in more detail below; the second approach for single-layer graphene oxidation is $O_2$ cracking using a hot STM or AFM tip. This method was pioneered by Hersam et al., and it results in the thermally reversible addition of epoxide groups to graphene at low to moderate coverage.[21] As yet, the product of this method is relatively uncharacterized in terms of surface properties.

## *Fluorination*

Fluorinated graphene has several useful properties from a surface engineering standpoint. The transformation of carbon-carbon $sp^2$ bonds to carbon-fluorine $sp^3$ bonds means that bond charge density is pulled away from graphene's basal plane, increasing the material's optical transparency, its electrical resistivity, and its hydrophobicity.

Graphene fluoride can be obtained via several routes. Direct fluorination of bulk graphite followed by exfoliation is conceptually the simplest method.[22–24] Graphite is unreactive to fluorine gas at room temperature but, above 300°C, fluorine molecules dissociate into atoms and intercalate between graphene layers, where they react nearly stoichiometrically with the graphite, giving $C_nF$ ($1.1 < n < 2$). This material ranges in color from dull gray to light yellow, depending on the extent of fluorination. Once the material is fluorinated, it is very stable and can be mechanically exfoliated in the same manner as unreacted graphite.[22] A major advantage of this method is that bulk graphite fluoride can be purchased commercially, so that researchers do not need to directly handle extremely reactive fluorine gas at high temperatures. However, because the starting material is graphite, the lateral size of the derived fluorinated graphene is limited by the graphite grain size, just as in the case of GO.

A safer and gentler fluorination method reacts single layer graphene with xenon difluoride ($XeF_2$), whose high chemical reactivity enables it to fluorinate carbon materials from aromatics to alkenes to carbon nanotubes.[25] An additional, practical advantage is that the vapor pressure of solid $XeF_2$ crystals allows commercial reactors to deliver precise concentrations of $XeF_2$ gas to the reaction chamber. Since 2010, numerous research groups have shown the utility of $XeF_2$ treatment as a robust route to modify graphene's surface properties through fluorination.[26,27] The reaction of $XeF_2$ gas with graphene is typically carried out at room temperature and at sub-atmospheric pressures, producing measurable fluorine concentrations after 20–30 seconds of exposure that saturate after only a few minutes.[26] A final useful feature of $XeF_2$ gas exposure is that fluorination only occurs where the graphene is exposed, allowing easy patterning of fluorination and the choice of fluorinating just the top of the graphene or both sides if a via is created.

At about the same time as the development of $XeF_2$ functionalization, researchers started exploring plasma-fluorination as a low pressure, gas-phase approach. Perhaps the earliest report of fluorinated monolayer graphene[28] used plasmas generated in Ar/$SF_6$ gas backgrounds. Subsequent works have employed a variety of fluorine-containing gas backgrounds including $SF_6$,[29,30] $F_2$,[31] $CF_4$,[32] and $CHF_3$.[33] Several experimental parameters in plasma processing can allow one to tune the F/C ratio; however, exposure times are typically minimized to avoid plasma damage, thereby limiting the fluorine uptake to 20–30%.

A fourth method of fluorination was introduced by Lee et al., based around the photodecomposition of a fluoropolymer.[34] Graphene was coated with a CYTOP fluoropolymer and various areas were exposed to a 488-nm laser. Lee et al. showed that the laser caused either photolytic or thermal decomposition of the polymer into fluorine and $CF_x$ radicals which then reacted with graphene at the precise points where the polymer was irradiated. This method allows photopatterning of graphene fluorination with at least 500 nm resolution, which might be useful for constructing lateral functionalized graphene heterostructures.

## Hydrogenation

Adding hydrogen to graphene modifies many of its properties. In common with other covalently bound functionalities, hydrogen adatoms reduce the conductivity of graphene.[35] Hydrogen is slightly electropositive compared with carbon, and as expected from this chemical consideration, hydrogen n-dopes graphene by donating electron density onto the lattice.[36] In addition, depending on the preparation method, hydrogen can add unpaired spins to the graphene lattice, inducing magnetic properties in a non-magnetic material.[37–39] Under suitable conditions, hydrogenation can also weaken the adhesion between graphene and substrate, allowing for smooth delamination of functionalized graphene as discussed below.[40] Finally, unlike typical organic chemical reactions, hydrogenation can activate graphene to further functionalization, especially by radical species such as chlorine and alkyl azo compounds.[41]

The earliest report of hydrogenated graphene by Elias et al. used plasma functionalization to add hydrogen atoms to graphene.[35] This method is still widely practiced today since it is fast and clean, with easily tunable parameters that make it amenable to electronics processing and lithography. Here again plasma processing can erode and damage the graphene and so hydrogen uptake must be balance against exposure times. Like plasma-fluorination, the extent of hydrogenation via plasma processing appears limited, as indicated by Raman spectroscopy and resistance measurements showing that maximum hydrogenation coverage is around 1 H per 5 nm$^2$.[35,42,43]

In contrast, hydrogenation via the Birch reduction is far more extensive and less erosive than plasma treatment.[44–46] The Birch reduction involves dissolving an alkali metal in liquid ammonia. The solvated electrons in this solution will transfer to the graphene sheet, enabling the subsequent extraction of a hydrogen from a proton donor such as ethanol or isopropanol. The method is extremely fast (complete reaction occurs in less than 1 minute) and chemically versatile: replacing proton donors with alkyl donors yields alkyl-functionalized graphene.[47] Birch hydrogenation also appears to be completely thermally reversible, as well as at least partially chemically reversible.[45,46,48] The parameters of the reduction can be tuned to produce either extensively hydrogenated graphene or partially hydrogenated graphene; however, control of the reaction is significantly less well-developed at the present time than for plasma hydrogenation. This drawback is mitigated if a very high level of hydrogenation is desired, as the Birch reduction is the only method that delivers high levels of hydrogenation without degrading graphene's basal plane.

Hydrogenated graphene is also produced less commonly via two other routes. First, like oxidation, hydrogen can be added to graphene locally via cracking of H$_2$ with a hot scanning probe tip.[49] This provides a material similar to that obtained by plasma processing, but on a much more local scale. The second method is electrochemical hydrogenation.[50] An electrochemical cell is prepared with graphene as the cathode. If the electrolyte is an acid, reduction of protons occurs at the graphene surface to form hydrogen atoms, which can then react with the graphene. Like the Birch reduction, this technique is not limited to proton reduction: if diazonium salts are added to a non-aqueous electrolyte, one can obtain aryl graphene derivatives.[51]

## Diazotization

Graphene reacts smoothly with aryl diazonium salts under mild conditions.[52] This method is particularly appealing since diazonium salts are organic groups, and as such, the full machinery of organic chemistry can be brought to bear on the problem of graphene surface engineering. One popular technique for biofunctionalization of single layer graphene is to graft diazonium with a functional group that can be coupled to a biological molecule.[53] Other work has focused on grafting organic groups that are sensitive to various stimuli, including light, which would induce a change in some property.[54–56] Diazonium functionalization is usually straightforward; although, it should be noted that some diazonium salts are quite reactive and will explode if handled improperly. However, for the most part, they are quite simple to work with. As an example, benzenediazonium tetrafluoroborate is a stable diazonium salt that is soluble in acetonitrile and that decomposes smoothly above 50°C. Reaction of single layer graphene with a solution of this material proceeds quickly, with monolayer coverage obtained within a few minutes. It was once assumed that the reaction was self-limiting to a monolayer; now, however, there is evidence that multilayer brushes can be built up on the graphene surface,[57] so some care is required if monolayer coverage is indeed the desired end-product of diazotization.

## Other functionalities

A number of additional graphene reactions could potentially prove useful for covalent attachment of functional groups and so are briefly mentioned here. Georgakilas et al.[58] and Quintana et al.[59] have demonstrated the cycloaddition reaction of azomethine ylides to graphene. Liu et al.[60] and Choi et al.[61] reported functionalization of graphene using azide molecules with minimal electronic disruption. Sarkar et al.[62] reported that graphene can undergo Diels-Alder reactions, behaving as both a diene or dienophile under various conditions. Huang et al.[63] and Yuan et al.[64] have also recently reported a two-step functionalization procedure, in which graphene is reacted with n-butyllithium, followed by attachment of a brominated target molecule. Hong and colleagues used photoreactive nitrene-based chemistry to covalently pattern planar graphene; the patterned areas serve as chemical handles from which atom transfer radical polymerization can be launched to add functionality, including design of surfaces that support cell growth.[65] While not all of these functionalization methods have resulted in specific applications, they offer several intriguing possibilities for future efforts. Ultimately, any functionalization strategy must strike a balance between the electronic and chemical properties of graphene to maximize overall surface performance.

## Plasma Processing of Graphene

Low temperature plasmas, a "gaseous soup" of ionized, reactive, and excited atoms and molecules, have long been an important tool for the synthesis and modification of advanced materials. Plasmas are produced by introducing an energy source that can ionize, dissociate, or excite the working gas, where gas composition and source intensity determine the amount and relative concentration of electrons, ions, excited neutrals, reactive neutral fragments (radicals), and photons delivered to materials exposed to the plasma. The energy and chemical reactivity delivered to surfaces via this mixture of species can etch, deposit on, or chemically modify the surface. This broad processing capability along with the ability to rapidly modify large areas (> 1000 cm$^2$) with precision down to a fraction of a micron make plasma processing attractive for chemically modifying graphene, especially where large scale modification is necessary.[66,67]

Many different plasma sources have been used to functionalize graphene and other atomically thin materials.[68,69] A primary difficulty with applying any plasma approach is the thinness and chemical stability of graphene. Essentially, the system needs sufficient power to generate and deliver the species needed to chemically modify the graphene, while minimizing the energy delivered to the surface so as not to remove a film that is only an atom thick. Consequently, the stated modes of operation have often been described as "low power"[70,71] or "remote".[72–74] While not quantitative, these adjectives clearly reflect that standard modes of operation fail to meet these requirements. More specifically, the kinetic energy of ions arriving at the surface are typically too high to effectively functionalize graphene. To understand the problem, one needs to consider the origin of the ion energy and the relevant properties of graphene.

In conventional discharge plasmas, electric fields energize the electron population so that some fraction of that population can sustain ionization of the gas. Consequently, the electrons have much higher average energy or "temperature" compared to the other species in the plasma. Typically, the heavy particles in the plasma (ions and neutrals) will remain near room temperature, whereas the electrons have temperatures exceeding 10,000 K. The temperature does not depend strongly on the input power but rather the *difference* between input power and energy losses within the system. The elevation in ion energies arises from the difference in mobility between lighter, more energetic electrons and the heavier, less energetic ions. To prevent the electrons from leaving at a higher rate than the ions, the plasma self-organizes to maintain a positive potential relative to adjacent surfaces such that electron losses are reduced. This positive potential accelerates ions such that they arrive at the substrate with energies well in excess of the electrons. In other words, the ion energy at adjacent surfaces scales with electron energy. In most plasmas, the ion energy is about 4–5 times the electron energy and in most discharges used for processing, the electron energy is a few eV.[75] Thus, an ion kinetic energy in excess of 15–20 eV is to be expected in most plasmas.

The energy transferred to a surface upon ion impact serves to break bonds, stimulate reactions, and heat the material. The relevant energies to consider in graphene are the energy required to remove carbon (sputtering), break carbon-carbon bonds, and stimulate reactions. For Ar$^+$ ion sputtering, the energy to remove carbon is about 30 eV,[76] the C-C bond energy is about 3 eV,[77] and the energy required to drive reactions will be similar. Although the sputtering threshold and the energy to drive reactions will depend on the specifics of the system,[78] they will be comparable. As such, careful control over ion flux and energy is required to achieve the desired chemical changes while minimizing damage and/or erosion. Given the expected energies noted above, this can be a challenge with conventional discharge plasma systems.

Reducing the source driving power and/or locating it remotely provides some control over the flux of ions and their energy; although, its limitations should be recognized. First, the input power strongly affects the production of species but has less influence on the electron temperature which also depends on energy sinks. For example, Lim and coworkers showed that an 8x increase in the power input (5 → 40W) produces a 25-fold increase in ion density ($4 \times 10^8$ → $10^{10}$ $cm^{-2}$) while dropping the electron temperature by about 15 % (3.5 → 3.0 eV).[79] In other words, decreasing the power reduces the flux of ions to the surface but not their energy. Operating in the low-power mode proved beneficial in the removal of residue from graphene-based devices. More recently, Narayanan et al. showed a strong correlation between increasing power and defect generation in an argon discharge.[80]

One strategy to avoid damage is the use of a remote plasma, where plasma generation is located relatively far from substrate so that gas phase reactions can cool hot species and/or reduce charged particle densities. In this system, increasing the pressure enhances these effects by decreasing the mean free path for collisions. The net effect of these geometries is to minimize (or eliminate) the flux of charged particles and reduce the ion energy at the substrate while still enjoying the production of reactive species associated with non-equilibrium, gas-phase chemistry. Remote reactors have been used to successfully incorporate oxygen and nitrogen groups,[81] as well as hydrogen[72] into graphene. Similar effects can be achieved using semi-transparent physical boundaries. Pham et al. utilized two screens between the source and the wafer chuck to remove charged particles from the flux of species incident to the surface, allowing them to dope graphene with chlorine without damage.[82]

### *Electron beam generated plasmas*

One way to the lower the ion energy is to generate the plasma with an electron beam which can provide large fluxes of very low energy ions.[83] For instance, the large area plasma processing system (LAPPS) employs magnetically-collimated, sheet-like electron beams to generate similarly sized plasma sheets for materials processing (Fig. 2).[84] The relatively simple configuration consists of an electron beam source, slotted anode, termination anode, sample holder, and magnetic field coils. Processing is effectively controlled by elevating the pressure which increases the dose of reactive species and energetic ions at the surface. A useful feature of this system is that the dose can be controlled independent of ion energy, with the coverage of functional groups on graphene increasing with increasing operating pressure (Fig. 3).[85,86] Here, the concentration of the particular functional group (as indicated in parentheses) is derived from the XPS spectra. While the degree of perturbation increases with increasing coverage,[86,87] no erosion or etching of the carbon backbone has been observed with this system.

Controlling the concentration of functional groups on the graphene enables one to engineer its properties. Figure 3 shows how nitrogen changes both the electrical properties and surface energy of the graphene. As the concentration of nitrogen moieties on the surface increases, the contact angle of water on the surface decreases,

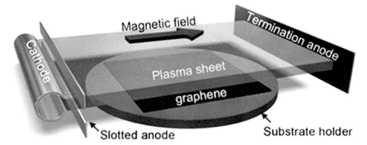

**Fig. 2.** The Large Area Plasma Processing System employs an electron beam generated plasma. Magnetically-collimated, sheet-like electron beams are used to generate similarly sized plasmas near the graphene surface. [Reproduced from Baraket 2012.[88]]

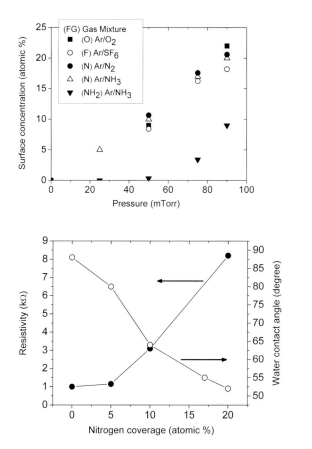

**Fig. 3.** (Top) The surface concentration of functional groups incorporated versus the operating pressure during beam-driven plasma processing. The legend indicates the operating background gas (e.g., $Ar/N_2$) and associated functional group (e.g., N), with the latter determined from XPS spectra. The results show how changing the operating environment (gas and pressure) can be used to control the surface chemistry. (Bottom) Typical changes in material properties after exposure to beam-driven plasmas, indicating and increase in both the resistivity and water contact angle with increasing nitrogen coverage. Water contact angle measurements are a quick measure of the free energy of a surface with lower values correlating with higher surface free energies.

indicating an increasing hydrophilicity which correlates with higher surface free energy. While plasma treatments in oxygen, hydrogen, and ammonia all increase graphene's hydrophilicity, the addition of fluorine has the opposite effect. On the other hand, the resistivity of graphene increases as more nitrogen is added. The increase in resistivity is independent of functional group although the magnitude will vary.

## *Patterning*

Plasma treatment combined with masking techniques is an attractive approach to patterning functional groups. Physical masks are useful since, compared to lithography approaches, they eliminate the solvent-based mask removal steps that can leave residue[79] or remove functional groups. When placed in intimate contact with the surface, physical masks can direct functional groups onto the surface so that they react only where desired. It follows from Fig. 3, that the resistivity and reactivity can thus be spatially tailored via the patterning of functional groups. For example, Hernandez and colleagues showed how a mask can create resistive and/or conductive channels by creating nitrogen-rich "lines".[87] They also demonstrated gold can be selectively grown, via electrodeposition, only on the regions functionalized with nitrogen, though the cause for this enhanced growth rate is not fully understood. It would be tempting to suggest that the increased wettability of the functionalized regions is the main driver. However, in subsequent works, the presence of fluorine was also found to improve the deposition of cadmium sulfide via chemical bath deposition[89] and hafnium oxide via atomic layer deposition.[90]

An alternate approach is to elevate the physical mask above the material to be patterned. Figure 4 shows how a cantilevered roof mask, raised slightly above the material to be patterned, can establish a chemical gradient by using the plasma's

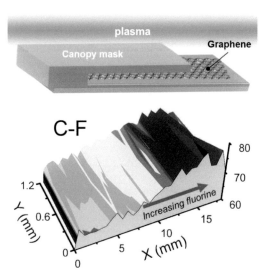

**Fig. 4.** A schematic of the canopy mask used in conjunction with plasma processing in an Ar/SF$_6$ background to develop a chemical gradient. XPS mapping of the surface shows a fluorine gradient that is uniform in the Y-direction but varies in the X-direction.

molding around the roof to vary the dose of reactive species with position.[85] A properly constructed mask can chemically grade the surface in one direction while leaving it homogeneous in the orthogonal direction, a useful property as will be discussed shortly.

## Transfer of Graphene Onto Different Substrates

Many methods have been developed for transferring graphene from one substrate to another, including polymer-supported transfer,[91] metal-supported transfer,[92] dry transfer that does not require any support,[93,94] and the original Scotch tape method.[95] Once transferred, the graphene can be chemically modified using the many techniques just outlined. This approach is not always viable since the receiving substrate must be sufficiently robust to withstand subsequent processing. For instance, graphene cannot be transferred onto living tissue and then subsequently functionalized, as the post-transfer processing would kill the cells. A more successful approach would transfer the graphene after it has been chemically modified; however, transfer of chemically modified graphene has generally proven challenging. Lee et al. observed that chemisorbed hydrogen on graphene was not retained when the hydrogenated graphene was transferred from its silicon oxide substrate via alkaline etching of the silica.[96] Similarly, attempts to transfer graphene fluoride have met with little success: Lyuksyutov et al. and Dubecky et al. both observed pH-dependent fluorine loss when graphene fluoride was immersed in aqueous solutions.[97,98]

The difficulty of functional group retention upon transfer stems from the large energy penalty associated with the out-of-plane distortion of graphene as it rehybridizes from the planar $sp^2$ to pyramidal $sp^3$ to form a covalent bond to an adsorbate. Paying this energy penalty most likely weakens the bond between graphene and the adsorbate and complicates retention of functional groups during the transfer process. This has ramifications for any transfer method that employs chemical etchants to release the graphene from its substrate.

The alternate strategy is to search for chemistries that both robustly bind to graphene and weaken its adhesion to the substrate. This combination should enable one to gently delaminate the graphene while leaving its chemical functionality intact. Highly hydrogenated graphene (HG) appears to have both attributes such that its weak adhesion to the substrate enables an intact HG sheet to be delaminated by simple immersion in water.[40] Notably, this is done without the use of polymer support or chemical etchants and is mild enough to co-transfer other covalently bonded functional groups such as methyl and diazonium-grafted aryl groups. Even extremely delicate properties such as ferromagnetism induced by partial hydrogenation are preserved during this hydrogen assisted transfer. Since graphene hydrogenation is thermally reversible,[44] the entire method can be considered as a polymer and etchant-free graphene transfer technique.

A possible mechanism for this adhesion weakening was considered by Wong et al., who performed DFT calculations on the effect of copper oxidation on graphene adhesion, finding that copper oxide formed at the edges of graphene significantly eases the delamination of the graphene.[99] The adhesion of graphene to its substrate is presumably dominated by a large metallic or plasmonic contribution to the van der Waals interaction. By hydrogenating the graphene, this interaction is minimized, and so are easily overcome during the delamination process by the force between water and the substrate.[40]

Finally, we point out that graphene oxide (GO) constitutes a distinct exception to the trend of losing chemical functionality during transfer. Oxidation of graphite via chemical methods is generally so extensive as to be irreversible. Even with vigorous reductive processing, reduced graphene oxide (rGO) remains sufficiently functionalized to retain many of GO's properties.[17]

## Challenges for Using Graphene for Surface Engineering

The use of two-dimensional materials for surface engineering has grown in the past few years; however, the field is not without its challenges. Perhaps the most notable challenge, and opportunity, is that the properties of a single sheet can be affected by the underlying surface. That is, the thinness of these materials and their close, conformal contact with the substrates causes them to exhibit a variety of substrate-dependent behaviors not normally seen with thicker films. Koehler et al. were one of the first groups to observe these types of effects by noting that diazonium aryl grafting takes place readily on single-layer graphene on silicon oxide, but proceeds much more slowly on bilayer graphene.[100] Later, Strano et al. explored this phenomenon in a more systematic way, showing that the aryl grafting reaction proceeds at different rates for SLG on a number of different surfaces, including hexagonal boron nitride (hBN), silicon oxide, and alkyl-terminated silicon oxide.[101] Substrate-dependent reactivity was also reported for the fluorination of graphene with $XeF_2$, which occurs at different rates and proceeds to different extents depending on whether the graphene is on silicon oxide, Cu, or Au metal.[102]

In general, oxide substrates such as $SiO_2$ induce greater reactivity in graphene than either metallic substrates or nonpolar insulating substrates such as polymers. This effect has been attributed to "charge puddles", inhomogeneities in surface charges or surface dipoles that make certain sites on graphene more locally reactive.[101] Charges on metallic surfaces distribute over the entire surface so that no single site is vastly more reactive than another, and nonpolar surfaces also do not tend to show a great deal of charge inhomogeneity. Metal oxides, however, often have charge dipoles at the surface that produce significant inhomogeneity over short distances. It should be noted that, for surface engineering purposes, most of these extrinsic effects diminish with the number of intervening layers and so can be alleviated by placing, for example, a spacer 2D material between graphene and the substrate.

Just as the chemical and physical nature of the substrate can affect the reactivity of 2D films, the converse is also true: the chemical nature of the 2D material can affect its interaction with its substrate. As mentioned before, graphene's adherence to the substrate decreases dramatically when it is hydrogenated.[40] This causes an obvious problem: weakening the graphene-substrate adhesion allows delamination of graphene and transfer onto a different substrate, but graphene's adhesion to this new substrate will be weakened as well. One answer is to thermally or chemically dehydrogenate HG to enhance adhesion.[41] However, the fact remains that engineering a robust surface chemistry from 2D materials involves examination not only of the materials themselves, but their interaction with the substrate. Indeed, in many cases one should consider the graphene and substrate as a hybrid system.

Another challenge in using 2D materials for surface engineering is the occasionally necessary tradeoff between desirable but incompatible chemical or physical properties. One example is the tradeoff between the coverage of functional groups on graphene and graphene's conductivity. Electrons in pristine graphene famously exhibit ballistic conduction which can be a desirable device characteristic. However, this conductivity diminishes with increasing coverage of covalently attached functional groups.[35,103] For applications where chemical versatility and conductivity are both needed, one must choose a concentration of functional groups that is dense enough to perform the requisite chemical functions but not so dense that the device becomes overly resistive. In certain cases, this problem has been solved: non-covalent pyrene linkers to graphene use π–π stacking to allow for a high density of functional groups at the graphene surface while only minimally disturbing the band structure of graphene.[104]

A further example of a tradeoff is that some functional groups destabilize at high surface coverages. Stine et al. showed that graphene fluoride with high fluorine concentrations spontaneously lost fluorine over a period of days to weeks before stabilizing at a lower concentration.[105] They performed reactivity and stability experiments using graphene on different supporting substrates (e.g., Au, $SiO_2$, Cu) and subsequently monitored the chemical composition of the samples over time when exposed to different ambients (e.g., air or dry $N_2$). For fresh samples with high coverages of fluorine, the fluorine content decreased by 50%–80% over the course of several days, regardless of whether the graphene was stored in air or $N_2$ ambients. Notably, these aged fluorinated graphene surfaces reacted to the same extent with ethylenediamine (EDA) as did freshly fluorinated samples, suggesting that only the nonvolatile F participate in alkylamine reactions. Although the fluoride coverage stabilized over a period of days to weeks, one might need to compromise on a lower-than-desired fluorine density to maintain stability of the system. Of course, if the functional group of interest is stable on graphene, this is less of a concern.

## Using Theory to Understand Chemically Modified Graphene

Surface chemical modification of graphene can change its fundamental properties at the atomic level, including its band gap, chemical reactivity and its electrical and optical properties. This can be done by surface chemical treatment and also by the choice of substrate. Surface engineering of graphene can involve modification of properties by creating vacancies,[106,107–109] doping graphene with heteroatoms,[26,110] or creating more elaborate organometallic complexes.[110,111] Theoretical work plays an important role in guiding and interpreting experiment for these complex systems. First principles calculations for a wide range of these systems have been made. This work generally involves density functional theory (DFT) or uses its results as inputs to physical models. The foundations of DFT were laid in two seminal papers by Hohenberg and Kohn[112] and by Kohn and Sham.[113] The former showed that the energy of an interacting many-electron system can be written as a unique functional of the electron density and the latter showed that one can map this complex interacting system onto a system of non-interacting particles with the same charge density. Using these two concepts,

DFT can compute the properties of materials without adjustable parameters. In the following section, we will discuss a several examples of this theoretical work on chemically functionalized systems.

Both hydrogen and fluorine adsorb directly above the C atoms of graphene, forming relatively simple bonds with the carbon $p_z$ orbitals oriented perpendicular to the surface. Thus they provide particularly good systems for beginning to understand its chemical functionalization. Experimental work has been done using $XeF_2$ to functionalize graphene grown on Cu and also for graphene transferred to $SiO_2$/Si substrates. Theory can give insights into properties such as the stability of the chemical groups and the impact of the arrangement of the functional groups on electronic properties. DFT calculations done as part of this work are illustrated in Fig. 5. The band gap of graphene in Fig. 5C was shown to increase with increasing F coverage. The binding energy per fluorine atom was found to peak at 25%. This coverage has F atoms at 3rd nearest neighbors, which results in isolated π resonances of the graphene valence states. More extensive experimental work on the properties fluorinated graphene has been done since this work.

Adsorbates may form ordered arrays, disordered arrays, or inhomogeneous arrangements of locally ordered adsorption geometries on graphene, and it is often difficult to characterize the geometries experimentally. First principles calculations can provide insights into the possible adsorption geometries for varying average coverages. Extensive calculations of the properties of F functionalized graphene have recently been done for dozens of arrangements of the F atoms.[114] A wide range of adsorption energies, lattice constants, bulk moduli, bond lengths, band gaps and magnetic properties were found as functions of coverage and of adsorbate arrangement. They include metallic

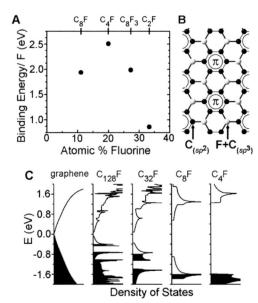

**Fig. 5.** (A) Calculated binding energy per F atom compared to the $F_2$ gas state. (B) Sketch of the calculated $C_4F$ configuration for the 25% coverage from (A). (C) Calculated total density of states (DOS) of single-side fluorinated graphene for several fluorine coverages.

systems and semiconducting systems with broadly varying band gaps as well as systems with local ferromagnetic, antiferromagnetic, and ferrimagnetic orderings.

Chemical functionalization of graphene is also of interest for modifying its mechanical and vibrational properties, including strength, internal stress, vibrational frequencies, and losses. Recent experimental work[115] used O and F functionalizations to demonstrate control of the vibrational properties of fairly complex multilayer graphene films. DFT calculations as a part of this work gave the dependences of the bulk moduli

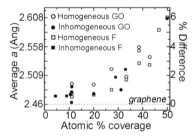

**Fig. 6.** Calculated average lattice constant versus O or F coverage. The right axis shows the percent deviation from the ideal graphene lattice constant. The top axis shows the atomic carbon ratio for a $C_xM_1$ material, where M is oxygen or fluorine.

and of the lattice constants of graphene films functionalized with O and with F, and they also considered the sensitivities of these properties to inhomogeneity in the adsorbate coverages. Results for the lattice constants as functions of coverage are shown in Fig. 6. They agree qualitatively with the overall trends in the experimental data, and they show that these properties should not depend sensitively on adsorbate inhomogeneity.

In order to form high quality films of functionalized graphene with desired properties, it is necessary to understand the basic processes by which the films are formed. However, even with relatively simple adsorbates, these processes are complex, involving dynamical interactions of the adsorbates with graphene and also interactions between adsorbates and interactions of adsorbates with defects in the system. These interactions are generally poorly understood. Recent theoretical work[106,107] has shed light on important aspects of these interactions. The primary amine $NH_2$, which is of interest as a linker between organic molecules and graphene, was considered as the adsorbate. DFT calculations[106] were made of the interactions of the $NH_2$ with B, N, F, H, $NH_2$ and OH adsorbates and of the interactions of these adsorbates with the graphene. Examples of the interactions between these adsorbates on surface are shown in Fig. 7, which gives the adsorption energies as functions of the separation between the two adsorbates on the surface. The interactions oscillate with the distance between adsorbates. These features arise from substrate mediated interactions between adsorbate through the $\pi$ valence electrons somewhat like RKKY interactions. DFT results are effective in describing relatively short-range interactions. However long-range interactions can be also important in surface formation. To address them, DFT results were used to construct detailed electronic tight-binding Hamiltonians, which included the interactions of the adsorbates with the $\pi$ valence electrons, Coulomb interactions of adsorbates, and lattice distortions.[107] This Hamiltonian was used to

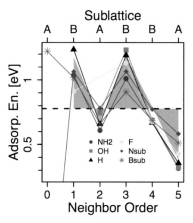

**Fig. 7.** Adsorption energies of the secondary $NH_2$ as a function of neighbor order from the primary defects listed in the Figure. The horizontal dashed line represents the adsorption energy of the first $NH_2$ molecule. The axis labeled "Sublattice" indicates the sublattice (A or B) of the C atom to which the secondary defect is attached.

determine long-range adsorbate-absorbate interactions, which were found to continue to oscillate over long distances. A changeover between 'weak' and 'strong' coupling regimes between adsorbates were found for certain parameters. It was shown that these interactions could be controlled with the system's chemical potential and by the initial adsorbate-substrate interaction, e.g., by the choice of adsorbate.

As described earlier in this chapter, low energy plasma processing can generate gradients of reactivity and wettability by spatially varying the gradients of functionalizations.[85] Measurements of the contact angles of droplets of water and of DMMP, a simulant for the nerve agent sarin, and also chemical force microscopy gave information about the chemical reactivities of the surfaces. For example, the O functionalized surfaces were found to be more hydrophilic than pristine graphene and F functionalized surfaces were more hydrophobic. Droplet motions in different directions over the surfaces with the chemical gradients of the O and F functionalizations were demonstrated. DFT calculations were made of the interactions of DMMP with O and F functionalized surfaces, and it was found that the adhesion energy with O rich surfaces was greater than that on pristine surfaces and that the adhesion with F rich surfaces was less than in the pristine case. These results are consistent with the contact angle and chemical force microscopy results and also with the directions of droplet motions on chemically graded graphene surfaces.

The properties of graphene systems can be affected by the substrate involved as well as by surface chemical functionalization. However, to date the role of the substrates is less well understood. Recent theoretical work has shed light on the role of substrates on graphene systems.[116] In this work, the magnetic properties of chemically-functionalized sheets of graphene provided sensitive measures of the electronic behavior of these systems. Of course, their magnetic properties are also of interest on their own, with several theoretical reports predicting defect-induced magnetism in graphene. However, experimental results for their magnetic properties often do not agree with one another nor with the theoretical results [see references in Dev et al.[116]]. Most theoretical work on magnetism in graphene uses free-standing graphene while experiments almost always involve substrates. Dev et al. showed that the magnetic properties of functionalized graphene depend strongly on its local environment, including the number of graphene layers and the substrate.[116] That work started with free standing graphene functionalized by fluorine or hydrogen such that there was an imbalance in the number of atoms on the two graphene sublattices. This imbalance results in net unpaired electrons and gives a net magnetic moment to the freestanding graphene. However, when functionalized graphene is placed on a copper substrate there is complete quenching of the magnetism (Fig. 8). This helps to explain why theoretical results for magnetic properties of free standing graphene systems are different from many experimental results for graphene on substrates and also why experimental results on different substrates may differ from one another. This behavior results from adsorbates changing graphene from an inert to a reactive surface, resulting in formation of metal-carbide bonds between the two sub-systems. This gives rise to three inter related effects that were shown to reduce/quench magnetism in graphene: (i) a transfer of charge between the substrate and graphene, (ii) distortion of the graphene, and (iii) most importantly, formation of interface states. This work also showed that if an additional layer of graphene is placed between the substrate and the

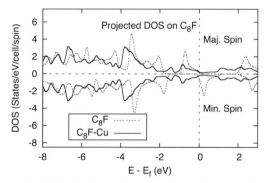

**Fig. 8.** DOS for freestanding $C_8F$ (dashed-curve), showing spin-splitting, and the projection of DOS on $C_8F$ in $C_8F$-Cu(111) structure (solid-curve), showing quenching of magnetism (reprinted from Dev et al.[116]).

functionalized graphene, the magnetic properties of the free standing functionalized graphene were maintained. That is, the additional layer of graphene effectively insulates the functionalized layer from the effects of the substrate. This is an example of how graphene can function as a controllable atomically thin 'veil' to insulate the substrate and surface properties from one another.

## Applications

### *Biofunctionalization*

Graphene, either in its pristine state or chemically modified, is a powerful tool for biofunctionalization.[117–119] There are several factors that contribute to its utility. First, the common use of carbon as a core building block of both graphene and biomolecules promotes biofunction because the surface is hospitable for small biomolecules, mid-sized enzymes, and full-sized cells alike. Moreover, biotoxicity ultimately becomes a function of geometry where carbon allotropes in non-natural shapes such as some carbon nanotubes can be problematic but natural products like graphene sheets are compatible. Biotoxicity can also appear when graphene is chemically modified with aggressively reactive terminal groups that lead to oxidation, phosphorylation, or even apoptosis of adsorbed biomaterial.[120,121] By taking into account these effects, biotoxicity can usually be anticipated and mitigated. Secondly, as noted earlier, graphene or its derivatives can be effectively coated onto nearly all substrates[122] and thereby induce compatibility in surfaces or materials that are otherwise inherently inhospitable to enzymes or toxic to cells.[123] This is crucial since it extends the range of materials available for biotechnology, bringing in many materials that lack inherent biocompatibility or whose surface chemistries are poorly developed. Thirdly, graphene's chemical flexibility enables it to be processed down multiple different reaction pathways. For instance, it can be functionalized with either non-covalent methods such as pyrene-based molecules[124,125] or covalent methods that leave terminal carboxyl groups or primary amines. In particular carboxyl and amine handles added to graphene open up almost limitless strategies for attaching biomolecules via commercial off-the-shelf products.

The first step in biofunctionalizing a new surface with graphene is typically to coat it with either spun-cast GO flakes or CVD graphene monolayers. GO-coated surfaces can be achieved by spin coating an aqueous/alcohol suspension of GO onto the substrate and then annealing to remove the solvent, leaving a nanometer-thick film of GO flakes that conformally coats the surface. The top layer of the GO displays several different oxygen-rich functional groups, including epoxy, carboxyl, and hydroxyl moieties, all of which are readily used in subsequent biofunctionalization pathways.[126] Alternately, CVD graphene can be coated onto a surface and then reacted to generate primary amines or other groups that are useful for subsequent binding. If transfer was assisted by hydrogen[40,48] or fluorine,[127] those chemical handles are immediately available for subsequent biofunctionalization. If the transfer was achieved by using the more conventional approach of coating with a support polymer followed by etching of the copper substrate, then the transferred graphene will need to be cleaned to expose the planar $sp^2$ carbon. In these cases, one typically biofunctionalizes the surface using $\pi$–$\pi$ stacking of aromatic molecules.

*Non-covalent biofunctionalization.* A common goal in biofunctionalizing graphene is to provide a sensor biochemical specificity through adding recognition elements such as antibodies, DNA, or peptides. Some sensor designs require that the graphene remain conductive. This makes covalent attachment a problem since such attachment necessarily disrupts graphene's $sp^2$ structure, altering its electronic properties. The requirement to both attach biomolecules and retain graphene's electronic structure spurred research into non-covalent biofunctionalization. Beginning with Ohno et al.[128] and Huang et al.,[125] most of these have taken advantage of physisorbed aromatic molecules that align with the graphene lattice through $\pi$–$\pi$ stacking.[129] The adsorbed aromatic molecule should, like 1-pyrenebutanoic acid, contain a functional group that can tether a biomolecular probe to the graphene surface. Building upon that work has led to multiple commercially available pyrene variants including amine and biotin functional groups for binding to commercial biomolecules. Another successful non-covalent attachment method uses adsorbed nanoparticles to tether the desired biomolecules.[130] As an example, gold nanoparticles were coated onto the graphene surface, and subsequently biofunctionalized through thiol chemistry. Another non-covalent functionalization has been reported by Cui and colleagues,[131,132] who specifically designed peptides that recognize and bind to graphene sheets. Phage display screened billions of potential binding peptides for affinity to graphene. The binding of the peptide to the graphene leaves intact the graphene structure because it uses only non-covalent van der Waals interactions, similar to an antibody/antigen interaction. Peptide linkers were used by Mannoor et al.[133] to add biological probes to a graphene BioFET for the detection of bacterial cells. Recently, another non-covalent functionalization scheme was reported by Park and colleagues,[134] where bilayer graphene is utilized for device production. Here, the top layer of graphene was covalently modified to include specific binding molecules, while the lower graphene sheet was left undisturbed to function as the transduction layer. The atomic thinness of the top graphene sheet thereby tethers the biomolecules in very close proximity to the lower graphene sheet, changing its electronic properties to transduce a signal.

*Covalent biofunctionalization.* While the ability of the non-covalent approaches to functionalize graphene without disrupting its bonding structure is attractive, they also have drawbacks specific to biofunctionalization. First, the portion of the molecule that anchors the biomolecule tends to be large and so spaces the biorecognition event farther away from the transduction element. This is important since many applications rapidly lose signal with distance; for example, Debye screening in bioFETs[135] or magnetoresistive sensors that capture magnetic particles above a sensor.[122] In the latter, the signal from a magnetic label to a sensor based on magnetoresistance drops quickly with separation distance (i.e., $1/d^6$). Secondly, non-covalently attached probe molecules are more likely to desorb,[134] not only lowering sensitivity but exposing the underlying graphene to possible non-specific adsorption of biomolecules, both of which can produce false results. As such, methods to covalently attach probe molecules to graphene have also been extensively explored. This was most easily accomplished for sensors using reduced GO,[136–138] exploiting the presence of epoxy and carboxyl groups that are already present in the GO flakes as attachment sites. Another method of covalent graphene functionalization is via the exposure to an electron beam driven plasma generated in an argon/ammonia background to form amine groups on the planar graphene sheet, which Baraket and colleagues[28,88] demonstrated for a DNA bioFET. Alternatively, the reaction of graphene with diazonium salts,[139–141] was utilized by Kasry and colleagues[142] to attach biotin for the subsequent detection of streptavidin.

*Applications of biofunctionalized graphene.* Applications of biofunctionalized graphene have secured a large footprint in the scientific literature and have been summarized in many excellent reviews.[117,119–121,143,144] Here, the focus is on two applications that rely on graphene's unique characteristics.

## Förster resonance energy transfer (FRET) assays

One of the most studied areas in the biofunctionalization of graphene—and other carbon allotropes—is its interaction with nucleic acids. Of particular interest is the

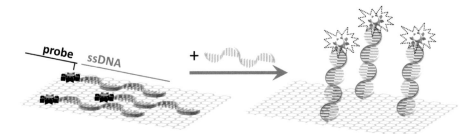

**Fig. 9.** Cartoon of FRET on graphene. Single stranded DNA with an attached fluorophore lies prone on a graphene surface with the graphene serving to quench the fluorescent signal. The addition of a complementary DNA results in hybridization, thereby lifting the double stranded DNA off the surface and restoring the fluorescence.

detection of nucleic acids based on Förster resonance energy transfer (FRET), where the fluorescence intensity of a fluorophore depends on its distance from the quencher. When the fluorophore and quencher are close, no light is observed, but, when separated, fluorescence reappears. Since graphene is an excellent quencher, many demonstrations of its use in FRET have been published (Fig. 9).[119,145–149] These publications also revealed the fascinating fact that single stranded DNA (ssDNA) lies prone on the graphene surface due to strong π–π stacking with the graphene. In contrast, double stranded DNA (dsDNA) lifts away from the graphene surface because the base pairing that constitutes hybridization precludes such stacking. The dramatic shift in orientation demonstrates that graphene is a complicated material for biofunctionalization, requiring researchers to be thoughtful in how they attach biomolecules to it. The interaction pathways only become more interesting when one considers protein attachment and the importance of tertiary structure to protein function. For example, Zhang and colleagues demonstrated that the spontaneous binding of several enzymes to GO directly correlates with the amount of oxygen present on the surface.[150] It becomes clear that there are complex interactions when biomolecules are attached to graphene and negative outcomes of biofouling or non-specific binding must be carefully managed. This active field of research is anticipated to grow in the coming years.

### Graphene Veil—A universal surface chemistry for biosensors based on graphene

Graphene could serve as a universal surface chemistry for biosensors.[122] The literature on biosensor surface preparation is vast—clear documentation of the time, effort, and resources poured into perfecting surface chemistries.[151–157] The breadth of the literature reflects that each combination of biomolecule and surface to be functionalized must be handled separately.[155–160] That is, a surface chemistry that works on gold typically does not work with silicon or gallium nitride or any of the other technologically relevant materials. Separating the issue of how a biomolecule was sensed from the problem of how to attach it to that sensor would fundamentally transform the field of biofunctionalization. For example, the attachment of a new protein to graphene need only be developed once before application to a range of materials. Consequently, all effort could be focused on producing a single biofunctionalization strategy that would then be shared widely. An ideal universal surface chemistry should yield a film thin enough to keep the biomolecules close to the sensing element, should be conformal, and should allow for multiple pathways to attach biomolecules. Importantly, the resulting coating must be sufficiently stable that it remains intact during washing steps and sensing and must not interfere with the transduction mechanism.

As described earlier, GO can conformally coat many surfaces with a film no more than a couple of nanometers thick, and these films activate several biofunctionalization pathways. Figure 10 shows the results for fluidic force discrimination (FFD) assays performed on five diverse sensor materials. FFD assays[161,162] are sandwich-style assays that capture the target molecule using the biorecognition element tethered to the graphene. A second capture agent tethered to a microparticle then binds to a second region of the target molecule to form the sandwich. Following biorecognition, a controlled fluidic force is applied to the bead labels that is strong enough to "push off" nonspecifically captured beads but too weak to disrupt the specific interactions. Consequently, the presence of the dark beads reflects the presence of the target molecule and the lack of beads means that the target was not present.

The first step in attaching biomolecules to GO begins by reacting ethylenediamine with the epoxy groups present on the GO surface. Amines will react spontaneously to relieve the epoxide ring strain, thereby covalently coupling an amine to the GO. From that point forward commercial products may be used, such as the two halves of the hydrazone reaction[122,163] to place neutravidin on the surface. Figure 10 shows the use of this chemistry to successfully produce both immunoassays and nucleic acid hybridization assays on a diverse set of substrates. Importantly, the assay for RCA atop silicon nitride was performed in beagle serum demonstrating compatibility with real world matrices. The slightly higher background signal due to matrix effects, ~ 10%, was expected and was consistent with previously published FFD assay results atop a glass slide using conventional surface passivation chemistries.[161,162] Finally, a graphene veil was successfully added to a flexible plastic substrate, polystyrene. For this experiment we demonstrated a different functionalization pathway. Neutravidin was covalently linked to the graphene veil through the native carboxyl groups using a carbodiimide crosslinker and a DNA hybridization assay was achieved. While this

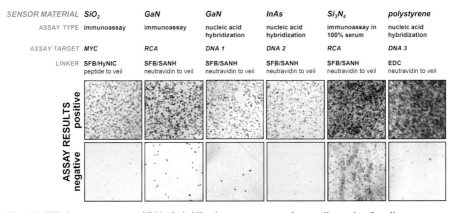

**Fig. 10.** FFD immunoassays and DNA hybridization assays on graphene veils coating five diverse sensor materials. (Reprinted from Mulvaney et al., *Biotechniques*, 2014.[122])

set of experiments is only a first demonstration, the work towards a universal surface chemistry marks one of the most promising uses of biofunctionalized graphene.

## *Inorganic films via chemical bath deposition*

Chemical bath deposition (CBD) has been widely used to deposit semiconducting thin films since it is scalable, cost-effective, non-destructive, and generally produces high quality films. In general, CBD is less constrained in the choice of substrate than other approaches since it requires only favorable physisorption to promote the deposition of the film. Yet not all substrates are amenable to CBD, and some substrates yield films whose properties could be improved. As an example of the latter, to improve the homogeneity of a cadmium sulfide (CdS) film deposited via CBD, Seo et al. coated glass with graphene, then treated it with UV-ozone to create defects and oxygen-rich functional groups that promoted the nucleation and growth of the CdS.[164] A more pronounced challenge is growing films on substrates, such as hydrophobic polymers, that provide poor adhesion to the deposited film.[165,166] Coating such substrates with graphene chemically modified to enhance binding can overcome these limitations. For instance, Lee et al. examined the use of films of graphene and chemically modified graphene to promote growth of CdS on polyethylene (PE), an otherwise inert substrate for CBD (Fig. 11).[167] Both visual inspection and atomic force microscopy show that, without a coating, PE will not support CdS growth. Even with a coating of pristine graphene, little CdS growth is observed. However, robust CdS growth was observed with coatings of chemically modified graphenes, such as fluorinated graphene (FG) or oxygenated graphene (OG). Controlling the extent of the fluorination enabled control over the nucleation and growth of CdS, while patterning the fluorination via e-beam

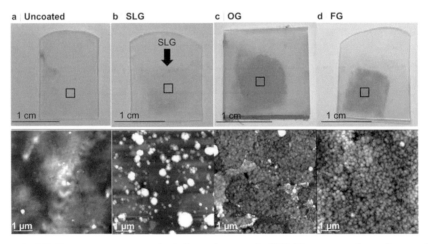

**Fig. 11.** CdS films were grown on polyethylene substrates via CBD. The top row shows photographs of PE substrates that were (a) uncoated, (b) graphene-coated, (c) OG-coated, and (d) FG-coated prior to CBD. The formation of CdS films on the PE substrate was strongly dependent upon the nature of the coating. The figures in the bottom row show the respective AFM topographic images from the PE samples, corresponding to the black boxes in the optical photographs. (Reprinted from Lee et al., *ACS Applied Materials & Interfaces*, 2017.[167])

plasma (described in a previous section) produced a patterned CdS film. Several other materials beyond CdS such as ZnO, ZnS, or PbS have been developed suggesting that modification of surfaces with chemically modified graphene is a generic strategy to promote the formation of semiconducting films via CBD.

## *Control of physisorption*

Common surface phenomena, such as the spreading of liquids or the adsorption of molecules, are dictated by interactions, such as van der Waals (vdW) and hydrogen bonding, active only at the surface. The short-range nature of these surface forces—vdW interactions extend to only around 20 nm above the surface[168] while hydrogen bonding extends less than 0.2 nm[169]—implies that they are completely determined by the chemical identity of the surface and, at most, the first few atomic layers of the solid. Thus, an ability to manipulate these surface layers should enable control over surface interactions and, by extension, the physisorption outcome. Atomic thinness and chemical diversity of monolayer materials, coupled with the methods of their transfer onto a solid, offer an unprecedented opportunity to fabricate surface layers from the bottom-up by sequentially stacking heterogeneous monolayers on the original solid. In this fashion, a stack of only a 10–20 nm thickness should impart pre-determined surface interactions to any solid. Indeed, because the individual monolayers are only 0.3–0.7 nm thick, the choice of monolayer and stacking sequence should enable extraordinarily fine tuning of the surface interactions. Graphene is particularly noteworthy in that it provides independent control over both components of surface energy, dispersive and non-dispersive.

An early attempt to control physisorption with monolayer materials was the examination of "wetting transparency", where changes in the water contact angle were monitored as graphene layers were added to a surface. If the water contact angle did not change when graphene was added, then graphene would be considered transparent to wetting. Whereas if the wetting was completely determined by graphene then it would be considered opaque. Initial experiments from the several groups investigating the effect on Si and $SiO_2$ generated contradictory conclusions: from complete wetting transparency[170] to partial transparency[171] to the wetting opacity.[172] Adding to the controversy, each study included molecular dynamics simulations supporting their respective conclusions. This controversy, which is still not fully resolved, highlights the challenges of reliable experimental measurements on transferred CVD monolayers. Notably, the water contact angle technique employed in the wetting transparency studies can present a serious experimental deficiency in that the ~ 1 μL droplets probe a surface area greater than 1 mm². Such a large area necessitates a graphene sample either on the metallic growth substrates or transferred onto a new substrate using polymer assisted transfer of CVD-grown graphene. However, that transfer process can tear or wrinkle the graphene and often leaves residual polymer contamination, all of which can impact the veracity of the water contact angle measurement. Indeed, the clean transfer of CVD graphene remains one of the primary challenges to surface engineering with the material.

The inhomogeneity of transferred CVD samples may be overcome by probing a smaller area of a cleaner surface. One can turn to atomic force microscopy (AFM) where the sharp tip of an AFM cantilever can directly probe surface forces.[173] Because the asperity of the tip has significantly smaller lateral dimensions (< 50 nm) than the water droplet, measurements on small but clean graphene samples (~ 1 µm laterally) fabricated by micromechanical exfoliation become possible. Further, prior to probing the surface interactions, the same tip can image the investigated surface with high spatial resolution (~ 10 nm), assuring its mechanical integrity and the absence of contamination. Last, but not least, this AFM-based approach allows the interactions to be measured as a function of separation between the surface and the tip, in contrast to the water contact angle measurement performed at the contact only. As we demonstrate later, this separation dependence becomes critical for analysis of interactions modified by the monolayers.

Figure 12A shows an AFM image of a $SiO_2$ substrate partially covered by exfoliated graphene. After the graphene-covered area is identified in the image, the tip is accurately positioned over it and surface interactions are measured, as schematically illustrated in Fig. 12B. Figure 12C presents the force acting on the tip as a function of separation recorded over areas covered with single-, double-, and triple-layer graphene. These force curves can be compared to that measured with the same tip from a separate highly oriented pyrolytic graphite (HOPG), which can be considered as an infinite stack of graphene layers.

The attractive force, indicated by its negative values, becomes monotonically stronger as the tip approaches the surface, consistent with the vdW interactions expected in the explored separation range (2 nm < d < 20 nm). Thus, the measured force is assigned to the vdW force acting between the sample and tip. We also note that the graphitic surfaces investigated are not expected to form hydrogen bonding at shorter separation. Because the interaction decays quickly with the separation (Fig. 12C), the force is effectively confined to the very asperity of the tip, allowing it to be exploited as a tiny nanoscale probe.

Examining Fig. 12C reveals that the force is the weakest over single-layer graphene, becomes progressively stronger over double- and triple-layer graphene and is the strongest over HOPG. Further, the interactions with the graphene-covered $SiO_2$ are qualitatively different from that with HOPG: the tip needs to approach closer than 7 nm from the surface to start detecting the force from the former, whereas an appreciable force already develops from the latter at a greater distance (~ 15 nm). As discussed later this trend indicates that the graphene strongly screens the vdW interactions of $SiO_2$.

The experimental data can be explained by a simple model of the vdW interactions between the AFM tip and the sample.[173] The tip asperity participating in the interaction is assumed to have a spherical shape with the curvature radius $R$. HOPG is treated as a homogeneous half-space with a perfectly flat surface, while the graphene-covered $SiO_2$ as the homogeneous half-space covered with a perfectly flat layer of thickness $t$. In the latter case, the interaction is split into two pairwise components between the tip and the graphene layer, and between the tip and the $SiO_2$ half-space. While the real vdW interactions are not pairwise,[168] such an assumption reflects the modelling simplicity, while the exact analytical expressions of the model are given in Tsoi et al.[173] The model provides a good fit to the experimental data yielding a product $A \cdot R$, where $A$ is a Hamaker constant characterizing the strength of vdW interaction.[169] Because

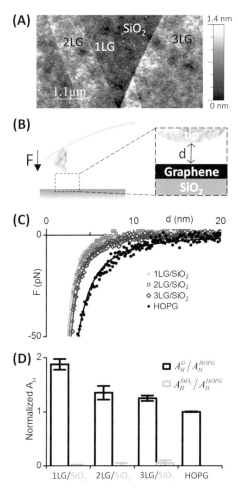

**Fig. 12.** (A) AFM topography image of graphene exfoliated on top of a SiO$_2$ substrate showing areas covered by varying numbers of graphene layers. 1LG = single-layer graphene, 2LG = double-layer graphene, 3LG = triple-layer graphene. (B) Schematic representation of the AFM force-separation measurement. Force F exerted on the tip by the sample is measured as a function of their separation d. (C) The F(d) dependence measured over areas covered by different numbers of graphene layers. Symbols are experimental data, lines are fits. In addition, the F(d) dependence measured with the same tip over a separate HOPG sample is included. (D) Hamaker coefficients determined from the fits for graphene and the underlying SiO$_2$ and normalized relative to HOPG. Reprinted from Tsoi et al.[173]

an independent measurement of a rather small $R$ may introduce a large uncertainty, we choose to normalize the products $A_{SiO2} \cdot R$ and $A_{Gr} \cdot R$, describing interaction of the tip with the SiO$_2$ substrate and graphene layer, respectively, to $A_{HOPG} \cdot R$ (Fig. 1D). The obtained results demonstrate two noteworthy trends. First, the contribution $A_{SiO2}$ is negligibly small in all samples, suggesting that even the single-layer graphene completely screens vdW interactions of the underlying SiO$_2$. In this manner, the tip experiences the vdW force from the atomically thin graphene alone and cannot detect the SiO$_2$ substrate. Because other probes, such as the water droplet, are expected to experience an identical interaction profile, this result clarifies the wetting transparency of graphene: the addition of graphene to $SiO_2$ should screen all of the vdW forces from $SiO_2$ as well as the hydrogen-bonding moieties of $SiO_2$ while provide minimal van der Waals interactions and no hydrogen bonding interactions. Thus, the single atomic layer should exhibit complete wetting opacity.

Note that the graphene screening of $SiO_2$ also explains the peculiar force curve shapes observed in Fig. 12C. The tip can detect an appreciable force from the atomically thin graphene alone only in its close proximity. In contrast, a large number of sub-surface graphene layers in HOPG combine to generate a non-negligible force even at greater separations. This is consistent with the general vdW principle that the separation between two interacting bodies determines how deep inside each other they probe.[168] Thus, probing the interaction as a function of separation can unambiguously uncover the vdW screening.

The second trend seen in Fig. 12D is a monotonic decrease in the vdW strength of graphene with number of layers, tending to that of HOPG. Remembering that the Hamaker constant defines interaction per unit volume,[169] the results in Fig. 12D indicate that each layer in the double-layer graphene exerts a weaker force on the tip than the single-layer graphene, however its two layers combine to generate a stronger net force than the single-layer graphene (Fig. 12C). Two aspects may contribute to this thickness-dependent Hamaker constant. The electronic band structure of graphene changes with the number of layers, tending to that of graphite.[174] In addition, doping of exfoliated graphene by the $SiO_2$ substrate depends on the number of layers.[175] Together, these two effects determine the electronic polarizability of graphene which is ultimately responsible for its vdW interactions.[168] Overall, the vdW screening and the thickness-dependence of the vdW interaction of graphene provide the first, simple demonstration of modified surface interactions with monolayer materials.

If graphene exhibits vdW screening, complete control over the surface forces also requires monolayers featuring full and partial vdW transparency. While the exact origin of the vdW screening is presently unknown, a likely explanation involves mobile carriers in graphene that can quickly rearrange spatially to screen electrodynamic waves exchanged in the vdW interactions between the $SiO_2$ and the tip.[173] In this context, replacing semi-metallic graphene with an insulating monolayer should result in vdW transparency. To this end, we treat the exfoliated graphene with $XeF_2$ gas, turning the uppermost layer into electrically insulating fluorinated graphene.[103,176]

Figure 13A shows an AFM image of fluorinated graphene on $SiO_2$. Prior to fluorination, the graphene flake contained a single-layer area on the right and a double-layer area on the left. Since fluorination affects only exposed graphene, it turns the former into fluorinated graphene and the latter into a fluorinated graphene/graphene stack (Fig. 13B), evident in the formation of stress-releasing wrinkles. Figure 13C shows force curves recorded over the two areas and compares them to that recorded with the same tip over HOPG. The vdW interaction over the fluorinated graphene is markedly different from that over the single-layer graphene shown in Fig. 12C and exhibits a close similarity in shape to the interaction with HOPG, suggesting that the sub surface $SiO_2$ exerts a significant vdW force across the fluorinated graphene on the tip. Indeed, fitting the experimental data with our simple model formally yields a zero contribution from the fluorinated graphene indicating that the entire force

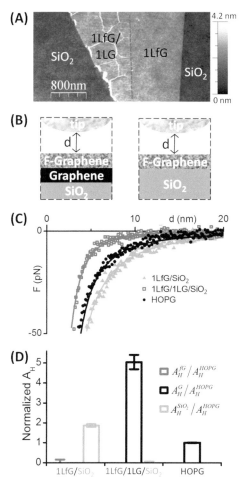

**Fig. 13.** (A) AFM topography image of graphene exfoliated on top of a SiO$_2$ substrate and treated with XeF$_2$ gas. 1LfG designates single-layer fluorinated graphene, 1LfG/1LG designates a double layer consisting of a top fluorinated graphene and underlying graphene. The gray dashed line denotes a boundary between 1LfG and 1LfG/1LG (B) Schematic representation of the AFM force measurement carried out over 1LfG/SiO$_2$ (right) and 1LfG/1LG/SiO$_2$ (left). (C) Representative *F(d)* dependences measured with an AFM tip over 1LfG/SiO$_2$, 1LfG/1LG/SiO$_2$, and a separate HOPG. Dots are experimental data, lines are fits. (D) Hamaker coefficients determined from the fits for the fluorinated graphene, underlying graphene and SiO$_2$ and normalized relative to HOPG. Reprinted from Tsoi et al.[173]

is supplied by SiO$_2$ (Fig. 13D). Thus, the fluorinated graphene exhibits the vdW transparency. Unfortunately, because the interaction from the exposed SiO$_2$ cannot be easily measured, the demonstrated transparency is not quantitative. That is, it cannot be determined whether it is full or partial.

The vdW interaction measured over the fluorinated graphene/graphene stack (Fig. 13B) provides a critical test to our early understanding of how monolayer materials affect surface interactions. The shape of the force curve is clearly different from HOPG and reminiscent of those reported for graphene in Fig. 12C. Fitting data with the model indicates that the entire force originates from the middle graphene layer. Thus, consistent with our developed intuition, graphene completely screens vdW interactions of the underlying $SiO_2$ and exerts a vdW force across the (fully or partially transparent) fluorinated graphene on the tip.

Tailoring physisorption with monolayer materials is attractive both from applied and basic perspectives. The former emphasizes the innovative procedure: an ultrathin (10–20 nm) heterogeneous stack of monolayers conformally applied to any pre-existing surface could impart to it a pre-determined physisorption outcome. From the basic point of view, this will become possible because surface interactions of the stack are tuned by its chemical composition. The potential finesse of the tuning is striking, given the atomic thinness and chemical diversity of the monolayer materials, as well as the possibility to arrange them into numerous combinations of hetero-stacks, all while preserving their chemical and electronic identity. Indeed, this approach should enable unprecedented surface interaction profiles, and with them, novel physisorption phenomena, such as the one described in the next section.

One can then use graphene and its chemical derivatives to control a molecule's interaction potential with a surface, tuning non-dispersive interactions (acid-base interactions, electrostatic double layer, hydrogen bonding) through chemical functionality and tuning dispersive ones (van der Waals) through the carrier density and band structure of the graphene. While the ability to do so on a wide range of substrates is technologically valuable, from a fundamental level the more interesting aspect is that it appears that all the dominant surface force may be controlled within a single system. These forces are enumerated in DLVO theory named after Boris Derjaguin, Lev Davidovich Landau, Evert Verwey, and Theodoor Overbeek who, in the 1940s, examined the root causes for the stability of colloids in a dispersion. DLVO theory's fundamental insight is that the attraction among colloids due to van der Waals forces are typically opposed by repulsive electrostatic charges on the colloids. The extended version of this model includes the other non-dispersive interactions to provide a fuller picture of the interaction.[177] This model is widely used since it can explain phenomena beyond colloids including ice crystal formation and the initiation of biofilms; however, it has remained hard to test since the opposing interactions, van der Waals vs. electrostatic, are hard to vary independently in most systems. With graphene surface engineering this limitation seems to be lifted, allowing the testing of some of the more interesting predictions of DLVO theory.

Perhaps one of the more interesting predictions is that of a secondary minimum. At very close separations between the two interacting bodies the van der Waals forces will always overwhelm the electrostatic repulsion, causing the two bodies to adhere. (Fig. 14). This is known as the primary minimum, after which there is typically a repulsive barrier due to the electrostatic forces. Critically, if the van der Waals force is sufficiently strong, then a secondary minimum appears after this barrier. While it is broad and shallow, it is expected to have a profound effect on the likelihood that the two bodies will bind together. Without secondary minimum, two bodies approaching each other generally have a single chance to cross over the barrier to reach the primary minimum and so permanently adhere. With a secondary minimum, the two bodies can become loosely associated and therefore repeatedly approach each other, providing multiple opportunities to cross the barrier. For this and other reasons, it is expected that

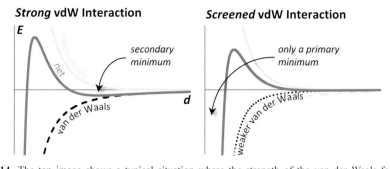

**Fig. 14.** The top image shows a typical situation where the strength of the van der Waals forces is sufficiently strong to form a secondary minimum, a shallow well that can help condense two bodies. The lower image shows that when the van der Waals forces are diminished, the secondary minimum disappears.

chemically modified graphene will find profound applications to modifying surfaces to control adhesion.

## Gradients in chemical functionality

As shown above, global chemical modification of surfaces clearly has many uses, so further control of the spatial distribution of functional groups should provide even greater functionality. For instance, a gradient of functional groups can control the flow of adsorbates. Previous methods for fabricating gradients relied on covalent attachment between SAM-forming molecules such as alkanethiols or silanes and the substrate.[3] With graphene, one could in principal establish a gradient on any surface. The previously discussed masking technique (Fig. 4) was used to produce chemical gradients on graphene that vary in total oxygen or fluorine content, with good lateral functional group uniformity (Fig. 15).[85] Oxygen gradients were shown to "pull" drops

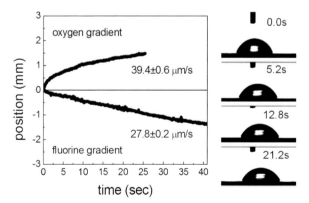

**Fig. 15.** When placed on graphene on which a chemical gradient has been built, a water droplet will be either pushed or pulled depending on the chemical functionality that is graded. The plot to the left shows a water droplet's position as a function of time on two different gradient. On the right are snapshots of one droplet on a fluorine gradient. The fluorinated graphene is more hydrophobic than graphene and so the water droplet, when placed on the region of high fluorine content, slowly moves toward the bare graphene.

of water and DMMP in the direction of increasing oxygen content, while fluorine gradients were shown to "push" such droplets in the direction of decreasing fluorine content. The direction of motion and droplet velocity can be broadly described in terms of increasing/decreasing hydrophilicity, which was correlated to high/low adhesion and binding energy determined using chemical force microscopy measurements and in density functional calculations. The plasma-based process of gradient production is flexible enough to accommodate a range of surface chemistries and coverages that could be used to further tailor performance.

# Conclusion

Graphene, and with it the emerging field of 2D materials, has given us powerful new tools to modify the physical and chemical properties of surfaces. In particular, the chemical flexibility of graphene enables the modification of surfaces to control their physical, chemical, electronic, and biochemical properties, among others. Moreover, new levels of control are available for heretofore poorly controllable properties such as van der Waals forces, which now may be readily tuned through chemical functionalization. Perhaps the most notable aspect is that this surface modification depends in only a limited way on the underlying substrate, a significant departure from traditional methods of creating thin films such as SAMs or Langmuir Blodgett films that require precise preparation of the underlying surface. It also opens up the possibility that a more universal chemistry can be envisioned where a surface treatment is developed once by creating it on a graphene sheet and then transferred onto a wide range of substrates like a chemical "sticky note". While some of these results remain to be accomplished, chemically modified graphene has already shown its use in tuning the wetting properties of different surfaces, controlling their biofunctionalization, and enabling the growth of subsequent films on otherwise inhospitable substrates.

# References

1. W. Kern, *Thin Film Processes II (Vol. 2)*, Academic press, 2012.
2. P. M. Martin, *Handbook of Deposition Technologies for Films and Coatings: Science, Applications and Technology*, William Andrew, 2009.
3. M. K. Chaudhury and G. M. Whitesides, *Science*, 1992, **256**, 1539.
4. S. P. Mulvaney, R. Stine, N. C. Long, C. R. Tamanaha and P. E. Sheehan, *Biotechniques*, 2014, **57**, 21.
5. E. H. Lock, S. C. Hernandez, T. J. Anderson, S. W. Schmucker, M. Laskoski, S. P. Mulvaney, F. J. Bezares, J. D. Caldwell, P. E. Sheehan, J. T. Robinson, B. N. Feygelson and S. G. Walton, *Surf. Coat Tech.*, 2014, **241**, 118.
6. J. S. Bunch, S. S. Verbridge, J. S. Alden, A. M. van der Zande, J. M. Parpia, H. G. Craighead and P. L. McEuen, *Nano Lett.*, 2008, **8**, 2458.
7. J.-H. Lee, E. K. Lee, W.-J. Joo, Y. Jang, B.-S. Kim, J. Y. Lim, S.-H. Choi, S. J. Ahn, J. R. Ahn, M.-H. Park, C.-W. Yang, B. L. Choi, S.-W. Hwang and D. Whang, *Science*, 2014, **344**, 286.
8. A. S. Mayorov, R. V. Gorbachev, S. V. Morozov, L. Britnell, R. Jalil, L. A. Ponomarenko, P. Blake, K. S. Novoselov, K. Watanabe, T. Taniguchi and A. K. Geim, *Nano Lett.*, 2011, **11**, 2396.
9. J. O. Sofo, A. S. Chaudhari and G. D. Barber, *Phys. Rev. B*, 2007, **75**, 153401.
10. W. K. Lee, K. E. Whitener, Jr., J. T. Robinson and P. E. Sheehan, *Adv. Mater*, 2015, **27**, 1774.
11. D. R. Dreyer, S. Park, C. W. Bielawski and R. S. Ruoff, *Chem. Soc. Rev.*, 2010, **39**, 228.
12. K. P. Loh, Q. Bao, P. K. Ang and J. Yang, *J. Mater. Chem.*, 2010, **20**, 2277.

13. C. K. Chua and M. Pumera, *Chem. Soc. Rev.*, 2013, **42**, 3222.
14. A. Criado, M. Melchionna, S. Marchesan and M. Prato, *Angew. Chem.-Int. Ed.*, 2015, **54**, 10734.
15. A. Hirsch, J. M. Englert and F. Hauke, *Acc. Chem. Res.*, 2013, **46**, 87.
16. J. Park and M. D. Yan, *Acc. Chem. Res.*, 2013, **46**, 181.
17. S. Stankovich, D. A. Dikin, R. D. Piner, K. A. Kohlhaas, A. Kleinhammes, Y. Jia, Y. Wu, S. T. Nguyen and R. S. Ruoff, *Carbon*, 2007, **45**, 1558.
18. W. S. Hummers and R. E. Offeman, *J. Am. Chem. Soc.*, 1958, **80**, 1339.
19. T. Enoki, M. Suzuki and M. Endo, *Graphite Intercalation Compounds and Applications*, Oxford University Press, New York, 2003.
20. K. E. Whitener Jr. and P. E. Sheehan, *Diamond Relat. Mater.*, 2014, **46**, 25.
21. M. Z. Hossain, J. E. Johns, K. H. Bevan, H. J. Karmel, Y. T. Liang, S. Yoshimoto, K. Mukai, T. Koitaya, J. Yoshinobu, M. Kawai, A. M. Lear, L. L. Kesmodel, S. L. Tait and M. C. Hersam, *Nat. Chem.*, 2012, **4**, 305.
22. R. Zboril, F. Karlicky, A. B. Bourlinos, T. A. Steriotis, A. K. Stubos, V. Georgakilas, K. Safarova, D. Jancik, C. Trapalis and M. Otyepka, *Small*, 2010, **6**, 2885.
23. S. H. Cheng, K. Zou, F. Okino, H. R. Gutierrez, A. Gupta, N. Shen, P. C. Eklund, J. O. Sofo and J. Zhu, *Phys. Rev. B*, 2010, **81**, 205435.
24. F. Withers, M. Dubois and A. K. Savchenko, *Phys. Rev. B*, 2010, **82**, 073403.
25. E. Unger, M. Liebau, G. S. Duesberg, A. P. Graham, F. Kreupl, R. Seidel and W. Hoenlein, *Chem. Phys. Lett.*, 2004, **399**, 280.
26. J. T. Robinson, J. S. Burgess, C. E. Junkermeier, S. C. Badescu, T. L. Reinecke, F. K. Perkins, M. K. Zalalutdniov, J. W. Baldwin, J. C. Culbertson, P. E. Sheehan and E. S. Snow, *Nano Lett.*, 2010, **10**, 3001.
27. R. R. Nair, W. Ren, R. Jalil, I. Riaz, V. G. Kravets, L. Britnell, P. Blake, F. Schedin, A. S. Mayorov, S. Yuan, M. I. Katsnelson, H. M. Cheng, W. Strupinski, L. G. Bulusheva, A. V. Okotrub, I. V. Grigorieva, A. N. Grigorenko, K. S. Novoselov and A. K. Geim, *Small*, 2010, **6**, 2877.
28. M. Baraket, S. G. Walton, E. H. Lock, J. T. Robinson and F. K. Perkins, *Appl.Phys Lett.*, 2010, **96**, 231501.
29. H. C. Yang, M. J. Chen, H. Q. Zhou, C. Y. Qiu, L. J. Hu, F. Yu, W. G. Chu, S. Q. Sun and L. F. Sun, *J. Phys. Chem. C*, 2011, **115**, 16844.
30. S. D. Sherpa, G. Levitin and D. W. Hess, *Appl. Phys. Lett.*, 2012, **101**, 111602.
31. K. Tahara, T. Iwasaki, A. Matsutani and M. Hatano, *Appl. Phys. Lett.*, 2012, **101**, 163105.
32. X. Hong, S. H. Cheng, C. Herding and J. Zhu, *Phys. Rev. B*, 2011, **83**, 085410.
33. M. Chen, H. Zhou, C. Qiu, H. Yang, F. Yu and L. Sun, *Nanotechnology*, 2012, **23**, 115706.
34. W. H. Lee, J. W. Suk, H. Chou, J. H. Lee, Y. F. Hao, Y. P. Wu, R. Piner, D. Aldnwande, K. S. Kim and R. S. Ruoff, *Nano Lett.*, 2012, **12**, 2374.
35. D. C. Elias, R. R. Nair, T. M. G. Mohiuddin, S. V. Morozov, P. Blake, M. P. Halsall, A. C. Ferrari, D. W. Boukhvalov, M. I. Katsnelson, A. K. Geim and K. S. Novoselov, *Science*, 2009, **323**, 610.
36. B. R. Matis, J. S. Burgess, F. A. Bulat, A. L. Friedman, B. H. Houston and J. W. Baldwin, *ACS Nano*, 2012, **6**, 17.
37. W.-K. Lee, K. E. Whitener, J. T. Robinson and P. E. Sheehan, *Adv. Mater.*, 2015, **27**, 1774.
38. H. Gonzalez-Herrero, J. M. Gomez-Rodriguez, P. Mallet, M. Moaied, J. J. Palacios, C. Salgado, M. M. Ugeda, J. Y. Veuillen, F. Yndurain and I. Brihuega, *Science*, 2016, **352**, 437.
39. A. J. M. Giesbers, K. Uhlirova, M. Konecny, E. C. Peters, M. Burghard, J. Aarts and C. F. J. Flipse, *Phys. Rev. Lett.*, 2013, **111**, 166101.
40. K. E. Whitener, W. K. Lee, N. D. Bassim, R. M. Stroud, J. T. Robinson and P. E. Sheehan, *Nano Lett.*, 2016, **16**, 1455.
41. K. E. Whitener, W. K. Lee, R. Stine, C. R. Tamanaha, D. A. Kidwell, J. T. Robinson and P. E. Sheehan, *RSC Adv.*, 2016, **6**, 93356.
42. M. M. Lucchese, F. Stavale, E. H. M. Ferreira, C. Vilani, M. V. O. Moutinho, R. B. Capaz, C. A. Achete and A. Jorio, *Carbon*, 2010, **48**, 1592.
43. E. H. Martins Ferreira, M. V. O. Moutinho, F. Stavale, M. M. Lucchese, R. B. Capaz, C. A. Achete and A. Jorio, *Phys. Rev. B*, 2010, **82**, 125429.
44. K. E. Whitener Jr, W. K. Lee, P. M. Campbell, J. T. Robinson and P. E. Sheehan, *Carbon*, 2014, **72**, 348.
45. R. A. Schafer, J. M. Englert, P. Wehrfritz, W. Bauer, F. Hauke, T. Seyller and A. Hirsch, *Angew. Chem.-Int. Edit.*, 2013, **52**, 754.

46. Z. Q. Yang, Y. Q. Sun, L. B. Alemany, T. N. Narayanan and W. E. Billups, *J. Am. Chem. Soc.*, 2012, **134**, 18689.
47. J. M. Englert, K. C. Knirsch, C. Dotzer, B. Butz, F. Hauke, E. Spiecker and A. Hirsch, *Chem. Commun.*, 2012, **48**, 5025.
48. K. E. Whitener, W. K. Lee, P. M. Campbell, J. T. Robinson and P. E. Sheehan, *Carbon*, 2014, **72**, 348.
49. J. E. Johns and M. C. Hersam, *Acc. Chem. Res.*, 2013, **46**, 77.
50. K. M. Daniels, B. K. Daas, N. Srivastava, C. Williams, R. M. Feenstra, T. S. Sudarshan and M. V. S. Chandrashekhar, *J. Appl. Phys.*, 2012, **111**, 7.
51. Z. Y. Xia, F. Leonardi, M. Gobbi, Y. Liu, V. Bellani, A. Liscio, A. Kovtun, R. J. Li, X. L. Feng, E. Orgiu, P. Samori, E. Treossi and V. Palermo, *ACS Nano*, 2016, **10**, 7125.
52. E. Bekyarova, M. E. Itkis, P. Ramesh, C. Berger, M. Sprinkle, W. A. de Heer and R. C. Haddon, *J. Am. Chem. Soc.*, 2009, **131**, 1336.
53. G. L. C. Paulus, Q. H. Wang and M. S. Strano, *Acc. Chem. Res.*, 2013, **46**, 160.
54. Z. Jin, T. P. McNicholas, C. J. Shih, Q. H. Wang, G. L. C. Paulus, A. Hilmer, S. Shimizu and M. S. Strano, *Chem. Mater*, 2011, **23**, 3362.
55. B. D. Ossonon and D. Belanger, *Carbon*, 2017, **111**, 83.
56. Y. Lu, M. B. Lerner, Z. J. Qi, J. J. M. Jr., J. H. Lim, B. M. Discher and A. T. C. Johnson, *Appl. Phys. Lett.*, 2012, **100**, 033110.
57. J. Greenwood, T. H. Phan, Y. Fujita, Z. Li, O. Lvasenko, W. Vanderlinden, H. Van Gorp, W. Frederickx, G. Lu, K. Tahara, Y. Tobe, H. Uji-i, S. F. L. Mertens and S. De Feyter, *ACS Nano*, 2015, **9**, 5520.
58. V. Georgakilas, A. B. Bourlinos, R. Zboril, T. A. Steriotis, P. Dallas, A. K. Stubos and C. Trapalis, *Chem. Commun.*, 2010, **46**, 1766.
59. M. Quintana, K. Spyrou, M. Grzelczak, W. R. Browne, P. Rudolf and M. Prato, *ACS Nano*, 2010, **4**, 3527.
60. L.-H. Liu and M. Yan, *Nano Lett.*, 2009, **9**, 3375.
61. J. Choi, K.-j. Kim, B. Kim, H. Lee and S. Kim, *J. Phys., Chem. C*, 2009, **113**, 9433.
62. S. Sarkar, E. Bekyarova, S. Niyogi and R. C. Haddon, *J. Am. Chem. Soc.*, 2011, **133**, 3324.
63. Y. Huang, W. Yan, Y. Xu, L. Huang and Y. Chen, *Macromol. Chem. and Phy.*, 2012, **213**, 1101.
64. J. Yuan, G. Chen, W. Weng and Y. Xu, *J. Mater. Chem.*, 2012, **22**, 7929.
65. D. Hong, K. Bae, D. Park, H. Kim, S. P. Hong, M. H. Kim, B. S. Lee, S. Ko, S. Jeon, X. Zheng, W. S. Yun, Y. G. Kim, I. S. Choi and J. K. Lee, *Chem. Asian J.*, 2015, **10**, 568.
66. Y. Li and N. Chopra, *JOM*, 2015, **67**, 34.
67. Y. Li and N. Chopra, *JOM*, 2015, **67**, 44.
68. A. Dey, A. Chroneos, N. S. Braithwaite, R. P. Gandhiraman and S. Krishnamurthy, *Appl. Phys. Rev.*, 2016, **3**, 021301.
69. K. Ostrikov, E. C. Neyts and M. Meyyappan, *Adv. Phys.*, 2013, **62**, 113.
70. J. A. Robinson, M. LaBella, M. Zhu, M. Hollander, R. Kasarda, Z. Hughes, K. Trumbull, R. Cavalero and D. Snyder, *Appl. Phys. Lett.*, 2011, **98**, 053103.
71. D. W. Yue, C. H. Ra, X. C. Liu, D. Y. Lee and W. J. Yoo, *Nanoscale*, 2015, **7**, 825.
72. D. C. Elias, R. R. Nair, T. M. Mohiuddin, S. V. Morozov, P. Blake, M. P. Halsall, A. C. Ferrari, D. W. Boukhvalov, M. I. Katsnelson, A. K. Geim and K. S. Novoselov, *Science*, 2009, **323**, 610.
73. N. Peltekis, S. Kumar, N. McEvoy, K. Lee, A. Weidlich and G. S. Duesberg, *Carbon*, 2012, **50**, 395.
74. G. Diankov, M. Neumann and D. Goldhaber-Gordon, *ACS Nano*, 2013, **7**, 1324.
75. M. A. Lieberman and A. J. Lichtenberg, *Principles of Plasma Discharges and Materials Processing*, John Wiley & Sons, 2005.
76. O. Lehtinen, J. Kotakoski, A. V. Krasheninnikov, A. Tolvanen, K. Nordlund and J. Keinonen, *Phys. Rev. B*, 2010, **81**, 153401.
77. M. S. Choi, S. H. Lee and W. J. Yoo, *J. Appl. Phys.*, 2011, **110**, 073305.
78. L. Delfour, A. Davydova, E. Despiau-Pujo, G. Cunge, D. B. Graves and L. Magaud, *J. Appl. Phys.*, 2016, **119**, 125309.
79. Y. D. Lim, D. Y. Lee, T. Z. Shen, C. H. Ra, J. Y. Choi and W. J. Yoo, *ACS Nano*, 2012, **6**, 4410.
80. R. Narayanan, H. Yamada, M. Karakaya, R. Podila, A. M. Rao and P. R. Bandaru, *Nano Lett.*, 2015, **15**, 3067.
81. N. McEvoy, H. Nolan, N. A. Kumar, T. Hallam and G. S. Duesberg, *Carbon*, 2013, **54**, 283.

82. V. P. Pham, K. H. Kim, M. H. Jeon, S. H. Lee, K. N. Kim and G. Y. Yeom, *Carbon*, 2015, **95**, 664.
83. S. G. Walton, D. R. Boris, S. C. Hernandez, E. H. Lock, T. B. Petrova, G. M. Petrov and R. F. Fernsler, *ECS J. Solid State Sci. Technol.*, 2015, **4**, N5033.
84. W. M. Manheimer, R. F. Fernsler, M. Lampe and R. A. Meger, *Plasma Sources Sci. Technol.*, 2000, **9**, 370.
85. S. C. Hernandez, C. J. C. Bennett, C. E. Junkermeier, S. D. Tsoi, F. J. Bezares, R. Stine, J. T. Robinson, E. H. Lock, D. R. Boris, B. D. Pate, J. D. Caldwell, T. L. Reinecke, P. E. Sheehan and S. G. Walton, *ACS Nano*, 2013, **7**, 4746.
86. S. C. Hernandez, V. D. Wheeler, M. S. Osofsky, G. G. Jernigan, V. Nagareddy, A. Nath, E. H. Lock, L. O. Nyakiti, R. L. Myers-Ward, K. Sridhara, A. B. Horsfall, C. R. Eddy, D. K. Gaskill and S. G. Walton, *Surf. Coat Tech.*, 2014, **241**, 8.
87. S. C. Hernández, F. J. Bezares, J. T. Robinson, J. D. Caldwell and S. G. Walton, *Carbon*, 2013, **60**, 84.
88. M. Baraket, R. Stine, W. K. Lee, J. T. Robinson, C. R. Tamanaha, P. E. Sheehan and S. G. Walton, *Appl. Phys. Lett.*, 2012, **100**, 233123.
89. W. K. Lee, S. C. Hernandez, J. T. Robinson, S. G. Walton and P. E. Sheehan, *ACS Appl. Mater Interfaces*, 2017, **9**, 677.
90. A. V. Jagtiani, H. Miyazoe, J. Chang, D. B. Farmer, M. Engel, D. Neumayer, S. J. Han, S. U. Engelmann, D. R. Boris, S. C. Hernandez, E. H. Lock, S. G. Walton and E. A. Joseph, *J. Vac. Sci. Technol. A*, 2016, **34**, 01B103.
91. A. Reina, X. Jia, J. Ho, D. Nezich, H. Son, V. Bulovic, M. S. Dresselhaus and J. Kong, *Nano Lett.*, 2009, **9**, 30.
92. J. Kim, H. Park, J. B. Hannon, S. W. Bedell, K. Fogel, D. K. Sadana and C. Dimitrakopoulos, *Science*, 2013, **342**, 833.
93. E. H. Lock, M. Baraket, M. Laskoski, S. P. Mulvaney, W. K. Lee, P. E. Sheehan, D. R. Hines, J. T. Robinson, J. Tosado, M. S. Fuhrer, S. C. Hernandez and S. G. Walton, *Nano Lett.*, 2012, **12**, 102.
94. L. G. Martins, Y. Song, T. Zeng, M. S. Dresselhaus, J. Kong and P. T. Araujo, *Proc. Natl. Acad. Sci. U.S.A.*, 2013, **110**, 17762.
95. K. S. Novoselov, A. K. Geim, S. V. Morozov, D. Jiang, Y. Zhang, S. V. Dubonos, I. V. Grigorieva and A. A. Firsov, *Science*, 2004, **306**, 666.
96. J. H. Lee, J. Avsar, J. Jung, J. Y. Tan, K. Watanabe, T. Taniguchi, S. Natarajan, G. Eda, S. Adam, A. H. Castro Neto and B. Özyilmaz, *Nano Lett.*, 2015, **15**, 319.
97. W. K. Lee, S. Tsoi, K. E. Whitener, R. Stine, J. T. Robinson, J. S. Tobin, A. Weerasinghe, P. E. Sheehan and S. F. Lyuksyutov, *Nano Res.*, 2013, **6**, 767.
98. M. Dubecký, E. Otyepková, P. Lazar, F. Karlický, M. Petr, K. Čépe, P. Banáš, R. Zbořil and M. Otyepka, *J. Phys. Chem. Lett.*, 2015, **6**, 1430.
99. K. Wong, S. J. Kang, C. W. Bielawski, R. S. Ruoff and S. K. Kwak, *J. Am. Chem. Soc.*, 2016, **138**, 10986.
100. F. M. Koehler, A. Jacobsen, K. Ensslin, C. Stampfer and W. J. Stark, *Small*, 2010, **6**, 1125.
101. Q. H. Wang, Z. Jin, K. K. Kim, A. J. Hilmer, G. L. C. Paulus, C.-J. Shih, M.-H. Ham, J. D. Sanchez-Yamagishi, K. Watanabe, T. Taniguchi, J. Kong, P. Jarillo-Herrero and M. S. Strano, *Nat. Chem.*, 2012, **4**, 724.
102. R. Stine, W.-K. Lee, K. E. Whitener, J. T. Robinson and P. E. Sheehan, *Nano Lett.*, 2013, **13**, 4311.
103. J. Robinson, J. Burgess, C. Junkermeier, S. Badescu, T. Reinecke, F. Perkins, M. Zalalutdniov, J. Baldwin, J. Culbertson and P. Sheehan, *Nano Lett.*, 2010, **10**, 3001.
104. Y. X. Xu, H. Bai, G. W. Lu, C. Li and G. Q. Shi, *J. Am. Chem. Soc.*, 2008, **130**, 5856.
105. R. Stine, W. K. Lee, K. E. Whitener, J. T. Robinson and P. E. Sheehan, *Nano Lett.*, 2013, **13**, 4311.
106. A. Hashimoto, K. Suenaga, A. Gloter, K. Urita and S. Iijima, *Nature*, 2004, **430**, 870.
107. J. Barzola-Quiquia, P. Esquinazi, M. Rothermel, D. Spemann, T. Butz and N. García, *Phys. Rev. B*, 2007, **76**, 161403.
108. O. V. Yazyev and L. Helm, *Phys. Rev. B*, 2007, **75**, 125408.
109. M. M. Ugeda, I. Brihuega, F. Guinea and J. M. Gómez-Rodríguez, *Phys. Rev. Lett.*, 2010, **104**, 096804.
110. P. Dev and T. L. Reinecke, *Phys. Rev. B*, 2015, **91**, 035436.
111. S. Sarkar, S. Niyogi, E. Bekyarova and R. C. Haddon, *Chem. Sci.*, 2011, **2**, 1326.
112. P. Hohenberg and W. Kohn, *Phys. Rev.*, 1964, **136**, B864.
113. W. Kohn and L. J. Sham, *Phys. Rev.*, 1965, **140**, A1133.

114. A. H. Castro Neto, F. Guinea, N. M. R. Peres, K. S. Novoselov and A. K. Geim, *Rev. Mod. Phys.*, 2009, **81**, 109.
115. H. Terrones, R. Lv, M. Terrones and M. S. Dresselhaus, *Rep. Prog. Phys.*, 2012, **75**, 062501.
116. P. Dev and T. L. Reinecke, *Phys. Rev. B*, 2014, **89**, 035404.
117. V. Palermo, *Chem. Commun.*, 2013, **49**, 2848.
118. C. Chung, Y. K. Kim, D. Shin, S. R. Ryoo, B. H. Hong and D. H. Min, *Acc. Chem. Res.*, 2013, **46**, 2211.
119. Y. Wang, Z. H. Li, J. Wang, J. H. Li and Y. H. Lin, *Trends Biotechnol.*, 2011, **29**, 205.
120. F. M. P. Tonelli, V. A. M. Goulart, K. N. Gomes, M. S. Ladeira, A. K. Santos, E. Lorencon, L. O. Ladeira and R. R. Resende, *Nanomedicine*, 2015, **10**, 2423.
121. C. Chung, Y.-K. Kim, D. Shin, S.-R. Ryoo, B. H. Hong and D.-H. Min, *Acc. Chem. Res.*, 2013, **46**, 2211.
122. S. P. Mulvaney, R. Stine, N. C. Long, C. R. Tamanaha and P. E. Sheehan, *BioTechniques*, 2014, **57**, 21.
123. A. Fabbro, D. Scaini, V. Leon, E. Vazquez, G. Cellot, G. Privitera, L. Lombardi, F. Torrisi, F. Tomarchio, S. Bonaccorso, S. Bosi, A. C. Ferrari, L. Ballerini and M. Prato, *ACS Nano*, 2016, **10**, 615.
124. Y. Ohno, K. Maehashi and K. Matsumoto, *Biosens. Bioelectron.*, 2010, **26**, 1727.
125. Y. X. Huang, X. C. Dong, Y. M. Shi, C. M. Li, L. J. Li and P. Chen, *Nanoscale*, 2010, **2**, 1485.
126. A. Bonanni, A. Ambrosi and M. Pumera, *Chem. Eur. J.*, 2012, **18**, 4541.
127. K. E. Whitener, R. Stine, J. T. Robinson and P. E. Sheehan, *J. Phys. Chem. C*, 2015, **119**, 10507.
128. Y. Ohno, K. Maehashi and K. Matsumoto, *J. Am. Chem. Soc.*, 2010, **132**, 18012.
129. Q. H. Wang and M. C. Hersam, *Nat. Chem.*, 2009, **1**, 206.
130. S. Mao, G. Lu, K. Yu, Z. Bo and J. Chen, *Adv. Mater.*, 2010, **22**, 3521.
131. Y. Cui, S. N. Kim, R. R. Naik and M. C. McAlpine, *Acc. Chem. Res.*, 2012, **45**, 696.
132. Y. Cui, S. N. Kim, S. E. Jones, L. L. Wissler, R. R. Naik and M. C. McAlpine, *Nano Lett.*, 2010, **10**, 4559.
133. M. S. Mannoor, H. Tao, J. D. Clayton, A. Sengupta, D. L. Kaplan, R. R. Naik, N. Verma, F. G. Omenetto and M. C. McAlpine, *Nat. Commun.*, 2012, **3**, 763.
134. S. J. Park, O. S. Kwon, S. H. Lee, H. S. Song, T. H. Park and J. Jang, *Nano Lett.*, 2012, **12**, 5082.
135. E. Stern, R. Wagner, F. J. Sigworth, R. Breaker, T. M. Fahmy and M. A. Reed, *Nano Lett.*, 2007, **7**, 3405.
136. N. Mohanty and V. Berry, *Nano Lett.*, 2008, **8**, 4469.
137. R. Stine, J. T. Robinson, P. E. Sheehan and C. R. Tamanaha, *Adv. Mater.*, 2010, **22**, 5297.
138. T. Kurkina, S. Sundaram, R. S. Sundaram, F. Re, M. Masserini, K. Kern and K. Balasubramanian, *ACS Nano*, 2012, **6**, 5514.
139. E. Bekyarova, M. E. Itkıs, P. Ramesh, C. Berger, M. Sprinkle, W. A. de Heer and R. C. Haddon, *J. Am. Chem. Soc.*, 2009, **131**, 1336.
140. D. B. Farmer, R. Golizadeh-Mojarad, V. Perebeinos, Y.-M. Lin, G. S. Tulevski, J. C. Tsang and P. Avouris, *Nano Lett.*, 2009, **9**, 388.
141. J. R. Lomeda, C. D. Doyle, D. V. Kosynkin, W.-F. Hwang and J. M. Tour, *J. Am. Chem. Soc.*, 2008, **130**, 16201.
142. A. Kasry, A. A. Afzali, S. Oida, S. J. Han, B. Menges and G. S. Tulevski, *Chem. Mater.*, 2011, **23**, 4879.
143. S. Liu and X. F. Guo, *NPG Asia Mater.*, 2012, **4**, 1.
144. R. Stine, S. P. Mulvaney, J. T. Robinson, C. R. Tamanaha and P. E. Sheehan, *Anal. Chem.*, 2013, **85**, 509.
145. J. Lee, J. Kim, S. Kim and D. H. Min, *Adv. Drug Deliv. Rev.*, 2016, **105**, 275.
146. H. J. Sun, J. S. Ren and X. G. Qu, *Acc. Chem. Res.*, 2016, **49**, 461.
147. S. W. Ding, A. A. Cargill, S. R. Das, I. L. Medintz and J. C. Claussen, *Sensors*, 2015, **15**, 14766.
148. J. Chen, W. H. Ding, Z. X. Luo, B. H. Loo and J. N. Yao, *J. Raman Spectrosc.*, 2016, **47**, 623.
149. L. Gao, C. Q. Lian, Y. Zhou, L. R. Yan, Q. Li, C. X. Zhang, L. Chen and K. P. Chen, *Biosens. Bioelectron.*, 2014, **60**, 22.
150. J. L. Zhang, F. Zhang, H. J. Yang, X. L. Huang, H. Liu, J. Y. Zhang and S. W. Guo, *Langmuir*, 2010, **26**, 6083.

151. K. E. Sapsford, W. R. Algar, L. Berti, K. B. Gemmill, B. J. Casey, E. Oh, M. H. Stewart and I. L. Medintz, *Chem. Rev.*, 2013, **113**, 1904.
152. S. Hvilsted, *Polym. Int.*, 2012, **61**, 485.
153. D. Mandler and S. Kraus-Ophir, *J. Solid State Electrochem.*, 2011, **15**, 1535.
154. R. Mout, D. F. Moyano, S. Rana and V. M. Rotello, *Chem. Soc. Rev.*, 2012, **41**, 2539.
155. Y. Zhou, C. W. Chiu and H. Liang, *Sensors*, 2012, **12**, 15036.
156. D. Samanta and A. Sarkar, *Chem. Soc. Rev.*, 2011, **40**, 2567.
157. X. L. Luo and J. J. Davis, *Chem. Soc. Rev.*, 2013, **42**, 5944.
158. E. Petryayeva and U. J. Krull, *Anal. Chim. Acta*, 2011, **706**, 8.
159. R. Stine, B. S. Simpkins, S. P. Mulvaney, L. J. Whitman and C. R. Tamanaha, *Appl. Surf. Sci.*, 2010, **256**, 4171.
160. S. Mariani and M. Minunni, *Anal. Bioanal. Chem*, 2014, **406**, 2303.
161. S. P. Mulvaney, K. M. Myers, P. E. Sheehan and L. J. Whitman, *Biosens. Bioelectron.*, 2009, **24**, 1109.
162. S. P. Mulvaney, C. L. Cole, M. D. Kniller, M. Malito, C. R. Tamanaha, J. C. Rife, M. W. Stanton and L. J. Whitman, *Biosens. Bioelectron.*, 2007, **23**, 191.
163. J. He, G. Z. Liu, S. P. Dou, S. Gupta, M. Rusckowski and D. Hnatowich, *Bioconjugate Chem.*, 2007, **18**, 983.
164. W. O. Seo, Y. Jung, J. Kim and D. Kim, *Appl. Phys. Lett.*, 2014, **104**, 133902.
165. M. H. Kunita, E. M. Girotto, E. Radovanovic, M. C. Goncalves, O. P. Ferreira, E. C. Muniz and A. F. Rubira, *Appl. Surf. Sci.*, 2002, **202**, 223.
166. D. Lim, J. Lee and W. Song, *Thin Solid Films*, 2013, **546**, 317.
167. W. K. Lee, S. C. Hernandez, J. T. Robinson, S. G. Walton and P. E. Sheehan, *ACS Appl. Mater. Interfaces*, 2017, **9**, 677.
168. V. A. Parsegian, *Van Der Waals Forces: A Handbook for Biologists, Chemists, Engineers, and Physicists*, Cambridge University Press, New York, 2006.
169. J. N. Israelachvili, *Intermolecular and Surface Forces: with Applications to Colloidal and Biological Systems*, Academic Press, London, 1991.
170. J. Rafiee, X. Mi, H. Gullapalli, A. V. Thomas, F. Yavari, Y. Shi, P. M. Ajayan and N. A. Koratkar, *Nat. Mater.*, 2012, **11**, 217.
171. C.-J. Shih, Q. H. Wang, S. Lin, K.-C. Park, Z. Jin, M. S. Strano and D. Blankschtein, *Phys. Rev. Lett.*, 2012, **109**, 176101.
172. R. Raj, Shalabh C. Maroo and E. N. Wang, *Nano Lett.*, 2013, **13**, 1509.
173. S. Tsoi, P. Dev, A. L. Friedman, R. Stine, J. T. Robinson, T. L. Reinecke and P. E. Sheehan, *ACS Nano*, 2014, **8**, 12410.
174. B. Partoens and F. M. Peeters, *Phys. Rev. B*, 2006, **74**, 075404.
175. D. Ziegler, P. Gava, J. Güttinger, F. Molitor, L. Wirtz, M. Lazzeri, A. M. Saitta, A. Stemmer, F. Mauri and C. Stampfer, *Phys. Rev. B*, 2011, **83**, 235434.
176. R. Stine, W.-K. Lee, K. E. Whitener, J. T. Robinson and P. E. Sheehan, *Nano Lett.*, 2013, **13**, 4311.
177. E. M. V. Hoek and G. K. Agarwal, *J. Colloid Interface Sci.*, 2006, **298**, 50.

# Index